About Island Press

Island Press is the only nonprofit organization in the United States whose principal purpose is the publication of books on environmental issues and natural resource management. We provide solutions-oriented information to professionals, public officials, business and community leaders, and concerned citizens who are shaping responses to environmental problems.

In 2002, Island Press celebrates its eighteenth anniversary as the leading provider of timely and practical books that take a multidisciplinary approach to critical environmental concerns. Our growing list of titles reflects our commitment to bringing the best of an expanding body of literature to the environmental community throughout North America and the world.

Support for Island Press is provided by The Nathan Cummings Foundation, Geraldine R. Dodge Foundation, Doris Duke Charitable Foundation, Educational Foundation of America, The Charles Engelhard Foundation, The Ford Foundation, The George Gund Foundation, The Vira I. Heinz Endowment, The William and Flora Hewlett Foundation, Henry Luce Foundation, The John D. and Catherine T. MacArthur Foundation, The Andrew W. Mellon Foundation, The Moriah Fund, The Curtis and Edith Munson Foundation, National Fish and Wildlife Foundation, The New-Land Foundation, Oak Foundation, The Overbrook Foundation, The David and Lucile Packard Foundation, The Pew Charitable Trusts, The Rockefeller Foundation, The Winslow Foundation, and other generous donors.

The opinions expressed in this book are those of the author(s) and do not necessarily reflect the views of these foundations.

About the Pacific Institute for Studies in Development, Environment, and Security

The Pacific Institute for Studies in Development, Environment, and Security, in Oakland, California, is an independent, nonprofit organization created in 1987 to conduct research and policy analysis in the areas of environmental protection, sustainable development, and international security. Underlying all of the Institute's work is the recognition that the urgent problems of environmental degradation, regional and global poverty, and political tension and conflict are fundamentally interrelated, and that long-term solutions dictate an interdisciplinary approach. Since 1987, we have produced more than sixty research studies, organized roundtable discussions, and held widespread briefings for policymakers and the public. The Institute has formulated a new vision for long-term water planning in California and internationally, developed a new approach for valuing well-being in the Sierra Nevada, worked on transborder environment and trade issues in North America, analyzed ISO 14000's role in global environmental protection, clarified key concepts and criteria for sustainable water use in the lower Colorado basin, offered recommendations for reducing conflicts over water in the Middle East and elsewhere, assessed the impacts of global warming on freshwater resources, and created programs to address environmental justice concerns in poor communities and communities of color.

For detailed information about the Institute's activities, visit www.pacinst.org, www.worldwater.org, and www.globalchange.org.

THE WORLD'S WATER
2002-2003

THE WORLD'S WATER

2002-2003

The Biennial Report on Freshwater Resources

Peter H. Gleick

with William C. G. Burns, Elizabeth L. Chalecki,
Michael Cohen, Katherine Kao Cushing, Amar Mann,
Rachel Reyes, Gary H. Wolff, and Arlene K. Wong

Pacific Institute for Studies in Development,
Environment, and Security
Oakland, California

ISLAND PRESS

Washington • Covelo • London

Library of Congress Card Catalog Number 98–24877
ISBN 1–55963–949–0
ISSN 1528–7165

Printed on recycled, acid-free paper ✪

Manufactured in the United States of America
09 08 07 06 05 04 03 02   8 7 6 5 4 3 2 1

To all our colleagues working on water problems around the world.

Contents

WATER BRIEFS *(continued)*

DATA SECTION   237

INDEX   329

Foreword

The importance of water to all life on our water planet is more obvious than ever before. Growing populations, dying ecosystems, political and military disputes, and unrelenting poverty are all intimately related to water or its absence. They require businesses, political leaders, teachers, communities, and citizens to address water problems with a new urgency. But to address a problem we must first understand it, and we must see a path that turns its challenges into opportunities.

One of the most valuable tools for finding that path is *The World's Water*. These books, produced every two years, offer insights into the wide array of global water problems and, one might say, aqueous solutions. Beginning with the first volume in 1998, *The World's Water* has systematically begun to lay out a new path, a "soft" path, for our water future. When I wrote *Soft Energy Paths* in 1977, there was an unshakable belief that large, centralized energy systems were the only way to meet unrelenting growth in demand, believed to be inextricably intertwined with growth in population and gross domestic production (GDP). Challenging that belief and offering a different path has helped the world begin the necessary transition to cheaper, cleaner, and more sustainable energy choices.

Interestingly, the level of U.S. energy demand shown in the "soft-path" graph I first published in *Foreign Affairs* in 1976—intensely controversial, even heretical, at the time—is now within a few percent of the actual level. Two-fifths of U.S. energy services are now provided not by new supply but by more productive end-use; yet we have barely scratched the surface of how much energy efficiency is available and worth buying. Almost unnoticed, too, the marketplace has shifted away from the once-universal centralized solutions, like power stations and grids, because they cost more, are less reliable and secure, and incur more financial risk than diverse, dispersed solutions matched to their tasks. The soft-path metaphor redefined the energy problem as not simply getting *more energy*—more, of any kind, from any source, at any price—but providing just the amount, quality, scale, and source of energy that will deliver each desired end-use service at least cost. Now the market too is following that logic wherever it is allowed to. The next step, now starting to emerge, is wherever "negawatts" (saved watts) are cheaper than megawatts, not only to market negawatts but also to make markets *in* them, so as to maximize competition in who saves and how, thus making the savings ever bigger, better, and cheaper.

The same kinds of choices in what questions we ask, and hence what answers we get, now apply to meeting water needs, from the village to the globe. Importantly, Peter

Gleick has begun to articulate these choices. His work on new visions for the future, alternative scenarios, water-use efficiency, and the proper application of economics and technology is showing us that a soft path for water is not only possible but also will work better, build faster, cost less, and provide more just and durable access to the vital water services that so many of our fellow-humans now shamefully lack.

This volume of *The World's Water* is coauthored by Gleick and colleagues at the Pacific Institute—the world's leading synthesists of new water thinking. It begins with a chapter outlining the ideas behind a soft-water path. Building on the concepts of rational economics, water-use productivity, and technology and decision making that are the right size for the job, Gary Wolff and Gleick argue that a soft path is not only preferable but also ultimately necessary. Two following chapters lay out issues and concerns in "the new economy of water"—the growing and worrisome trend toward private control over water systems and efforts to turn water resources into commodities and trade goods that may not sufficiently protect the public interest. In these chapters, Gleick, Wolff, Elizabeth L. Chalecki, and Rachel Reyes offer balanced insights into the volatile arguments on both sides of these issues and suggest a sensible course for public and private policy makers.

Each of the previous volumes has also tackled various problems associated with the "hard-water path," such as the consequences of building big dams. This new book reviews the findings of the recently completed report from the World Commission on Dams and offers a case study of the impacts of the major dam construction project in southeastern Turkey on irreplaceable archeological sites. Other chapters address indicators of well-being associated with water, the discovery and importance of water in other parts of our own solar system and beyond, efforts to restore the Colorado River Delta, and much more.

Also building on previous work, the book revisits the implications of climate change for water resources. Gleick has published extensively on this subject, including the lead water report of the National Assessment in the United States. In this volume, William Burns takes a detailed look at the vulnerability of the water resources of island nations to climate change—and provides a depressing picture of the perhaps inevitable destruction of several unique cultures and societies.

As usual, the book includes a carefully selected set of data on water and a newly updated chronology of water-related conflicts. This striking chronology, compiled over the past decade and stretching back hundreds of years, is the authoritative listing of international and regional violence over shared water resources. It has become an invaluable source to researchers and policy makers in security studies, and it strongly reinforces the importance of a soft-water path in avoiding future conflict, terrorism, and injustice. Many security scholars fear that major conflicts and disputes will be over water. This need not be. Modern understanding of how to use water productively, how to provide it at the right scale and quality for each task, and to enable water efficiency to compete fairly with water supply can stretch existing supplies to provide enough— as the South African Constitution says, "for all, for ever."

Taken alone, this volume provides critical new insights into new solutions to both old and new water problems. Taken together, the series *The World's Water* is an invaluable resource. Nearly as precious as water itself. Drink of it deeply and often—and then act.

Amory B. Lovins
Rocky Mountain Institute

Acknowledgments

Producing a new book on critical water issues every two years is a challenging task, though there is no shortage of issues that need to be addressed. This year, for the first time, *The World's Water* is a multiauthored book, with contributions from the staff of the Pacific Institute. The broader insights and voices of these contributors will help to keep the series timely and useful. As always, this book has benefited from feedback, conversations, reviews, data, critiques, and suggestions from many people. We would especially like to thank the Legal Services Department of the Author's Guild, William Arrocha, Maude Barlow, Shirley Brown, Stefan Deconinck, Navroz K. Dubash, Mairi Dupar, Nabil El-Khodari, Anthony Falkland, Ed Glenn, John Hay, Christine Henges-Jeck, Jippe Hoogeveen, Roger Jones, Larry Karp, Sanjeev Khagram, Smitu Kothari, Tundu Lissu, Amory Lovins, Marcus Moench, Carlos Munoz, Jennifer Pitt, Sandra Postel, Jerry delli Priscoli, John Roberts (for his insights into water units), Gautum Sethi, Jamie Skinner, Shiv Someshwar, Caroline Sullivan, Bob Wilkinson, Aaron Wolf, Peggy Wu, and probably many others for their insights, information, and helpful comments. Peggy Wu provided several of the maps and offered advice on other graphics. Any errors are, of course, our own.

Support for *The World's Water 2002–2003* was provided by grants to the Pacific Institute for Studies in Development, Environment, and Security from the William and Flora Hewlett, Horace W. Goldsmith, Henry Luce, Rockefeller, and John D. and Catherine T. MacArthur Foundations. I thank them for their continued willingness to support the research and policy work these books represent.

Finally, as always, I thank my wife, Nicki Norman, for her continued enthusiasm for my work, if not my travel.

Peter H. Gleick

Introduction

Welcome to the third volume of *The World's Water*. What started as an effort to explore a variety of critical water issues and to disseminate water data and information has now expanded into a larger endeavor. In recent years, attention to water problems has grown as our understanding of the connections between this vital resource and human and ecological health has improved. Now, more than ever, policy makers and the public need useful information and knowledge about both water problems and their potential solutions. The goal of *The World's Water* is to fill at least part of that need. In this volume, with new authors and new data, we continue to try to raise important questions and offer new insights and approaches.

In 2002, the world will come together to review progress since the Earth Summit in Rio de Janeiro in 1992, including improvements in protecting and managing our freshwater resources. In 2003, Japan will host the Third World Water Forum; Spain will host a major World Water Congress. Tens of millions of dollars will be spent on these efforts, and tens of thousands of people will participate. Among the fundamental questions they will be asking is whether all of the efforts of the past decade have produced any real progress. Has there been any? More than a billion people still lack access to adequate safe drinking water. Nearly two and a half billion people lack access to improved sanitation. Thousands still die every day from water-related diseases. New instances of water-related conflict and violence are reported weekly. Climate change and impacts on water seems to be accelerating.

Despite the long way to go, progress *is* being made. The world is refocusing attention on critical unmet water needs, and for the first time since the 1980s has set an explicit goal—to reduce by half the proportion of people unable to reach or afford safe drinking water. A new set of principles and criteria have been developed by the World Commission on Dams to guide the construction of dams and other water infrastructure. New voices are beginning to be heard, including those of business and labor, local communities and representatives of the environment, and new scientists from many disciplines. Even senior policy makers are taking note: more than 110 Ministers from around the world attended the Second World Water Forum in The Hague in 2000 and issued a joint statement. Another ministerial statement on water was issued at the International Freshwater Conference in Bonn in December 2001. (Both statements are reproduced in full in the In-Brief section of this book.) Such statements do not provide water to poor people or save ecosystems, but they do lay the foundation for real actions and educate policy makers about the need for such actions.

The three volumes of this series are meant to be viewed as a compendium—a consistent and evolving collection of discussions, information, and raw data on the changing nature of the world's water. Earlier volumes of *The World's Water* have reported on various aspects of the "changing paradigm" of water management, development, and planning. This volume goes further in the chapter by Gary Wolff and Peter Gleick, describing the transition in terms of a choice of paths we can take. People do not want to "use" water. They want to drink and bathe, swim, produce goods and services, grow food, and otherwise meet human needs and desires. Achieving these ends can be done in different ways, often with radically different implications for water. One path—what we call the "hard" path—relies almost exclusively on centralized infrastructure and decision making. The second path—the "soft" path—also relies on centralized infrastructure, but complements it with extensive investment in decentralized facilities, efficient technologies, and human capital. It strives to improve the overall productivity of water use rather than seek endless sources of new supply. It delivers diverse water services matched to the users' needs, and works with water users at local and community scales. And the soft path acknowledges the importance of protecting our natural environment and meeting basic human needs for water.

This volume also tackles perhaps the most controversial issue in the water community today: the "new economy of water." As policy makers have broadened their search for solutions to water problems, new efforts are underway to expand international trade in water and to encourage private companies to play more of a role in water provision and management. These efforts are enormously controversial. Proposals to "privatize" water systems and to include water resources in global trading agreements have led to growing disputes and even violence. Many unanswered questions remain about the true implications and consequences of treating water as an economic good and whether these new approaches can effectively, equitably, and adequately serve human and environmental needs. Should we price and sell a resource as fundamental and vital as water? Will private companies protect ecosystems, water quality, and access to water for the poor? There is little doubt that the headlong rush toward open trading markets and privatization of public water systems has failed to address important issues and concerns. We tackle these questions in Chapters 2 and 3 by Gleick, Wolff, Elizabeth L. Chalecki, and Rachel Reyes, and we offer some fundamental principles that we believe are necessary to prevent inequitable, uneconomic, and environmentally damaging privatization agreements.

Another theme running throughout the series is the issue of measuring water scarcity and other aspects of water problems. In this volume, Gleick, Chalecki, and Arlene Wong explore attempts to define different "indicators" and "indices" of water well-being. The use of water-related indicators to measure the vulnerability of water systems, the quality of human or ecosystem health, or the level of development, has greatly accelerated in recent years. In part, this is the result of growing interest in human development. It also results from a growing understanding of the limitations of the traditional measures, such as gross national product, used during the twentieth century to track quality of life. Recently, new indicators have been proposed to address water-related issues, such as water availability, access to clean water and safe sanitation, the time and effort required to collect domestic water, cost and price, quality, vulnerability of water systems to climate change, and more. Such efforts are useful but fraught with limitations: this chapter discusses appropriate design and development of such indices.

The connections between water and climate change were explored in the first volume. Since then, new reports from the Intergovernmental Panel on Climate Change and the National Assessment in the United States have raised even greater concerns about the likelihood and severity of the consequences of climate change for the world's water resources. In this volume, William Burns reviews a particular set of impacts—the consequences of climate changes for the vulnerable water resources of 30,000 islands in the Pacific Ocean—home to more than 7 million people and 22 separate political entities. While the popular press has focused on the threat of inundation of island coastal areas by rising sea levels, perhaps the most critical near- and long-term threat to these islands is the possible impacts of climate change on their freshwater quality and availability.

A critical aspect of sustainable water management and use must be the protection of natural ecosystems and their water resources. The earlier volumes laid out principles for ecosystem protection and described the efforts of some nations, such as South Africa, to legally commit water to the environment. In some places, efforts are now underway to restore natural ecosystems that have been destroyed or damaged by human activities and water withdrawals, such as in the Everglades, Florida, and the Sacramento–San Joaquin Delta in California. Other ecosystems are not as fortunate and remain under threat or seriously degraded. One class of threatened ecosystems is river deltas—around the world the rich marshes, wetlands, and ecosystems at the mouths of river systems are being threatened and destroyed by both water withdrawals and water contamination. Examples include the Nile, the Yellow, the Ganges-Brahamaputra, the Jordan, the Tigris-Euphrates, and many more. Chapter 6 by Michael Cohen describes one such threatened delta—at the mouth of the Colorado River shared by the United States and Mexico—and efforts by non-governmental organizations and government agencies to try to restore regular flows there.

The first book reviewed the state of the world's dams and challenged the old paradigm of relying on ever-larger numbers of dams to meet human needs. The second volume explored new efforts to take out or decommission dams that either no longer serve a useful purpose or have caused egregious ecological impacts. In November 2000, the World Commission on Dams, headquartered in South Africa, concluded a groundbreaking two-year study on the sustainability of large dams. The work of the Commission is considered by many to be the most comprehensive assessment of dams yet undertaken. In this volume Katherine Kao Cushing summarizes the Commission's objectives, work products, and processes, and describes the varied and controversial responses to the Commission's work. In a related "In Brief," Amar Mann describes one of the consequences of failing to consider all of the consequences of major dam projects—in this case the implications of the massive Southern Anatolia Project for the rich and varied archaeology of ancient Turkey.

In late October 2001, the Mars Odyssey spacecraft reached our nearest planetary neighbor and safely entered orbit. Within a very short period of time, its instruments reported strong evidence for the presence of Martian water resources. In "Water and Space," Chalecki offers us a summary of the fascinating search for water in outer space, including the moon, Mars, the moons of Jupiter, and far more distant places. The presence of water in space has important implications for the search for extraterrestrial life and provides a critical resource necessary for space exploration.

Some of the other information found in the first two volumes of *The World's Water* is again updated here. The Chronology on water-related conflicts has proven to be an invaluable resource for both researchers looking at the security risks of shared water resources and for policy makers seeking to reduce those risks. This Chronology has been modified and expanded, as has the section on water-related Internet sites. Both of these resources are also available at the web site associated with this book: http://www.worldwater.org.

Finally, we once again offer a substantial Data Section—designed to be a regular feature. Some of the data sets from the first volumes are updated here, such as the country data on water availability and water use and information on national and global access to basic water services such as clean drinking water and improved sanitation. A wide range of new data is added, however, including tables on water and economics, and data on dams worldwide. Downloadable selections from all three volumes of data tables are posted at http://www.worldwater.org. Thanks to all the readers who have offered feedback on these books—your suggestions and encouragement make it worth the effort.

Peter H. Gleick

The Soft Path for Water

Gary Wolff and Peter H. Gleick

> *. . . I shall be telling this with a sigh*
> *Somewhere ages and ages hence:*
> *Two roads diverged in a wood, and I,*
> *I took the one less traveled by,*
> *And that has made all the difference.*
> ROBERT FROST (1874–1963),
> "THE ROAD NOT TAKEN"

The world is in the midst of a major transition in the way we think about—and manage—our vital and limited freshwater resources. Earlier volumes of *The World's Water* have described different aspects of this transition as the "changing water paradigm." This chapter goes further, describing the transition in terms of a choice of paths we can take. People do not want to "use" water. People want to drink and bathe, swim, produce goods and services, grow food, and otherwise meet human needs and desires. Achieving these ends can be done in different ways, often with radically different implications for water.

There are two primary ways of meeting water-related needs, or more poetically, two paths. One path—the "hard" path—relies almost exclusively on centralized infrastructure and decision making: dams and reservoirs, pipelines and treatment plants, water departments and agencies. It delivers water, mostly of potable quality, and takes away wastewater. The second path—the "soft" path—may also rely on centralized infrastructure, but complements it with extensive investment in decentralized facilities, efficient technologies, and human capital.[1] It strives to improve the overall productivity of water use rather than seek endless sources of new supply. It delivers diverse water services matched to the users' needs and works with water users at local and community scales.

---

1. Amory Lovins (1977) originally coined the term "soft path" for energy use. We warmly acknowledge his paternity in the terminology and many of the concepts discussed here. The Rocky Mountain Institute's definition and discussion of the soft path for water is available at www.rmi.org/sitepages/pid278.php.

This chapter tells the tale of these paths up to the present. Decisions made today, and actions of future generations, will write the conclusion of the story.

Water is a critical and essential resource. In the past, water policy has typically revolved around the idea that regular additions to supply were the only viable options for meeting anticipated growth in population and the economy. This idea led to the construction of pipelines and aqueducts that bring water to many towns and cities and to the irrigation canals that bring water to dry but fertile soils. Once "easy" sources of raw water are captured, however, this path leads to more and more ambitious, intrusive, and capital-intensive projects that capture and store water far from where the water is needed, culminating in the massive water facilities that dominate parts of our landscape.

The traditional approach to water supply led to enormous benefits. The history of human civilization is intertwined with the history of the ways humans have learned to manipulate and use water resources. The earliest agricultural communities arose where crops could be grown with dependable rainfall and perennial rivers. Irrigation canals permitted greater crop production and longer growing seasons in dry areas, and sewer systems fostered larger population centers.

During the industrial revolution and population explosion of the nineteenth and twentieth centuries, the demand for water rose dramatically. Unprecedented construction of tens of thousands of monumental engineering projects designed to control floods, protect clean water supplies, and provide water for irrigation or hydropower brought great benefits to hundreds of millions of people. Thanks to improved sewer systems, cholera, typhoid, and other water-related diseases, once endemic throughout the world, have largely been conquered in the more industrialized nations. Vast cities, incapable of surviving on their local resources, have bloomed in the desert with water brought from hundreds and even thousands of miles away. Food production has kept pace with soaring populations largely because of the expansion of artificial irrigation systems that now produce 40 percent of the world's food. Nearly one-fifth of all of the electricity generated worldwide is produced by turbines spun by the power of falling water.

On the other hand, half the world's population still suffers with water services inferior to those available to the ancient Greeks and Romans. According to the World Health Organization's most recent study, more than 1 billion people lack access to clean drinking water, and nearly 2.5 billion people do not have improved sanitation services (World Health Organization 2000). Preventable water-related diseases kill an estimated 10,000 to 20,000 children each day, and the latest evidence suggests that we are falling behind in efforts to solve these problems. There were new, massive outbreaks of cholera in the 1990s in Latin America, Africa, and Asia. The number of cases of dengue fever—a mosquito-borne disease—doubled in Latin America between 1997 and 1999. Millions of people in Bangladesh and India are drinking water contaminated with arsenic. And population growth throughout the developing world is increasing the pressure on limited water supplies.

The effects of our water policies extend beyond jeopardizing human health. Tens of millions of people have been forced to move from their homes—often with little warning or compensation—to make way for the reservoirs behind dams. Certain irrigation practices degrade soil quality and reduce agricultural productivity, threatening to bring an end to the Green Revolution. Groundwater aquifers are being pumped down faster than they are naturally replenished in parts of India, China, the United States, and elsewhere. And disputes over shared water resources have led to violence and continue to raise local, national, and even international tensions (see the Water and Conflict Chronology in the Water Briefs section of this volume).

Negative impacts on natural habitat are also significant. More than 20 percent of all freshwater fish species are now threatened or endangered because dams and water withdrawals have destroyed the free-flowing river ecosystems where they thrive (Ricciardi and Rasmussen 1999). On the Columbia and Snake Rivers in the northwestern United States, 95 percent of the juvenile salmon trying to reach the ocean do not survive passage through the numerous dams and reservoirs that block their way. More than 900 dams that block almost all New England and European rivers keep Atlantic salmon from their spawning grounds, and their populations have fallen to less than 1 percent of historic levels. Perhaps most infamously, the Aral Sea in central Asia is disappearing because water from the Amu Darya and Syr Darya Rivers that once sustained it has been diverted to grow cotton. Twenty-four species of fish formerly found in the sea and nowhere else are now thought to be extinct.

A Better Way

In the twenty-first century we can no longer ignore these costs and concerns. The old water development path—successful as it was in some ways—is increasingly recognized as inadequate for the water challenges that face humanity. We must now find a new path with new discussions, ideas, and participants. In the 1990s, progress was made in building bridges among competing water interests and in expanding the venues for discussing and resolving disputes. In time, these efforts will lead to better ways of meeting diverse water needs in a sustainable and equitable manner. But the process of defining, developing, and implementing an alternative path has not yet been completed. Moreover, powerful groups in the water sector continue to claim that decentralized investments in capital and people cannot and will not be effective. They believe that bigger and bigger centralized facilities are just fine. They want to continue on the hard path, and they claim it is still the best way to meet global water needs.

We refer to the traditional path as the "hard path" and the newer, alternative path as the "soft path." The adjective *soft* refers to the nonstructural components of a comprehensive approach to sustainable water management and use, including equitable access to water, proper application and use of economics, incentives for efficient use, social objectives for water quality and delivery reliability, public participation in decision making, and more.

The soft path can be defined in terms of its differences from the hard path. The two paths differ in at least six ways:

1. The soft path redirects government agencies, private companies, and individuals to work to meet the water-related *needs* of people and businesses, rather than merely to supply water. For example, people want to be clean or to clean their clothes or produce certain goods and services using convenient, cost-effective, and socially acceptable means. They do not fundamentally care how much water is used, and may not care whether water is used at all (Box 1.1). Water utilities on the soft path work to identify and satisfy their customers' demands for water-based services. Since they are not concerned with selling water per se, promoting water-

Box 1.1 Meeting People's "Needs": How Much Water Is Really Needed?

People do not "need" water, except to drink and cook meals that maintain basic human health. People need services and goods such as food crops or waste disposal. Food production will always require some amount of water, but far less than is currently being used in most agricultural settings. Waste disposal does not require any water, although using some amount of water for this purpose may be appropriate.

Agriculture is the largest consumer of water worldwide: often 80 percent of water use in a country or region. The potential for water efficiency improvements from techniques such as furrow diking, land leveling, direct seeding, drip irrigation, micro-sprinklers, and water accounting is large. For example, micro-irrigation systems (primarily drip and micro-sprinklers) often achieve efficiencies in excess of 95 percent, as compared with flood irrigation efficiencies of 60 percent or less. As of 1991 (Postel 1997), however, only 0.7 percent of irrigated farmland worldwide was being micro-irrigated.

Toilet flushing is the largest indoor use of water in Western, nonconserving, single-family homes. New technology, however, can manage human wastes without water. Electrically mixed, heated, and ventilated composting toilets have no odors or insect problems and produce a finished compost that does not endanger public health (Del Porto and Steinfeld 1999). These devices safely and effectively biodegrade human wastes into water, carbon dioxide, and a soil-like residue. Although they use about 500 kilowatt-hours of electricity per year, they can displace an equal or greater amount of electricity currently used to deliver water and treat wastewater. At present, electric-assisted composting toilets cost much more than conventional flush toilets.

Whether water-efficient technology will become socially acceptable as it becomes economically competitive is an important question. Soft-path planners believe that farmers want to grow crops rather than use water and will implement any water-conserving technology that makes economic and social sense. Soft-path planners believe that most people want human wastes managed in a convenient, cost-effective, hygienic way, and that they will accept alternatives that use little or no water if these criteria are met. Hard-path planners assume that future choices will look like current choices, with very slow technological progress. This is because hard-path planners erroneously equate water supply with the underlying needs that must be satisfied.

use efficiency becomes an essential task rather than a way of responding to pressure from environmentalists. The hard path, in contrast, fosters organizations and solutions that make a profit or fulfill their public objectives by delivering water—and the more the better.

2. The soft path leads to water systems that supply water of various qualities, with higher quality water reserved for those uses that require higher quality. For example, storm runoff, gray water, and reclaimed wastewater are explicitly recognized as water supplies suited for landscape irrigation and other nonpotable uses. This is almost never the case in traditional water planning: all future water demand in urban areas is implicitly assumed to require potable water. This practice exaggerates the amount of water actually needed and inflates the overall cost of providing it. The soft path recognizes that single-pipe distribution networks and once-through consumptive-use appliances are no longer the only cost-effective and practical technologies. The hard path, in contrast, discounts new technology, and over-emphasizes the importance of economies of scale and the behavioral simplicity of one-pipe, one-quality-of-water, once-through patterns of use.

3. The soft path recognizes that investments in decentralized solutions can be just as cost-effective as investments in large, centralized options. For example, there is nothing inherently more reliable or cost-effective about providing irrigation water from centralized rather than decentralized rainwater capture and storage facilities, despite claims by hard-path advocates to the contrary. Decentralized investments are highly reliable when they include adequate investment in human capital, that is, in the people who use the facilities. And they can be cost-effective when the easiest opportunities for centralized rainwater capture and storage have been exhausted. In contrast, the hard path assumes that water users—even with extensive training and ongoing public education—are unable or unwilling to participate effectively in water-system management, operations, and maintenance.

4. The soft path requires water agency or company personnel to interact closely with water users and to effectively engage community groups in water management. Users need help determining how much water of various qualities they need, and neighbors may need to work together to capture low-cost opportunities (Box 1.2). In contrast, the hard path is governed by an engineering mentality that is accustomed to meeting generic needs.

5. The soft path recognizes that ecological health and the activities that depend on it (e.g., fishing, swimming, tourism, delivery of clean raw water to downstream users) are water-based services demanded, at least in part, by their customers, not just third parties. Water that is not abstracted, treated, and distributed is being used productively to meet these demands. Water is part of a natural infrastructure that stores and uses water in productive ways. The hard path, by ignoring this natural infrastructure, often *reduces* the amount and quality of water available for

Box 1.2 The Benefits of User Participation:
Condominial Sewers

Most cities in less-developed parts of the world are surrounded by neighborhoods that have minimal or no piped water or wastewater services. Hundreds of millions of people live in these peri-urban areas, often as squatters in great poverty but also as legal homeowners with modest income. Traditional sanitary sewers are composed of a lateral line from each home to a trunk sewer line in or along the street that eventually connects to large-diameter sewer mains that service relatively large areas. Although main and trunk lines are usually built and maintained by a government agency, lateral lines and in-house plumbing are usually built and maintained by each household.

Brazilian engineer Carlos de Melo, working with residents of peri-urban areas in northeast Brazil, developed an alternative sanitary sewer system that cost residents only about one-third as much as the traditional system. His innovation was to replace conventional, deep, individual-house lateral sewers with shallow laterals that provide service to a series of homes, often passing through back and side yards in ways that reduce the total length of sewer required. This innovation would not be possible without the cooperation of groups of neighbors who share each "condominial" sewer lateral.

Involving users of the system has had advantages other than cost. For example, the system has built-in incentives for maintenance. If one household drain blocks, neighbors quickly bring it to the attention of the household and the problem is fixed rapidly. Less solid waste is disposed inappropriately down the sewers because blockages have immediate effects in the neighborhood. And the operations and maintenance cost for the formal sewer agency declines. Condominial sewer systems—a soft-path wastewater technology made possible by water user cooperation—are increasingly being used around the world.

Sources: World Bank 1992, Wright 1997.

use. The hard path defines infrastructure as built structures, rather than separating it into built (gray) and natural (green) components.[2]

6. The soft path recognizes the complexity of water economics, including the power of economies of scope. The hard path looks at projects, revenues, and economies of scale. An economy of scope exists when a combined decision-making process would allow specific services to be delivered at lower cost than would result from separate decision-making

---

2. "Gray infrastructure" is a term apparently coined by Lovins, though Falkenmark uses "blue" and "green" to refer to separate human and natural water uses.

processes. For example, water suppliers, flood control districts, and land-use authorities (e.g., local government) can often reduce the total cost of services to their customers by accounting for the interactions that none of the authorities can account for alone. This requires thinking about land-use patterns, flood control, and water demands in an integrated, not isolated, way.

Dominance of the Hard Path in the Twentieth Century

The water-supply hard path has dominated water development in the twentieth century, at least in part for the legitimate reasons presented below. But rationales often outlive their usefulness. Furthermore, the choice of development paths is not an absolute choice between hard and soft paths. It is a choice about the path of water development after a basic built water infrastructure has been provided. Should we try to supply more and more water, or is it time to shift our focus from new physical supply to reassessing how fixed supplies of water can be better used to meet ongoing water-related needs? Should water managers stick with the kinds of projects and techniques they know and continue to fail to meet the water-related needs of some people and many ecosystems? Or is it time to emphasize new approaches that seem more likely to meet these needs?

Shared Water Sources versus Individual Water Use

Natural *sources* of water are typically shared. A "water hole" for every household is not realistic in most situations. Even when groundwater is the primary source of water supply and each household has a well, use or abuse of underground water by each household can affect neighboring households. Some degree of agreement and cooperation among users of a shared resource is desirable unless water is abundant. As a result, over hundreds and even thousands of years, people have created social, cultural, and legal rules and built facilities that govern and physically manage shared sources of water.

In contrast, *end use* of water is not shared. There is no compelling reason for users to discuss or improve end-use practices when water is abundant. Even when water is not abundant, end-use practices tend to be viewed as personal rather than community issues. As long as water is allocated in a way that is acceptable to the group, why should the group care how water is used? The belief that the customer is always right once water has been delivered to them—even when the customer is ignorant or wasteful[3]—is deeply embedded in the hard-path paradigm. The soft-path paradigm is emerging only as more efficient water use by individual customers is recognized to be a legitimate social concern because of the growing social (and environmental, economic, and cultural) costs of the failure to be efficient.

---

3. Water law in California and other places requires (in theory) that water use be reasonable and beneficial; that is, extremely wasteful practices are prohibited by society. But other than extreme situations, the difference between beneficial use and waste is an individual—not social—choice under current laws and customs in most parts of the world.

Economies of Scale in Collection and Distribution

Another reason for the dominance of the hard path is the economy of scale that often exists—or is believed to exist[4]—in water collection (e.g., reservoirs), treatment, and distribution systems. Each person carrying water from a well or river takes much more time and effort than building, operating, and maintaining a shared system of water pipes. And once a system is constructed, devices that increase use efficiency are of little value so long as water is abundant. In such circumstances, the marginal cost of water from the piped system is much lower than the marginal cost of water conservation efforts.

Economies of scale exist by definition when the cost per unit of water declines as the total amount of water that is managed increases. When shared water sources are difficult to expand further, the cost per unit of water will increase if the system is expanded. Once economies of scale have been exhausted, the marginal cost (or cost of each additional liter) of water from piped systems will (sooner or later) become higher than the marginal cost of water conservation efforts. For example, low-flow toilets were not a difficult technological advance and are inexpensive to purchase and install. But they became widely cost-effective in the United States only after the seemingly cheapest sources of surface and groundwater had been exploited. This is why the emergence of the soft water development path is an appropriate response to history as well as a change from historical practices.

Simple End-Use Technology When Water or Energy Is Abundant

Wherever fresh water was abundant historically, end-use technologies were simple. Washbasins, with or without running water, or pipes located at the proper height over well-drained surfaces were adequate for drinking, cooking, bathing, and clothes washing. Machinery that used water—such as electrically powered clothes washers—were designed much later to replace human labor, not to use water more efficiently. Sophisticated and technically efficient water-measurement and -use devices, a key component of soft-path water systems, were not necessary and did not develop as rapidly as did water collection and distribution technologies (e.g., dams, pipes, pumps).

Where water was not abundant, numerous soft-path techniques were developed. The choice of crops is a good example: people need nutritional food but can choose among a variety of crops to meet that need. Olives are an important part of Middle Eastern and Mediterranean cultures because they are well adapted to semi-arid regions. Rice was not grown in most arid regions prior to the availability of inexpensive

---

4. Conventional accounting is often criticized for failing to include social and environmental costs that do not require money expenditures by the project developer. For example, the cost of the Aswan High Dam did not include many environmental and social costs that were identified after the dam began operation. Consequently, we caution readers to carefully examine claimed economies of scale. Some economies of scale are real; others are not real but seem to be due to a poor choice of analytical method. It is also useful to think of the size of the "gap" between narrow and broader cost estimates as depending on the knowledge level and relative political power of those who bear the "unaccounted for" costs. Economist James Boyce points out that unaccounted for environmental and social costs (so-called external costs) will be larger, in theory, when the costs are imposed on as yet unborn future generations, living people who are unaware of the costs being imposed on them, or living people who are aware but are not politically powerful enough to force a fuller cost accounting.

energy, unless farmland was located downhill from water sources. Both crop choices are examples of simple end-use "technology" decisions when water is scarce and energy to bring water from distant places is expensive.

Myths about the Soft Path

Because the emergence of the soft path diminishes the power and influence of entrenched interests, there is resistance to it among practitioners of hard-path planning. This has led to myths—widely believed falsehoods—about the soft path. The next section explains and debunks the six most common myths about the soft path.

Myth: Efficiency Opportunities Are Small

Traditional water planners and managers often believe that opportunities for improving water-use efficiency are small. A side effect of the initial focus on structures to capture and deliver water was that water planning agencies and companies came to be dominated by engineers.[5] Engineers are understandably very cautious in their decision making and designs. After all, lives and livelihoods often depend on the engineer being right. But because efficiency improvements depend on the behavior of water users, rather than professional personnel of the water agency or company, engineers often have little experience with the potential for efficiency improvements. Even when improved end-use devices are involved, such as low-flow toilets, many water agencies and companies believe that reduced water use due to installation of such devices will be relatively small or transient. Without professional supervision, so the thinking goes, these devices will break down over time and use more water than when new, or customers may defeat the water-saving potential through misuse (e.g., double flushing).

Myth: Water Demand Is Relatively Unaffected by Market Forces

Water demand has traditionally been projected, erroneously, as independent of costs, prices, subsidy considerations, and market forces. This means that the price elasticity of demand has been implicitly assumed to be equal to zero.[6] Given this mindset, it is not surprising that most historical estimates of future water demand have greatly over-estimated actual demand (Figure 1.1). In places where an effort has been made to identify and capture improvements in water-use efficiency, water demands have been cut by 20, 30, 40 percent, or more (Gleick et al. forthcoming).

---

5. On a personal note, both authors of this chapter have formal engineering degrees, and one (Wolff) is a licensed professional engineer.

6. Because water is an essential and basic good, some argue that the price elasticity of water demand is small. That is the case in some instances, or over short time periods; but even then the price elasticity is certainly not zero. On the other hand, there are numerous examples of water users responding to price increases by significantly changing their behavior or investing in or inventing devices that reduce water use dramatically without a reduction in the final, water-based service that is desired. The price elasticity of water demand is high in many instances or over long enough time periods.

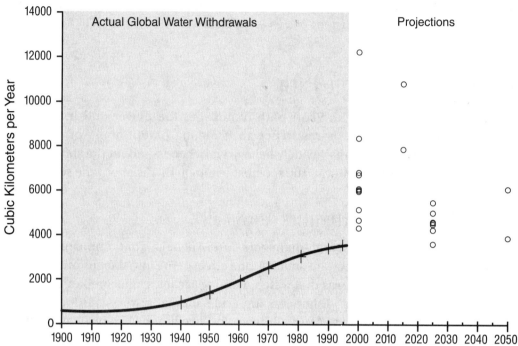

FIGURE 1.1 Many "projections" have been made of future water use. This figure shows actual global water withdrawals and projections made over the past 40 years. Almost all projections overestimated future water use—often substantially.
Source: Gleick 2000

Myth: Conserved Water Is Not "Real"

Another result of the difference in mindset between the hard and soft paradigms is the fundamental misunderstanding—or myth—that efficiency improvements involving nonconsumptive water uses are not real. Consumptive uses of water prevent that water from being reused. These include evaporative loss, contamination that cannot be removed by treatment, or discharge to a salt sink such as the ocean. Hard-path planners understand that greater efficiency in consumptive uses makes real, physical water available to other users. But they have been confused about nonconsumptive uses of water: uses that permit reuse (e.g., showers or runoff water from excess irrigation).

Consequently, a considerable number of confusing terms have appeared in the water conservation literature in the past few years. Among the terms used to describe the "kinds" of water that might be saved are *real* water, *paper* water, *applied* water, and *new* water. Some of these distinctions have been valuable in identifying where and when conservation is most beneficial, allowing planners to focus on the improvements in water-use efficiency that are the most appropriate and valuable (Keller and Keller 1995, Molden 1997, Seckler 1996, Gleick and Haasz 1998). Some of these terminological distinctions, however, have been misleading or have misrepresented the reality of efficiency improvements.

For example, water accounts for the Nile River (Molden 1997) indicate that only 20–30 percent of irrigation water diverted from the Nile is evaporated or transpired by crops. The remainder is return water that is reused downstream. Some have claimed that upstream reductions in nonconsumptive use are relatively unimportant because water loss is not the same as water waste (Palacios-Velez 1994). Of course this is strictly correct: irrigation return flows have not been "wasted." But upstream conservation would allow a larger upstream area to be irrigated, or would improve water quality for downstream natural systems or irrigators, or might improve the navigability of the Nile. These types of benefits need to be weighed against the costs of conserving water that is nonconsumptively used at present.

The California Department of Water Resources (CDWR) made a similar claim in the water plans prepared in the 1990s: upstream reductions in nonconsumptive use do not produce "new" water (CDWR 1998). Sacramento, under this line of thought, should not bother to install residential water meters or retrofit houses with low-flow showerheads or toilets since water conserved in Sacramento will be used downstream anyway. This error led state water planners to ignore improvements in urban water-use efficiency in inland regions (CDWR 1998). That, in turn, led to an overestimate of future urban demand in California in 2020 by more than a billion cubic meters (Gleick and Haasz 1998).

Of course, CDWR was correct in saying that "new" water was not produced. But the water savings from efficiency improvements in Sacramento would be real, nonetheless, because they allow new customers and new demands in Sacramento to be supplied with real water without taking more water from nature, or they allow higher quality flows of water for some distance downstream of Sacramento. Higher quality downstream flows would enhance natural systems that provide recreation, fishing, tourism, and other benefits. Whether the conserved water is "new" or not is irrelevant to the key issue: Does conservation of nonconsumptively used water create benefits that exceed the costs of conservation?

Myth: Efficiency Improvements Are Speculative and Risky

Even when hard-path water planners believe that the technical potential for water efficiency improvements is large, and the water saved is "real," they also tend to believe that relying on efficiency improvements is risky. Traditional water planners are happy to base water policy on assumptions about the probability a that there will be b people in the future using c amounts of water per person requiring d new reservoirs and treatment plants. Yet the same planners have historically rejected as speculation that economic or technical strategy w will lead to the installation of x high-efficiency toilets and showerheads that each save y cubic meters of water capable in total of eliminating the need for z new reservoirs and treatment plants. Hard-path water planners argue that our well-being and economic health rely on proven approaches rather than speculative policies that may not work.

Of course there is a germ of truth in the fear that decentralized efforts to curb demand may be unreliable. Left entirely alone, water customers may fail to notice that efficient water-using devices are no longer operating efficiently or may fail to continue their initially water-conscious and efficient behavior. One need only think about how often lights are left on in empty rooms when the public thinks that electricity is inexpensive and abundant. On the other hand, aggregate water-use patterns—such as seasonal and daily variations—are so predictable and reliable that water planners use them routinely. One wonders if rigorous statistical analysis would find that conservation is any less reliable than, for example, surface water supply in areas that experience periodic droughts.[7]

Furthermore, reliable and permanent efficiency improvements are achievable. They require training, public education, and on-going feedback to customers from the water supplier. Just as centralized water systems are unreliable when they are improperly designed, operated, or maintained, decentralized systems will be unreliable unless water users are reminded to behave in a consistent way and are rewarded or punished for failure to do so.

Achieving reliable reductions in water demand requires new "soft-path customer relations." Customers will need more information than in the past. For example, they need to know when their water use is higher than the average for similar users (e.g., single-family residences with similar floor size) or similar functions (e.g., irrigating a 100-square-meter garden). Whether their usage is higher or lower than average, they also need to be told when water-efficiency investments are financially desirable for them to make, taking into account savings other than on their water bill (e.g., energy bill savings from heating less water). Some customer classes (e.g., small businesses) may need help financing or obtaining credit with which to finance such investments.

Just as fire departments reorganized dramatically—becoming more specialized and professional—as buildings increased in height, the emergence of water scarcity is not just causing changes in water-use patterns (e.g., precision irrigation) but also causing water suppliers to rethink their roles and objectives. If supplying physical water is the only objective, demand management will seem speculative and unreliable. But when working closely with customers to meet their water-related needs becomes an equally important objective, demand management will be seen as a reliable way to stretch existing water supply.

Myth: Efficiency Improvements Are Not Cost-Effective

As mentioned above, centralized water facilities were historically lower cost—within a somewhat narrow accounting methodology—than decentralized investments in efficiency. This may still be the case in some circumstances. But the belief that efficiency improvements are more expensive than new or expanded centralized water supply is incorrect in most circumstances.

On the water-supply side of the economic comparison, recent centralized projects have cost considerably more than initially envisioned (Box 1.3) even without accounting for the environmental and social costs of new dams and reservoirs, which are usually neglected. Dams and reservoirs routinely end up costing two, three, or even more times their originally projected capital costs, and new facilities are many times more expensive than the first ones built in a region or watershed.

On the conservation side of the economic comparison, the cost of efficiency improvements has trended downward as technologies mature and comprehensive efforts are implemented. For example, some brands of early low-flow toilets (e.g., 1.6 gallons per flush), as well as a few still sold today, occasionally require double-flushing to provide equivalent service to older, more water-intensive toilets (e.g., 3.5 or 6 gallons per flush models). But the best low-flow toilets no longer have this problem, and they

---

7. We credit Bob Wilkinson for this penetrating question.

Box 1.3   Perennial Cost Overruns on the Hard Path

There are many anecdotal and documented stories of water-supply project cost overruns. According to international data, however, cost overruns are the rule rather than the exception. World Bank statistics on dam construction projects over the last four decades suggest that construction cost overruns averaged 30 percent on the 70 hydropower projects funded by the Bank since the 1960s. Another World Bank study found that three-quarters of the 80 hydro projects completed in the 1970s and 1980s had costs in excess of their budgets, and almost one-third of the projects studied had actual costs that exceeded estimates by 50 percent or more.

There are many reasons for cost overruns, including delays, design errors, poor quality construction, or corruption of project advocates and managers. For example, the Chixoy Dam in Guatemala was delayed for nine years by the collapse of some poorly designed tunnels. Its final cost of $1.2 billion was more than five times its initial cost estimate. The Yacyreta Dam on the Parana River between Argentina and Paraguay became known as a "monument to corruption" as the cost of the project increased to $8 billion from an original estimate of $1.6 billion. Cost overruns are not restricted to projects in less-developed countries: the recently completed Diamond Valley Reservoir in southern California had actual costs considerably higher than originally estimated. And last year, an economist at the U.S. Army Corps of Engineers blew the whistle on biased cost-benefit studies for projects along the Missouri and Mississippi Rivers, leading to calls for reform of the cost estimating and evaluation process.

Sources: World Commission on Dams 2000, International Rivers Network 2001

---

cost no more on average than inadequate low-flow models or the older wasteful models. As water suppliers have learned which models reliably save water, and provided this information to their customers, the cost of conserving water via low-flow toilet installation has fallen. This trend will continue as the technology for pressure-flush toilets (around 0.6 gallons per flush) and waterless toilets and urinals improves.

Another reason for the belief that efficiency improvements are too expensive is that hard-path water planners usually compare the cost of efficiency improvements without accounting for secondary benefits. Box 1.4 illustrates this problem with some common indoor residential water- and energy-using appliances. Without accounting for the avoided energy expense for water heating, only toilets and showerheads are cost-effective compared with water-supply augmentation at $325 or more per acre foot. After accounting for the energy benefit, washing machines and dishwashers are also highly cost-effective, even if new water supply could be obtained for free.[8]

---

8. This analysis was done assuming natural gas costs, but similar or better results would be obtained for electric water heating or other sources of heat.

Benefits of water-use efficiency, other than reduced water or energy bills, that are increasingly being discussed (Dziegielewski 1999) or quantified include the following:

- Reductions in *peak water system loads*. Peak loads determine the size of capital facilities required, hence capital costs. Lower peak loads mean

---

Box 1.4 The Importance of Secondary Benefits

One can estimate the cost of conserved water by dividing the amortized *additional* cost of a new high-efficiency water-using device (e.g., a new showerhead) by the average annual reduction in water use that the device will produce over its service lifetime, as compared with an average "conventional" water-using device. When high-efficiency devices are installed only when older devices would be replaced naturally (due to wear or remodeling), or are installed during new construction, the cost of installation is not part of the *additional* cost of high-efficiency devices. This is called "natural replacement." The "amortization divided by water savings" method, however, assumes that operation and maintenance costs other than water expenses are the same for both devices. This is a reasonable assumption for some devices, for example, toilets.

It is unreasonable for devices that use heated water. The following comparison of the cost of conserved water from four indoor residential water-using devices shows how important accounting for just one secondary benefit can be for water users in California. The only benefit other than water savings that we quantified was the reduced energy bill that results from using less heated water, assuming that natural gas is used for water heating. Electric water heating would only make these efficiency improvements more attractive. The negative cost estimates mean that the annual financial benefits other than reduced water expenses would justify the conservation investment *even if water were free.*

Device Installed (Natural Replacement)	Cost of Conserved Water (in Dollars per Acre-Foot)	
	Without Energy Benefit	**With Energy Benefit**
1.6 gallon per flush toilet	$45	$45
2.5 gallon per minute showerhead	$324	-$736
Average high-efficiency clothes washer (includes vertical and horizontal axis machines)	$865	-$177
Average high-efficiency dishwasher	$862	-$102

Source: Gleick et al. forthcoming

that existing capital facilities can serve more customers, avoiding or reducing the expense of these facilities.

- Reductions in *peak energy demands.* Energy and water-supply networks are similar in many ways. Reduction of peak energy demand caused by less water pumping, treatment, or heating will similarly allow energy utilities to serve more customers with existing capital facilities, avoiding or reducing capital expenses that are ultimately paid by energy purchasers.

- Reductions in *wastewater treatment expenses,* both operational and for expansion of existing sewers or treatment facilities.

- Reductions in *environmental damage* from water withdrawals or waste-water discharges in environmentally sensitive locations.

- Increases in *employment;* for example, by increasing the rate at which appliances or irrigation systems are monitored, serviced, or replaced. Investments in large, centralized capital facilities increase employment during construction but use relatively little labor once construction is complete.

The myth that efficiency improvements are not economically competitive with expansion of centralized supplies is slowly being overcome. Water planners are beginning to realize that the cost escalation, construction delays, and interest charges that so often plague large capital-intensive water projects do not have the same effect on conservation programs. It is an inherent characteristic of small-scale water-efficiency efforts that their lead times are substantially shorter than those of conventional big systems. This has been seen over and over again: for example, in Santa Barbara, California, a severe drought in the late 1970s stimulated local residents to support the construction of a large desalination plant, as well as a pipeline to connect to the centralized state water project. When the very high economic costs of those hard-path options were passed on to consumers, reductions in demand occurred so fast that the hard-path options became unnecessary. The desalination plant has not been put into routine operation and is currently in the process of being "mothballed" (partially decommissioned). If effective pricing programs, education, and community planning had been done first, the expense of these facilities could have been long delayed, perhaps completely avoided.

Whether in development, distribution, installation, or repair, producing small and technically simple systems is faster than designing, permitting, financing, and constructing large-scale reservoirs. As Lovins (1977) noted for the energy industry, the industrial dynamics of the soft path are very different from the hard path, the technical risks are smaller, and the dollars invested are far more diversified, reducing financial risk.

Myth: Demand Management Is Too Complicated

One of the reasons that *efficiency approaches* are difficult for traditional water agencies to adopt is that they *shift the burden from engineering logistics to social ones.* Traditional water agencies are usually dominated by engineering experts who know how to design and build large structures that can serve a million people. But these same experts are unfamiliar with methods for designing and implementing efficiency programs that reach a million individual customers. It is not surprising that working with individual customers and coordinating among many customers seems too complicated to engineers. These tasks are complicated, but no more so than engineering projects. But different professional skills and training are required, so demand management appears too complicated when one is not trained in people management, human capital investment (e.g., educators), and related disciplines.

It is also true that the types of information and technical assistance that water suppliers must offer to customers on the soft path would not have been as manageable a decade or two ago. There have been rapid advances in information processing, water-use monitoring, and more efficient end-use devices and practices. Even those who are skilled practitioners within the field may find it difficult to keep up with the newest information. This reinforces the perception that demand management is too complicated. As the demand management field matures, however, this perception—and myth—will weaken. Indeed, growing numbers of water agencies now have water conservation departments, and professional societies are adding water conservation experts and groups.

One Dimension of the Soft Path: Efficiency of Use

In this section we describe the water-use efficiency dimension of the soft path. Other dimensions of the soft path (e.g., matching water quality with type of need, public participation, etc.) are not described in detail in this chapter due to space limitations. Lens, Zeeman, and Lettinga (2001) and Wright (1997) are excellent introductions to the wastewater dimension of the soft water development path. Gleick et al. (1995) present a detailed quantitative assessment of future soft-path water development in California, and "Moving Toward a Sustainable Vision for the Earth's Fresh Water," Chapter 7 of *The World's Water 1998–1999,* included a qualitative global vision of a sustainable water path. We also expect to revisit this issue in future volumes of this biennial report.

Definitions and Concepts

The concept of integrating nonstructural water-management approaches into water planning goes back many decades. In 1950, the Water Resources Policy Commission of the United States published *A Water Policy for the American People,* which noted:

> We can no longer be wasteful and careless in our attitude towards our water resources. Not only in the West, where the crucial value of water has long been recognized, but in every part of the country, we must manage and conserve water if we are to make the best use of it for future development. (WRPC 1950)

In the early 1960s, Gilbert White called for broadening the range of alternatives examined by water managers who had previously only focused on structural solutions to water problems (White 1961). Under White's approach, managers should consider both structural and nonstructural alternatives, including zoning, land-use planning, and changing water-use patterns. Unfortunately, traditional water management has, in general, continued to concentrate heavily on the construction of physical infrastructure.

One of the first challenges along the soft path is to define conservation and water-use efficiency. Baumann et al. (1980) defined water conservation using a benefit-cost approach: "the socially beneficial reduction of water use or water loss." Under this definition, water conservation involves trade-offs between the benefits and costs of water-management options. This leads to economically efficient outcomes, by finding the level of conservation where the incremental cost of demand reduction is the same as the incremental cost of supply augmentation, taking into account all costs of water conservation and supply augmentation, including environmental and social costs. The advantage of this definition is that it focuses on comprehensive demand and supply management with the goal of increasing overall well-being, not curtailing water use.

In contrast, the term *water conservation* is presently used in most instances to refer to reducing water use by any amount or any means. *Technical efficiency* is a measure of water conservation: how much water is actually used for a specific purpose compared to the minimum amount necessary to satisfy that purpose.[9] *Water-use efficiency* is synonymous with technical efficiency. Under these definitions, the theoretical maximum water-use efficiency occurs when society actually uses the minimum amount of water necessary to do something. In reality, however, this theoretical maximum efficiency is rarely, if ever, achieved because the technology is not available or commercialized, because the economic cost is too high, or because societal or cultural preferences rule out particular approaches.

While technical efficiency and water-use efficiency can be useful concepts, they offer little guidance as to how much reduction in water use is enough (Dziegielewski 1999). In theory, a society could conserve too much water, expending some resources on water conservation that would be better spent on other goods or services. Consequently, the best use for numerical measures of technical efficiency is for comparison over time or between locations. Just as the speedometer in a car tells the driver how fast he is going, but not whether that is too fast or too slow, measures of technical efficiency help to know where we are relative to a social objective but are not adequate to establish social objectives.

Establishing social objectives requires a balance between costs and benefits, broadly defined. Because all costs and benefits cannot be quantified in most cases, social objectives for water conservation and efficiency improvements are best made through a democratic political process. Box 1.5 presents five terms that are useful when social objectives are being discussed and selected.

Finally, the concepts of *water productivity* and *water intensity* are also useful to soft-path planners. Unlike water efficiency, which is a percentage, water productivity is the amount of measurable output per unit of water that is used. The units of output can be physical (e.g., tons of wheat) or economic (e.g., the dollar value of the good or service produced). Figure 1.2 shows water productivity for the U.S. economy from 1900 through 1996. Productivity was relatively constant until the 1970s, when a combination of factors (such as rising environmental awareness, advances in technology, and the shift toward a service economy) caused water productivity to rise steadily.

Water intensity is the inverse of water productivity: the amount of water needed to produce a unit of output. For example, the 1996 data point in Figure 1.2 shows U.S.

---

9. The numerical measure of technical efficiency is calculated by dividing minimum use by actual use. Since actual use is larger than or equal to minimum use, reducing actual use will increase the ratio up to a maximum of 1.0 (100 percent efficiency). This measure has one mathematical oddity, however. When the minimum use is zero, the ratio will always be 0 (0 percent efficiency), no matter what actual water use is! Should actual water use fall to zero, the ratio will become undefined, since 0 divided by 0 is undefined.

Box 1.5   Some Water Efficiency Definitions

Best available technology (BAT): The most water-efficient, technically proven, commercially available technology for reducing water use. A good example is the electric-assisted composting toilet, capable of meeting all disposal needs without the use of water. These toilets are proven and commercially available. BAT is useful for quantifying a maximum savings *technically* available. This is an objective assessment of potential, independent of cost or social acceptability. Thus, the BAT for toilets uses no water.

Maximum available savings (MAS): For a given agency or region, MAS is an estimate of the maximum amount of water than can be saved under full implementation of BAT, independent of current costs.

Best practical technology (BPT): The best technology available for reducing water use that meets current legislative and societal norms. This definition involves subjective judgments of social acceptability and will change over time, but it defines a more realistic estimate of maximum practical technical potential, independent of cost. For example, the current BPT for toilets in the United States is the 1.6-gallon-per-flush ultra-low-flow toilet, though this value can continue to decrease over time.

Maximum practical savings (MPS): For a given agency or region, MPS is an estimate of the maximum amount of water that can be saved under full implementation of BPT, independent of current costs.

Maximum cost-effective savings (MCES): For a given agency or region, the MCES is the maximum amount of water conservation where the marginal cost of conservation is less than or equal to the marginal cost of developing new water supply. Any lesser amount of conservation will force customers to pay more for water-related services than they would pay with a little more conservation and a little less water.

---

water productivity to be $13.85 of gross national product per cubic meter of water. Therefore, overall water intensity in the U.S. economy in 1996 was about 0.07 cubic meters of water per dollar of gross national product (1/13.85 = 0.0722). Increases in water productivity imply decreases in water intensity.

Water productivity and intensity can be measured in a variety of ways. Figure 1.2 uses data on total U.S. water withdrawals from natural systems. These data capture both direct and indirect (embodied) uses of water.[10] *Direct use* is water used at the point of production. *Indirect use* is water needed to produce nonwater inputs used at the point of production. For example, the direct water use in producing a box of

---

10. Because some imports are used to produce economic output in the United States, some amount of indirect (embodied) water use is not accounted for in the Figure. However, we believe that the contribution of imports to overall water demands is very modest.

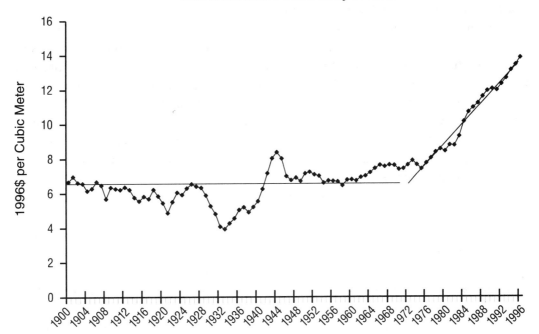

FIGURE 1.2 The economic "productivity" of water in the United States was relatively constant between 1900 and 1970 at around $6.50 of gross domestic product (GDP) per cubic meter of water withdrawn. After 1970, however, the economic productivity of water began to rise dramatically and now approaches $15 per cubic meter. All dollar figures are in constant 1996 dollars.
Source: Gleick 2000

breakfast cereal may be quite small, while the indirect water use embodied in the grains used to produce breakfast cereal is much larger.

This distinction means that measures of water productivity and water intensity can be based on direct water-use data or direct plus indirect water-use data. Direct water-use data, alone, may be misleading, as the breakfast cereal example shows. The best measures of water productivity or water intensity include both direct and indirect water use.

In the following section we provide some brief examples of the concepts and definitions. The examples only begin to demonstrate the numerous successes and opportunities along the soft water development path. Readers are referred to Gleick et al. (1995), Postel (1997, 1999), Vickers (1999), Owens-Viani, Wong, and Gleick (1999), Wilkinson (1999), and Vickers (2001) for additional or more comprehensive examples.

Efficiency improvements in agriculture. As noted previously, agriculture is the largest user of water worldwide: often 80 percent of water use in a country or region. Agriculture's place as the largest water-use sector implies that each 1 percent decline in agricultural water use reduces demand for water much more than each 1 percent decline in municipal, commercial, institutional, or industrial water use. For example, a 25 percent reduction in agricultural water use in an area where agricultural use is 80 percent of total use makes possible a doubling (100 percent increase) of water use in all other sectors combined.

The potential for water efficiency improvements in agriculture is large. Efficiency improving techniques include furrow diking, land leveling, direct seeding, drip irrigation, low-energy precision application (LEPA) sprinklers, low-pressure sprinklers, water accounting, and many others. Postel (1999) provides examples from around the world of the gains from these techniques.

One category of techniques—micro-irrigation systems (primarily drip and micro-sprinklers)—often achieves efficiencies in excess of 95 percent as compared with flood irrigation efficiencies of 60 percent or less, as noted in Box 1.1. Postel (1999) presents water productivity gains for a wide range of crops from shifting to drip from conventional surface irrigation in India. Although water use declined as much as 65 percent (sugarcane), water productivity increased as much as 255 percent (cotton). This is because more precise application of water can both reduce total water use and increase yields. Sugarcane and cotton yields increased 20 and 27 percent, respectively, along with substantial reductions in water use.

As of 1991, however, only 0.7 percent of irrigated farmland worldwide was being micro-irrigated (Postel 1997). While we expect this number has increased since 1991, more recent comprehensive data on global irrigation technology are not available. Israel and Cypress had demonstrated by 1991, however, that high percentages of irrigated land could be managed under these techniques (48.7 and 71.4 percent of the irrigated land in these countries, respectively). Because micro-irrigation has been expensive, historically, it has been used mostly on higher valued crops. But systems that cost only a fraction the cost of conventional precision systems are now available (Postel 1999), and will probably become even cheaper as time passes.

Another example is laser leveling of fields. This technique causes water to be distributed more uniformly, reducing the water required to ensure that all parts of the field are irrigated adequately. Recent experience growing wheat, alfalfa, and cotton in the Welton-Mohawk Valley of Arizona (Vickers 2001) found that water use declined between 20 and 32 percent as a result of laser leveling, and yields increased from 12 to 22 percent. This practice requires that land be leveled every two to five years at a relatively modest cost: about $40 per acre for each leveling.

Efficiency improvements at the municipal scale. A number of municipal water suppliers have implemented aggressive water conservation programs. Municipal conservation programs that are fully integrated have shown impressive successes. Box 1.6 lists elements of comprehensive programs. Postel (1997) includes an excellent summary of successful municipal programs in Jerusalem, Israel; Mexico City, Mexico; Los Angeles, California; Beijing, China; Singapore; Boston, Massachusetts; Waterloo, Canada; Bogor, Indonesia; and Melbourne, Australia. Reductions in water demand varied from 10 to 30 percent.

Vickers updates the results from the Massachusetts Water Resources Authority, serving the Boston area, and presents recent data for the city of Albuquerque, New Mexico (Vickers 1999). These municipalities have reported reductions of 25 and 18 percent, respectively. Similarly, Owens-Viani (1999b) presents results from the Marin Municipal Water District (MMWD) in northern California. The MMWD conservation management plan led to a reduction in demand of about 15 percent in the first 10 years of implementation, despite a 7.5 percent increase in the district's population. However, this is only about half of their 20-year target of up to a 32 percent reduction in absolute demand despite increases in population. After adjusting for population growth, this 20-year target—already half achieved—amounts to a reduction in water use of around 45 percent per capita.

Wilkinson (1999) provides a story that puts a human face, and some dollars and cents, on this type of soft-path success. The 815-acre University of California at Santa

---

Box 1.6 Elements of Integrated Water-Management Plans

Elements of fully integrated water-management plans can include

- monitoring and evaluation of water use;
- indoor and outdoor efficiency standards established and enforced under local ordinances;
- water budgets for nonresidential customers;
- conservation rate structures;
- a technical assistance program;
- landscape seminars;
- demonstration gardens and garden contests;
- rebates for decentralized investments that displace or delay centralized investments;
- ongoing public information campaign and school education program; and
- full economic integration of efficiency improvements efforts with development of new water supplies (including recycled water).

Barbara (UCSB) campus is like a small town, with its own fire department, police station, post office, medical clinic, food service facilities, offices, laboratories, housing for approximately 4,200 students and 150 faculty families, recreational and laundry facilities, and extensive landscaping. Total campus water use was reduced by nearly 50 percent between 1987 and 1994, even as the campus population served increased slightly. Total cost savings to the campus from 1987 to 1996 were $3.7 million, excluding energy and maintenance savings. Capital costs for water-efficiency measures during this period were estimated to be less than $1 million, including significant equipment replacement that would have taken place anyway for maintenance reasons.

A good example of the type of innovation that emerges on the soft path but is neglected on the hard path was the discovery that pool filter backwash water could be cost-effectively discharged into tank trucks that clean the sewer lines with high-pressure water. The UCSB program also demonstrates the human element of changing past practices. Maintenance staff, for example, claimed that sewers would back up if toilets were changed to 1.6 gallon per flush models, and that the landscape would have to become all cactus and rocks to reduce the quantity of irrigation water. The University responded by purchasing examples of nearly every available model of low-flow toilets and showerheads and asked skeptical maintenance staff to test them according to simple performance criteria. Preference tests were then conducted with campus residents to ensure satisfaction.

Installation was complemented by a strong educational program that communicated what was being done and why. Wilkinson notes that the direct involvement of maintenance staff and end users, and the pre-testing of technology, were major factors in the success of the program. Sewer problems actually decreased, in part because of an improved preventive maintenance program. And irrigation water savings were achieved without planting any cacti.

Efficiency improvements at businesses. There are many opportunities for commercial, industrial, and institutional (CII) customers to use water more effectively. Pike (1997) evaluates opportunities for commercial and institutional water users in the United States and finds that average potential savings vary from 9 to 31 percent within 18 categories of users (e.g., eating and drinking places, vehicle dealers and services). Similar statistics are present in Vickers (2001), including examples from outside the United States. Potential reductions in industrial water use are often larger. Studies in the San Francisco Bay Area and urban portions of southern California (EBMUD 1990; Brown and Caldwell 1990; ERI Services 1997; Gleick et al. 1995; Hagler Bailly Services 1997; Sweeten and Chaput 1997; U.S. EPA 1997; Wilkinson 1999; Owens-Viani 1999a, 1999b; Wilkinson, Wong, and Owens-Viani 1999) have found typical opportunities or actual successes (averaged by business or institution type) in the range of 16 to 54 percent.

Most of these opportunities are cost-effective and widely applicable. The Brown and Caldwell analysis, for example, found that typical reductions of 30–40 percent usually had estimated payback periods of less than one year and estimated average savings of $50,000 per year. The report concluded, "The cost effective water conservation measures successfully used at the case study facilities can readily be adopted by other facilities and other industries."

Conservation opportunities are most often expressed as percentages of pre-conservation water use, as in the studies above. Water-use efficiency measures, however, show how close or far current use is from minimum water use without loss of particular water-based services. They allow comparison of conservation successes since water-use efficiency is defined as the minimum water use necessary with an analyst-defined "benchmark" technology divided by actual water use. Finally, water productivity figures show how much water is used per unit of output. For example, Border Foods in New Mexico, one of the largest green chile and jalapeño pepper producers in the world, increased water productivity from about 1.4 pounds of product per gallon used to about 2 pounds per gallon in the 1992–95 time period (Vickers 2001).

The Oberti Olives processing plant in Madera, California, provides an excellent example of the alternative measures that can be used to evaluate efficiency changes (Owens-Viani 1999a). The Oberti plant processes 128 tons of olives per day, washing, curing, storing, and packaging. This amounts to about 600 cans of olives per minute. The plant also produces between 40,000 and 80,000 gallons of olive oil per year. Oberti obtains fresh water from private wells. Water use prior to recovery of in-plant wastewater for in-plant reuse was about 1.3 million gallons per day. After installation of a membrane filtration system in 1997, water use was reduced to an average of 110,000 gallons per day. Table 1.1 shows three different ways of measuring water-use improvements for the Oberti plant.

The estimates of water-use efficiency assume that membrane filtration is the best available technology (BAT) for water management at the Oberti plant. Recall that BAT is defined as the technology that would use the least water to perform a specific list of functions. "Best" in BAT is not necessarily best for the water user. For example, Oberti took action because it was required by the state of California to construct double-liners in its wastewater ponds or manage its wastewater differently. It investigated both evap-

TABLE 1.1   Three Measures of Progress in Water Use at the Oberti Olive Plant

	Before 1997	After 1997
Water conserved	Not applicable	A 91 percent reduction (1,190,000 gallons per day)
Water-use efficiency	8.5 percent	100 percent (*)
Water intensity	10,156 gallons per ton of product	859 gallons per ton of product

* Assuming new technology was "best available technology" or BAT.
Source: Owens-Viani 1999a

oration and membrane filtration systems for treating wastewater. The energy cost of the evaporation system caused it to be inferior to the membrane filtration system, but one can readily imagine that something other than membrane filtration might have been the best alternative for Oberti.

An oddity of the water-use efficiency measure, both before and after, is that it depends on the definition of BAT. Since BAT may change over time, water-use efficiency measures are useful only when BAT is defined and applied consistently across all users and times that are compared. Nonetheless, water-use efficiency is probably the most useful of the measures in identifying the possibility for future reductions in water use. Progress is not possible beyond 100 percent unless BAT improves.

Many water planners still believe that using less water somehow means a loss of prosperity. The traditional assumption, repeated over and over in water plans and discussions about the risk of future water shortages, is that continued improvements in well-being require continued increases in water use. This might be true in the absence of technological progress, but with technological improvements there is enormous room for economic growth without growth in water use. For example, producing a ton of steel before World War II required 60 to 100 tons of water. Today, each ton of steel can be produced with less than 6 tons of water: a ten-fold improvement in water productivity (and a ten-fold reduction in water intensity). Further, because a ton of aluminum can be produced using only 1.5 tons of water, replacing the use of steel with aluminum, as has been happening for many years in the automobile industry, can further lower water use without reducing economic activity (Gleick and Haasz 1998). And telecommuting from home can save the hundreds of liters of water required to produce, deliver, and sell a liter of gasoline, even accounting for the water required to manufacture computers and telephone systems.

The reality is that the link between water use and economic well-being (often measured as some form of gross domestic product) is not immutable. It can be modified and even broken, as it already has in the United States. This is shown clearly in Figure 1.3, which presents the data behind Figure 1.2 (water productivity in the United States) in another way. Figure 1.3 shows water withdrawals and gross domestic product in 1996 dollars for the United States from 1900 to 1996. From 1900 to 1980, these curves rose in lockstep—increases in national income were matched by similar increases in water withdrawals. Then, in 1980, this relationship was broken, with continued rapid increases in national income, but a leveling off—even a decrease—in total water withdrawals.

Similar patterns are emerging around the world. For example, Japan used nearly 50 million liters of water to produce a million dollars of commercial output in 1965; by 1989 this had dropped to 13 million liters per million real (inflation-adjusted) dollars of commercial output—almost a quadrupling of water productivity. Data from Hong Kong

U.S. GNP and Water Withdrawals

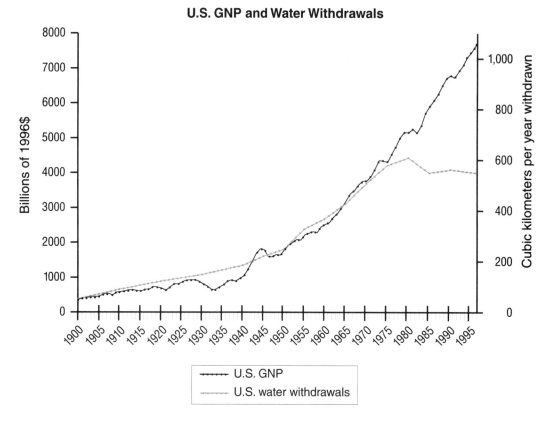

FIGURE 1.3 U.S. GNP (in billions of 1996 dollars) is plotted here with total U.S. water withdrawals (in cubic kilometers per year) from 1900 to 1998. Until the mid–1970s, these two curves rose exponentially together. By the late 1970s, however, total U.S. water withdrawals began to level off and then decline, even while total U.S. economic productivity grew.
Source: Gleick 2000

(Figure 1.4) exhibit this pattern as well. If data are available and progress has been sustained, similar figures can be developed for any successful conservation program, such as those described previously in Boston, Albuquerque, Marin County, and the University of California at Santa Barbara. Time-series data on water conservation, water-use efficiency, water productivity, and economic output versus water input are perspectives on a single phenomenon: the success of the soft path of water development.

An example of the MPS and MCES measures. Opportunities for further progress on the soft path are most practically identified by comparing two measures of efficiency potential: maximum practical savings (MPS) and maximum cost-effective savings (MCES). Figure 1.5 (Gleick et al. forthcoming) shows that the future potential is large even in a region and sector that has already conserved a considerable amount of water. The upper line in the figure is the water-use path of "business as usual" in the indoor residential sector of California through 2020. The hard path requires that the state develop new water supplies to meet this projected "need." The lower line in the figure is the MPS from implementation of the current best practical technologies (BPT) for indoor, residential water use. The lower line is also the MCES for indoor, residential water use. MCES is equal to MPS because we have estimated that all practical indoor residential conservation technologies are cost-effective to implement (see Box 1.4).

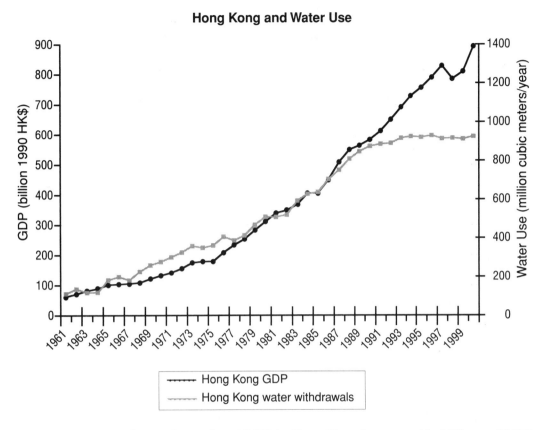

FIGURE 1.4 Gross domestic product (GDP) in Hong Kong (measured in billions of 1990 Hong Kong dollars) is plotted here with total Hong Kong water withdrawals (in cubic kilometers per year), from 1961 to 2000. Until the late 1980s, these two curves rose exponentially together. By the early 1990s, however, total water withdrawals in Hong Kong began to level off and then decline, even while total economic productivity grew. *Source:* David Yongqin Chen, Chinese University of Hong Kong, 2001

The soft path as shown moves California from the uppermost line to the lowermost line over a 20-year time period. While the hard path would require about a 45 percent increase (of 900,000 af/yr, or over 1.1 billion cubic meters per year) in water supply (for indoor residential purposes) by the year 2020, the soft path actually *reduces* total indoor residential demand by about 25 percent (that is, by about 500,000 af/yr, or over 600 million cubic meters) despite population growth. This means that the soft path for indoor residential use can cost-effectively conserve about 1,400,000 af/yr (1.7 billion cubic meters) by the year 2020, or around half the water hard-path planners would say is "needed" for indoor residential purposes by 2020. This simple example, for a single sector of California water use, shows the dramatic gains possible by shifting effort from the supply side to the demand side.

Moving Forward on the Soft Path

There are many pieces that must be put together to move along a soft water path. Below we describe four that we consider necessary for improving water-use efficiency,

**Indoor Residential Water Use in California with
Various Cost-Effective Actions**

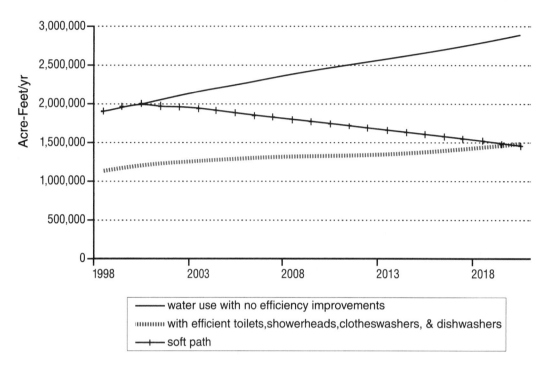

FIGURE 1.5   Nearly 2 million acre-feet of water is used in California homes today. This figure shows the amount of water residents would use with no improvement in the efficiency of toilets, washing machines, dishwashers, and showerheads (the solid line), the amount of water required if all homes were fitted with currently available, cost-effective efficient water fixtures (the lower broken line), and a "soft path" transition from inefficient use to efficient use.
Source: Pacific Institute 2002

but others are necessary for reforming institutions, opening up decision-making processes, and so on. We hope to elaborate on these in coming works.

The first step is identifying the physical and economic potential for improving water-use efficiency and allocation. The second is identifying the institutional, economic, and technological barriers that impede these improvements. The third is to make a social choice about how much efficiency is desirable. Finally, government agencies or water suppliers must implement comprehensive, integrated economic, educational, and regulatory policies that remove the barriers and achieve the socially desirable level of water savings.

Step 1. Identifying the Potential

Identifying the actual potential for improving water-use efficiency (however defined) requires that water planners ask and answer some basic questions:

   a. Who is going to require water?
   b. For what purpose or goal is water needed?
   c. What kind of water is necessary to meet specific goals?
   d. How much water of a particular quality is needed to meet any given goal?

Surprisingly, the first three questions have rarely been asked or answered with sufficient detail.

a. Who is going to require water? This question is typically addressed in a rudimentary way by identifying user categories such as urban and agricultural users. Urban users are often broken into residential, industrial, commercial, and institutional classes. But a detailed analysis of the diverse kinds of water users within these classes is rarely provided. Even more rare is inclusion of environmental or ecological users of *in situ* waters at or downstream of the sources of raw water.

b. For what purpose or goal is water needed? This question gets immediately to the heart of the hard-path/soft-path choice. Water is but a means to ends and is useful only insofar as it helps to meet those ends. What is desired is not to use a certain amount of water but to achieve certain goals and purposes: to remove wastes, produce objects, grow food, generate energy, provide recreation. Without understanding what we want to do, it is impossible to evaluate the water needed to accomplish our goals. And yet, no country, state, water district, or water company seems to have developed a complete inventory of the tasks required to meet the specific water-related needs or desires of their customers.

Answering this question helps to make explicit deeper factors that drive water use and supposed future "needs." Is using water to dispose of human wastes always necessary? Can a computer chip be made without water? Does it really make economic sense to grow cotton (or as much cotton) rather than some other crop in California or Egypt?

c. What kind of water is necessary to meet specific goals? Different water demands can be met with waters of differing quality. Traditionally in industrialized nations, water delivered to a home or business is treated to drinking-water standards. Only a tiny fraction of domestic and business water use, however, is used for drinking, cooking, or other health-sensitive functions. Landscape irrigation and car washes can use lower-quality water. Similarly, some high-technology businesses require water that is higher quality than drinking water. Until the pattern of quality needs is understood, one cannot say—as hard-path planners do—that single-pipe, single quality, once-through patterns of water distribution and use are the only economical patterns.

d. How much water of a particular quality is needed to meet any given goal? As we have noted earlier, getting rid of human wastes in toilets can take 23, or 13, or 6 liters of water, or even no water at all, depending on the technology used. Growing a ton of cotton can take 800, or 1,200, or even 3,000 tons of water per year, depending on the climate, soil, irrigation technology, crop variety, and efforts of the farmer. Extensive research and data are available telling us how much water is used to flush a toilet, or produce a computer chip, or grow cotton, but very little research has been done to evaluate the minimum amount of water required to do these things.

Step 2. Identifying Barriers

Data and information gaps are among the most common and frustrating barriers to identification of the potential for water-use efficiency. In many cases, even the most fundamental data do not exist. Most countries do not collect national water-use data. In other cases, the data that do exist are incomplete, inconsistent, or from poorly documented original sources. In order to make intelligent decisions, however, the right data and information are essential. For example, every municipal water agency in the world reports that more water is treated and put into their distribution systems than can be accounted for in water-use surveys or billings. Yet every traditional measure of aggregate water use comes from estimates of gross water withdrawal, not from point-of-use surveys.

A few representative examples of data or information gaps that would be useful to fill are:

- irrigated area, by crop type and irrigation method, by region, watershed, and country;
- irrigated and unirrigated residential landscape areas as a function of lot size, house size, income, climate zone, or other demographic data;
- the distribution of residential water-using appliances in cities or urban regions, by type and use (e.g., the distribution of various liters per flush toilets and other devices for management of human waste);
- regional measures of agricultural water use separated into evaporation and transpiration components;
- water delivered to categories of businesses with similar functions (e.g., all retail stores) (data are sometimes available from a water agency or company, but the data are rarely collated and summarized);
- industrial water use with detailed information on water required to produce categories of goods, measures of economic output, or employees; and
- summaries of the actual cost of conservation and efficiency measures in a variety of locales, estimated by a consistent methodology.

Great uncertainties still remain about the physical, practical, and economic potential of soft-path management techniques. The magnitude of this potential depends on water prices, rate designs and structures, existing and developing technology, public opinion and preferences, and policies pursued by water agencies and managers. Nonetheless, many of the uncertainties associated with water-use efficiency programs can be reduced with modest investments in data collection and analysis.

There are numerous other barriers to estimation of different measures of water-use efficiency, such as MAS, MPS, and MCES. For example, the attitudes described in the section on "Myths about the Soft Path" are barriers to estimating efficiency and conservation potential. These attitudinal barriers might be summarized as "lack of political or managerial will" to pursue the soft path. The failure of some California cities to fully implement residential metering of water use, despite information (available for decades) that shows that such metering can reduce water demand by 15 to 20 percent (Flack 1981), is evidence of the depth of the institutional and attitudinal barriers to the soft path.

Step 3. Making Social Choices

Many water policy makers believe that only a limited amount of conservation and efficiency is economical, despite a lack of essential information with which to test this belief (or even in the face of conflicting evidence), and that going beyond modest conservation objectives will lead to unacceptable changes in lifestyle or highly restrictive regulatory requirements. The twin demons of government intrusion in the market and social engineering are raised and reviled surprisingly often, despite the fact that most hard-path water projects backed by powerful constituencies have benefited from vast government subsidies and often lead to substantial impacts to local populations.

Rational discussion requires that these concerns be considered and analyzed. It is certainly true that the soft path is not easy to follow. It requires institutional changes, new management skills, and greater direct participation by water users. Fully branching off of the hard path onto the soft path is not easy, but it is increasingly apparent that continuing on the hard path as physical water scarcity increases causes equal or greater difficulties.

Because social choices that involve significant change are difficult to make and even harder to implement, Vickers (1999) concludes:

> No U.S. water supplier has yet to exploit water efficiency's full capabilities to optimize customer water demands. . . . Despite the great promise of water conservation to enable the public and nonresidential customers to live within their water means, in reality few utilities have made significant investments in conservation programs to make much of a difference.

This assessment is accurate, although some U.S. and international water suppliers (e.g., the examples provided above) seem to be moving toward this objective. As the soft path is investigated and matures, even greater levels of efficiency will become possible. The soft path requires a social choice to invest in the people, businesses, and cooperative arrangements that are needed for the maximum cost-effective water savings to become reality.

Step 4. Implementing Comprehensive Demand-Management Programs

Some examples of soft-path successes were provided above. But what does a water supplier need to do to implement comprehensive and successful demand-management programs? In essence, the supplier needs to fully integrate a variety of new tools with older tools. Successful water-use efficiency programs inevitably include combinations of regulations, economic incentives, technological changes, and public education. Considerable experience in every sector of the economy suggests that the most effective water-use efficiency programs include combinations of all of these approaches (Gleick et al. 1995; Owens-Viani, Wong, and Gleick 1999; Vickers 2001).

Regulatory tools include government policies to encourage water conservation and efficiency improvements, including appliance efficiency standards, landscape ordinances, and efficient building codes. Economic incentives include marginal cost volumetric water prices, rebates for water-saving end-use devices or practices, low-cost loans or assistance in obtaining credit for capital investments by customers, environmental fees or surcharges that allow compensation for those damaged by additional water withdrawals (e.g., fishermen), and trading of water rights or water-use permits.

Technological tools include all of the devices that use water more efficiently than in the past, and advanced media and informational techniques to communicate with and persuade water users to behave in ways that are socially desirable (i.e., achieve the social objectives chosen in each locale).

Education of the public means ensuring that information on options, costs, technology, and regulations are fully available to water users. Smart choices will only be made when those choices are known and understood.

Finally, unless demand management is fully integrated with water-supply planning, it will remain an underused and misunderstood part of our water future.

Conclusions

We have reached a fork in the road. We must now make a choice about which water path to take. We know where the hard path leads—to a diminished natural world, concentrated decision making, and higher economic costs. The soft path leads to more productive use of water, transparent and open decision making, and acceptance of the ecological values of water.

In most past water-planning efforts, demand has been projected independently of changes in technology, costs, prices, customer preferences, and market forces. It is not surprising that so many past demand projections greatly overestimated the level of demand that eventually materialized. This bias in demand projections has caused large capital-intensive supply augmentation projects to appear economically desirable. Actual experience, however, is that these projects have much larger costs than initially estimated. Improved end-use efficiency technologies are now, and are likely to remain, very cost-effective. That they are not widely recognized as such is a shame.

There is no single estimate of the potential for improving the efficiency or productivity of water use. Each situation comes with a different set of assumptions, social conditions, physical structures, technological opportunities and constraints, and costs. These characteristics will determine, ultimately, how much future demand for water can be reduced or modified, and how much reduction in future demands for water is socially desirable.

Examples of soft-path successes are becoming more common. Some of them have been touched on in this chapter, but there are many more. Despite this, efforts by water agencies to implement comprehensive water-efficiency programs are few and far between. Even in developed countries, where some of the best funded and most promising water conservation programs exist, the majority of urban water agencies rely on voluntary practices and programs that are not comprehensive, are incompletely implemented, and are inadequately monitored.

Soft-path water approaches are inherently more democratic than large centralized capital projects, and require a much wider range of professional skills to implement. The soft path requires institutional change, not just better technology. But most institutional changes, historically, take place only with concerted social action or after crises force the changes to occur. Humanity in the twenty-first century will either choose the soft water path or pay dearly—in both money and a diminished natural world—for clinging to the more familiar hard path.

REFERENCES

Brown and Caldwell, Inc. 1990. *Case Studies of Industrial Water Conservation in the San Jose Area.* Brown and Caldwell, Pleasant Hill, California.

California Department of Water Resources (CDWR). 1998. *The California Water Plan Update, Bulletin 160–98.* California Department of Water Resources, Sacramento.

Del Porto, D., and C. Steinfeld. 1999. *The Composting System Toilet Book.* The Center for Ecological Pollution Prevention, Concord, Massachusetts.

Dziegielewski, B. 1999. "Management of water demand: Unresolved issues." *Water Resources Update,* Issue 114, pp. 1–7.

EBMUD. 1990. *Industrial Water Conservation.* East Bay Municipal Utility District (EBMUD), Oakland, California.

ERI Services, Inc. 1997. *Commercial, Industrial, and Institutional Water Conservation Program, 1991–1996, Prepared for the Metropolitan Water District of Southern California.* ERI Services, Inc., Los Angeles.

Flack, J.E. 1981. "Residential water conservation." *Journal of the Water Resources Planning and Management Division*. American Society of Civil Engineers, Proceedings Paper 16080, pp. 85–95.

Gleick, P.H. 1998. *The World's Water 1998–1999*. Island Press, Covelo, California.

Gleick, P.H., et al. 2000. *Water: The Potential Consequences of Climate Variability and Change*. A Report of the National Water Assessment Group, U.S. Global Change Research Program, U.S. Geological Survey, U.S. Department of the Interior and the Pacific Institute for Studies in Development, Environment, and Security. Oakland, California.

Gleick, P.H., and D. Haasz. 1998. *Review of the CalFed Water-Use Efficiency Component Technical Appendix*. Working Paper of the Pacific Institute for Studies in Development, Environment, and Security. Oakland, California.

Gleick, P.H., P. Loh, S.V. Gomez, and J. Morrison. 1995. *California Water 2020: A Sustainable Vision*. A report of the Pacific Institute for Studies in Development, Environment, and Security. Oakland, California.

Gleick, P.H., G.H. Wolff, D. Haasz, and A. Mann. Forthcoming. *Untapped Opportunities: Residential Water Conservation in California*. Pacific Institute for Studies in Development, Environment, and Security, Oakland, California.

Hagler Bailly Services. 1997. *Evaluation of the MWD CII Survey Database, Prepared for the Metropolitan Water District of Southern California*. Hagler Bailey Services, San Francisco.

International Rivers Network. 2001. "When the rivers run dry—the World Bank, dams and the quest for reparations." Available at http://irn.org/programs/finance/damfacts.html. 8–13–2001.

Keller, A., and J. Keller. 1995. *Effective Efficiency: A Water Use Efficiency Concept for Allocating Freshwater Resources*. Water Resources and Irrigation Division Discussion Paper, No. 22. Center for Economic Policy Studies. Arlington, Virginia.

Lens P., G. Zeeman, and G. Lettinga (eds.). 2001. *Decentralised Sanitation and Reuse: Concepts, Systems and Implementation*. IWA Publishing, London.

Lovins, A. 1977. *Soft Energy Paths: Toward a Durable Peace*. Harper Colophon Books.

Molden, D. 1997. *Accounting for Water Use and Productivity*. System-wide Initiative on Water Management (SWIM), No. 1. International Irrigation Management Institute. Colombo, Sri Lanka.

Owens-Viani, L. 1999a. "Reducing water use and solving wastewater problems with membrane filtration: Oberti Olives." In L. Owens-Viani, A.K. Wong, and P.H. Gleick (eds.), *Sustainable Use of Water: California Success Stories*. Report of the Pacific Institute for Studies in Development, Environment, and Security. Oakland, California. Pp. 113–119.

Owens-Viani, L. 1999b. *Marin Municipal Water District's Innovative Integrated Resource Management Program*. In L. Owens-Viani, A.K. Wong, and P.H. Gleick (eds.), *Sustainable Use of Water: California Success Stories*. Report of the Pacific Institute for Studies in Development, Environment, and Security. Oakland, California. Pp. 11–26.

Owens-Viani, L., A.K. Wong, and P.H. Gleick (eds.). 1999. *Sustainable Use of Water: California Success Stories*. Report of the Pacific Institute for Studies in Development, Environment, and Security. Oakland, California.

Palacios-Velez, E. 1994. "Water use efficiency in irrigation districts." In H. Garduno and F. Arreguin-Cortez (eds.), *Efficient Water Use*. UNESCO, Montevideo, Uruguay.

Pike, C. 1997. "Some implications of the 1997 California food processor survey." Paper presented at the 214th American Chemical Society National Meeting, September 9, Las Vegas, Nevada.

Postel, S. 1997. *Last Oasis: Facing Water Scarcity*. W. W. Norton. New York.

Postel, S. 1999. *Pillar of Sand: Can the Irrigation Miracle Last?* W. W. Norton. New York.

Ricciardi, A., and J.B. Rasmussen. 1999. Extinction rates of North American freshwater fauna. *Conservation Biology*, pp. 1,220–22.

Seckler, D. 1996. *The New Era of Water Resources Management: From Dry to Wet Water Savings*. Research Reports No. 1. International Irrigation Management Institute. Colombo, Sri Lanka.

Sweeten, J., and B. Chaput. 1997. "Identifying the conservation opportunities in the commercial, industrial, and institutional sector." Paper presented to the American Water Works Association annual meeting, June, Atlanta, Georgia.

U.S. Environmental Protection Agency (U.S. EPA). 1997. *Study of Potential Water Efficiency Improvements in Commercial Businesses: Final Report*. United States Environmental Protection Agency with the State of California Department of Water Resources Grant No. CX 823643–01–0. April.

Vickers, A.L. 1999. "The future of water conservation: Challenges ahead." *Water Resources Update*, Issue No. 114, pp. 49–51.

Vickers, A.L. 2001. *Handbook of Water Use and Conservation.* WaterPlow Press, Amherst, Massachusetts.

Water Resources Policy Commission. 1950. *A Water Policy for the American People: The Report of the President's Water Resources Policy Commission,* Vol. 1. U.S. Government Printing Office, Washington, D.C.

White, G.F. 1961. "The choices of use in resource management." *Natural Resources Journal,* Vol. 1, pp. 23–40.

Wilkinson, R. 1999. "Increasing institutional water-use efficiencies: University of California, Santa Barbara program." In L. Owens-Viani, A.K. Wong, and P.H. Gleick (eds.), *Sustainable Use of Water: California Success Stories.* Report of the Pacific Institute for Studies in Development, Environment, and Security. Oakland, California. Pp. 99–105.

Wilkinson, R., A.K. Wong, and L. Owens-Viani. 1999. "An overview of water efficiency potential in the CII Sector." In L. Owens-Viani, A.K. Wong, and P.H. Gleick (eds.), *Sustainable Use of Water: California Success Stories.* Report of the Pacific Institute for Studies in Development, Environment, and Security. Oakland, California. Pp. 77–84.

World Bank. 1992. *World Development Report 1992: Development and the Environment.* Oxford University Press, New York.

World Commission on Dams. 2000. *Dams and Development: A New Framework for Decision-Making: The Report of the World Commission on Dams.* Available at http://www.dams.org/report/wcd_overview.htm.

World Health Organization. 2000. *Global Water Supply and Sanitation Assessment 2000.* The World Health Organization, Geneva, Switzerland.

Wright, A. 1997. *Toward a Strategic Sanitation Approach.* United Nations Development Program–World Bank Water and Sanitation Program, Washington, D.C.

Globalization and International Trade of Water

Peter H. Gleick, Gary Wolff, Elizabeth L. Chalecki, and Rachel Reyes

Water has an economic value in all its competing uses and should be recognized as an economic good.
THE DUBLIN PRINCIPLES, 1992

[A]ccess to basic water and sanitation are universal rights, and cannot therefore be negotiated as commodities.
NGO STATEMENT AT THE HAGUE, MARCH 2000

Among the most powerful and controversial new approaches to water policy is the idea that water should be considered an "economic good," increasingly subject to the rules and power of markets, prices, and international trading regimes. In the last decade, this idea has been put into practice in many ways. Prices have been set for water previously provided for free. Commercial trade in bottled water has boomed. Proposals have been floated to transfer large quantities of fresh water across international borders, and even across oceans.

These ideas and trends have generated enormous controversy. Many unanswered questions remain about the true implications and consequences of treating water as an economic good and whether these new approaches can effectively, equitably, and adequately serve human and environmental needs. Debate is growing about how— and even whether—to price and sell a resource as fundamental and vital as water. Controversy is building about how to protect ecosystem quality and access to water when private actors set the rules. Concern has been raised about how fresh water should be defined and treated by sweeping new international trade agreements. This chapter addresses these issues; the next chapter looks at the comparable controversy over efforts to privatize water systems around the world—a different but related concern.

In some places and in some circumstances, treating water as an economic good offers some major potential advantages in the battle to provide all humans with their basic water requirements while protecting natural ecosystems. But there is little doubt that the headlong rush toward open trading markets and large-scale international transfers has failed to address some of the most important issues and concerns about water. In particular, water has vital social, cultural, and ecological roles to play that cannot be protected by purely market forces. In addition, certain management goals and social values appear to require direct and strong government support and protection. As a result, any efforts to turn water into an economic commodity must be accompanied by guarantees to respect certain principles and support specific social objectives. Among these are the need to provide for basic human and ecosystem water requirements, permit equitable access to water for poor populations, include affected parties in decision making, and increase reliance on water-use efficiency and productivity improvements. Openness, transparency, and strong public regulatory oversight are fundamental requirements in any efforts to trade water as a good. These principles are defined and summarized here.

The Nature and Economics of Water: Background and Definitions

Because some often-used words and phrases—*globalization, privatization, commodification, water as a social good, water as an economic good*—are critical to the discussion and analysis in this chapter, we explicitly define these five terms and discuss them in the context of current water-management questions. We also define public and private goods in Box 2.1.

Globalization

A "global" economy is being created by the intensification of trading across national boundaries and the transnational character of large corporations, creating interdependencies among the national economies of each region and of the world. Over the past few years, new rules and processes governing trade in goods and services have been developed, leading to an expanding influence of multinational corporations and to a series of international agreements with broad implications for consumers, governments, and the natural environment.

Box 2.1   Private and Public Goods

Economists define private goods as those for which consumption (or use) by one person prevents consumption (or use) by another. Public goods are those that can be used by one person without diminishing the opportunity for use by others. Water-supply systems are public goods because, in most circumstances, delivery of water to one household does not prevent delivery of water to another household. The economic definitions of private and public goods should not be confused with public or private ownership of goods. A private good can be publicly owned.

These rules and processes have come to be known as *globalization,* defined here as the process of integrating and opening markets across national borders. The entire process of globalization is highly controversial, raising great concern about national sovereignty, corporate responsibility, equity for the world's poorest people, and the protection of the environment. The controversy extends to proposals to encourage large-scale trading of fresh water across borders. Indeed, among the most controversial water issues today are questions about how to implement—indeed, *whether* to implement—international water trading and sales.

Privatization

Privatization in the water sector involves transferring some or all of the assets or services of public water systems into private hands. This is separate from, but related to, globalization. There are numerous ways to privatize water, such as the transfer of the responsibility to operate a water delivery or treatment system, a more complete transfer of system ownership and operation responsibilities, or even the sale of publicly owned water rights to private companies. Alternatively, various mixes are possible, such as soliciting private investment in the development of new facilities, with transfer of those facilities to public ownership after investors have been repaid. Increasingly, offers to privatize water services are coming from large, multinational corporations, and as these efforts intensify, so does opposition at local, regional, and international levels.

When the service being privatized has "public good" characteristics, like water, government regulation or oversight has traditionally been applied. Economists and others argue that goods and services previously provided by public officials or agencies may become less vulnerable to political manipulation when privatized, but private entities may also become less responsive to public interests. Such public interests include protection of water quality, commitment to efficiency improvements that reduce the volume of water sold, maintenance of basic service levels, transparent prices and billing practices, and investments in water reclamation or additional sources of water supply.

Commodification

Commodification is the process of converting a good or service formerly subject to nonmarket social rules into one subject to market rules. Even with today's sophisticated economies, many goods and services are still traded or exchanged outside of markets. For example, villages often have complex informal social arrangements that govern access to common water supplies, rationing of these supplies during drought, and so forth. Water exchanges within the village community or between communities may require commitments to return water in the future or other social commitments, rather than or in addition to payment in currency. The processes of globalization and privatization tend to require that water (and water services) be treated as commodities, subject to the rules of marketplaces and free of traditional cultural rules. Of course, water in some forms has long been considered a commodity, particularly bottled water of various types. In recent years, however, the sales of different forms of water have boomed, including flavored waters, glacier water, distilled and partially distilled waters, and other "designer" waters. This has led entrepreneurs to begin to explore the possibility of large-scale movements of waters for commercial, rather than purely community, purposes.

Water as a Social Good

There is no single, universally accepted definition of social goods and services.[1] One widely used definition is that social goods are those that have significant "spillover" benefits or costs. Literacy is a social good, for example, because it benefits not just literate individuals but also makes possible a higher level of civilization for all members of a society. Widespread availability of clean and affordable water is a social good under this definition because such availability improves both individual and social well-being. Improvements in water quality for one individual mean better water quality for all individuals who share that water-supply system. But social goods can have private-good characteristics as well: more water for one individual can mean less water for other individuals who share a water-supply system.

Access to clean water is fundamental to survival and critical for reducing the prevalence of many water-related diseases (UN 1997). Indeed, piped water is typically one of the first community services people seek as communities develop, even before electricity, sanitation, or other basic services. Ensuring that the public receives an adequate supply of social goods requires some level of governmental action, since completely "free" markets often do not find it profitable to provide social goods. For example, water quality affects public health, in both the short term and the long term. However, private water sellers have no incentive to mitigate long-term water-quality issues that do not affect the salability of the water (i.e., carcinogens that do not affect taste, odor, or appearance). Similarly, improvements in water-use efficiency and productivity are often economically beneficial to society as a whole but may reduce revenues to water sellers. Completely "free" markets will not encourage private sellers to either improve water quality or water-use efficiency.

Other dimensions of water supply also have a social good character and therefore require governmental action, oversight, or regulation. Collection, storage, treatment, and distribution of water often require large capital facilities that exhibit economies of scale. Since privately owned and operated monopolies will maximize their profits by selling less of their product than is efficient (and thus artificially raising its price), government action to control prices, quantities of water supplied, and capital investments may be appropriate and desirable.

Modern societies usually recognize that markets will be more efficient and effective if social goods are regulated to some degree by government, and in some instances provided directly by government (e.g., energy, communications, transportation, education, criminal and civil courts, police and military forces). Furthermore, many development economists and theorists urge widespread provision of at least some social goods as a prerequisite for the transformation of poorer economies into highly productive, modern economies.

Because water is important to the process of economic development, essential for life and health, and has cultural or religious significance, it has often been provided at subsidized prices or for free in many situations. In theory, this makes water available to even the poorest segments of society. This is politically popular but brings with it a financial burden because society must pay for the subsidy. It can also encourage wasteful use of water, and the perverse result that many of the poor do not have access to clean water at reasonable prices because those who have access use more

---

1. Economists often mean "goods and services" when they say only "goods." We also use this convention.

water than they need. Balancing these public and private benefits is the challenge discussed below.

Water as an Economic Good

Frustration over the failure to meet basic needs for water has been growing over the dozen years after the massive effort of the International Drinking Water Supply and Sanitation Decade (1981–90). Despite an impressive increase in the number of people with access to clean water, the number without access remains unacceptably high. During the 1990s, mobilization of the financial, engineering, and physical resources required to supply clean water to those without it was recognized to be infeasible without more efficient use of water and a rethinking of national and international water priorities and policies. Among these was the potential value of applying economic tools and principles. Consequently, the International Conference on Water and Environment, held in Dublin, Ireland, in January 1992, included the following principle among the four so-called Dublin Principles: "Water has an economic value in all its competing uses and should be recognized as an economic good" (ICWE 1992).

Of the four principles enunciated in Dublin, this one has stirred the most debate and confusion. Water is essential for human life. Treating it solely as a commodity governed by the rules of the market implies that those who cannot afford clean water must suffer the many ills associated with its absence. However, making it available at subsidized prices can lead to inefficient use and short supply. The "needle to be threaded" in water management is how to get the most value from water that is available, while not depriving people of sufficient clean water to meet their basic needs.[2] The complete commodification of water, however, is not a necessary consequence of the movement toward management of water as an economic good.

What does recognition of water as an economic good mean?[3] Among other things, it means that water has value in competing uses. Managing water *as an economic good,* broadly defined, means that water will be allocated across competing uses in a way that maximizes its value to society. Such allocation can take place through markets, through other means (e.g., democratic or bureaucratic allocations), or through combinations of market and nonmarket processes.

It is important to note, however, that a broad economic approach to water management does not inevitably lead to management of water as if it were a commodity in *all* its aspects. For example, water pricing that subsidizes the fixed charge portion (for the physical water connection) of water rates but imposes a volumetric charge (for actual water used) that reflects the highest value use of water, treats each unit of water consumed as a commodity but treats the piped connection itself as a social good. This

---

2. Gleick (1996) discusses the concept of "basic needs" for water in the context of international statements and fundamental human requirements. He estimates a "basic water requirement" for domestic uses and argues that these uses should be considered essential social goods. He also notes that most people can afford to pay for basic water needs, but that when they cannot, governments should subsidize the small amounts of water involved.

3. Much has been written on this subject (see, for example, Rogers et al. 1998, McNeill 1998, Briscoe 1997). Rogers et al. (1998) discuss this issue at length, including examples from Thailand and India. Their discussion emphasizes estimation of the costs and benefits of ecological, cultural, and social factors under current conditions. Costs and benefits can change, perhaps significantly, if property rights and rules, social preferences, technology, or institutions change.

pricing scheme could allow the poor to satisfy their basic water needs but also reduce wasteful and inefficient use of water.

Water Managed as Both a Social and Economic Good

Following the Dublin meeting, the United Nations Conference on Environment and Development (held in Rio in 1992) clearly recognized that water should be managed as both a social and economic good: "Integrated water resources management is based on the perception of water as an integral part of the ecosystem, a natural resource, and a social and economic good . . ." (UN Agenda 21, Chapter 18.8).

The theory of allocating water across its conflicting uses, however, often conflicts with practice when there is no way to measure, or capture, all its costs and benefits. Some argue that attempting to place a market value on the social good aspects of water (i.e., ecosystem and cultural values) may result in uncertain and misleading dollar value estimates. Managing water solely as an economic good may ultimately cause poor people or small businesses to be priced out of the market, leaving them without enough of an essential social good.

Critics have pointed out that rapid implementation of private-public partnerships for water supply has involved, in too many cases, blatant disregard for the needs of the poor. That privatization or globalization of water management can harm people, however, does not imply that they *must* harm people. It is also true that the rush toward a global economy and management of water as an economic good has neglected the ecological value of water (e.g., from the ecosystem services provided by healthy river and coastal habitat) and its cultural value as well (e.g., value associated with local control, preservation of traditional practices, or the religious significance of water).

Those who approach water management from a narrow economic perspective argue that the mistakes of the past can be addressed by more complete cost/benefit analyses, and other quantitative tools, that account for such values. Having quantified environmental, cultural, and distributional impacts, so they argue, monetary gains from managing water as an economic good must be weighed against any adverse social impacts of managing water as an economic good. Most of those who take this approach see the economic and social good characteristics of water as incompatible.

A broader economic approach, however, recognizes that some significant benefits and costs—especially some types of cultural and ecological benefits and costs—cannot be quantified in practice. Consequently, the results of cost/benefit analysis are almost always incomplete and therefore inadequate as the sole basis for water-management decisions. The broader approach involves quantifying costs and benefits when doing so is feasible and affordable, but more important, it seeks to put into place stakeholder participation processes that can lead to the changes in institutions (e.g., formal or informal property rights and rules or organizations) that will allow water to be allocated to its highest value uses (which maximizes social benefits less social costs, even those that have not been measured) through processes that are accepted as fair and equitable by stakeholders.

Once the possibility of extensive stakeholder participation leading to institutional change becomes part of the economic analysis, more opportunities to manage water as both a social and an economic good are available. In the final section of this chapter we present some principles for water-management policy that can lead to simultaneous management of water as an economic and social good.

Water as Both a Renewable and Non-renewable Resource

Many natural waters are renewable resources, made available by the hydrologic cycle of the coupled atmospheric-oceanic-terrestrial system. In this sense, continued flows of water are not affected by withdrawals and use of water. Unlike non-renewable resources, the amount of water available for use in a basin in the future is not necessarily altered by past withdrawals of water in that basin. In contrast, the withdrawal and use of non-renewable resources, such as coal or oil, reduces the amount that can subsequently be used.

Not all natural waters are renewable, however, and some that are renewable can be made non-renewable through human actions. Some groundwater basins and lakes, for example, have extremely slow rates of recharge and inflow. Water extracted from these basins or bodies of water in excess of the natural recharge or inflow rate is, therefore, equivalent to pumping oil—it reduces the total stock available for later use—and hence is non-renewable and exhaustible. Contamination of a groundwater stock, similarly, can make a renewable resource into a non-renewable resource. Finally, human actions to modify watersheds, such as cutting forests or paving land, can affect the overall hydrologic balance, reducing recharge or flow characteristics and altering timing, availability, and renewability of water. In extreme cases, this can exhaust a formerly renewable resource.

Whether a particular water resource is renewable or exhaustible is important for international trade discussions. All natural waters, therefore, cannot be treated alike. In fact, how the World Trade Organization (WTO) treats them will depend on their classification as exhaustible or renewable resources—and if renewable, on the minimum flows required to sustain animal and plant life or human health. All exhaustible stocks of water may qualify as non-renewable mineral resources for exemption under GATT (General Agreement on Tariffs and Trade) Article XX(g). Some renewable flows of water may qualify for an exemption under GATT Article XX(b): those that are "necessary to protect human, animal or plant life or health." These issues are discussed below.

Water Provides Other Benefits

Fresh water is vital to protect and maintain human, animal, and plant health, yet these benefits are rarely protected by private financial markets or trading systems. Water bodies provide habitat for aquatic life. Riparian systems provide moisture for vegetation and terrestrial biota, nutrient transport between one ecosystem and another, recreational and transportation opportunities, and aesthetic benefits. Larger systems such as the Great Lakes provide broad regional climate and weather services. Reducing water quantity or the quality of a water body by means of large-scale withdrawals or transfers may significantly alter these *in situ* benefits. Changing the timing of flows in a river, even when quality and total quantity remain unchanged, may also alter ecological conditions.

Diversions or transfers of water from watersheds to other regions have led to many ecological and human health disasters. The diversion of water from the Amu Darya and Syr Darya Rivers in central Asia has caused the destruction of the Aral Sea ecosystem, the extinction of the sea's endemic fish populations, the dramatic shrinking of the sea itself, and widespread local health problems associated with the exposure and atmospheric transport of salts. See Figures 2.1 and 2.2. Withdrawals of water from many rivers and streams in North America and Europe have led to reductions and extinctions in many fish populations, particularly anadromous fish, which are born in freshwater rivers, migrate out to the open ocean, and then return to fresh water to spawn. Depletion of river flows have severely damaged river deltas and local communities, such as in the Sacramento/San Joaquin delta in California, the Nile River delta in Egypt, and the Colorado River delta in Mexico, where the local Cocopa Indian communities have been affected (Morrison, Postel, and Gleick 1996).

The transfer of water from one ecosystem or ecoregion to another may support economic development, but it also runs the risk of contributing to or accelerating the loss of ecosystem integrity (Linton 1993) or causing adverse economic or social effects in the area of origin. Measuring or quantifying all benefits of water in economic terms would be necessary to incorporate them into decisions to trade, market, or manage water, yet as discussed above, such measurements are complex and often incomplete and inaccurate.

Water Has Moral, Cultural, and Religious Dimensions

Water has more than economic and ecological importance; it has cultural or symbolic importance as well. It figures prominently in religious rituals such as baptism and ritual bathing, and in the national identities of many native peoples (Graz 1998). Since water is fundamental to life in all forms, deep-seated feelings may be relevant to water-management decisions. For example, strong concerns about what is fair or just may arise when water supplied to urban dwellers decreases water availability in rural areas, or when water supply to urban residents whose basic need for water is not being met is blocked to protect rural economies or natural systems.

Moral dimensions of water management intersect with the property rights issues that underlie economically efficient allocation of water. If local people "own" or have a right to water in its natural place, they must be persuaded to accept removal of water from its natural place for the reallocation of water to be efficient. Even when outsiders are willing to pay very large sums for water—perhaps enough to make locals extremely rich in money terms—locals may be unwilling to accept such trade. That is, when water *in situ* or from a particular source has cultural or symbolic significance, as well as its usual uses, it may have very different value to people of different cultures.

In practice, those who want to remove water from an ecosystem and sell it may work hard to confuse the property rights situation, or to obtain the right to water *in situ* for themselves. In the former case, what is economically efficient is unclear because economic efficiency depends on a clear definition of property rights. In the latter case, what is economically efficient in a simple market situation is different from what is economically efficient when the property right is assigned to those who are unwilling to accept loss of *in situ* water. That is, more than one economically efficient outcome is possible; indeed, a different efficient outcome may exist for each different ownership arrangement. This little-known and -understood economic "fact" is an intersection point between water as a social good and water as an economic good. Successfully managing water as both an economic and social good requires that the property rights issues associated with water in its natural state be investigated thoroughly and made as explicit as possible in the laws, regulations, and informal rules that affect water management.

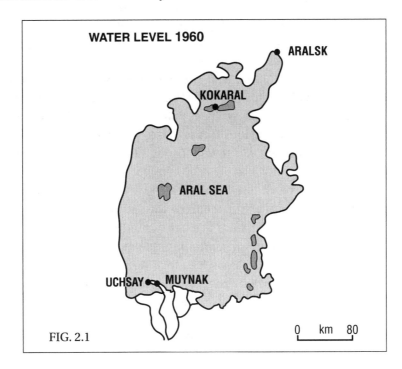

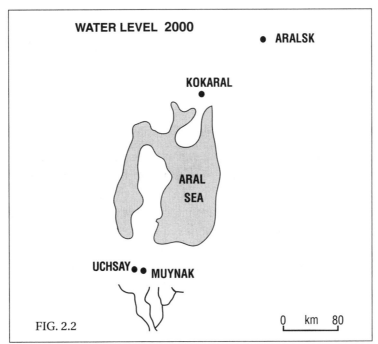

FIGURES 2.1 AND 2.2 Aral Sea extent in 1960 and 2000.
Source: http://visearth.ucsd.edu/VisE_Int/aralsea/aralanim.html

The Globalization of Water: International Trade

The world's water is unevenly distributed, with great natural variations in abundance. Indeed, the complex and expensive water systems that have been built over the past few centuries have been designed to capture water in wet periods for use in droughts and to move water from water-rich regions to water-poor regions. As domestic, industrial, and agricultural demands for fresh water have grown, entrepreneurs have created a wide range of markets for water, leading to various forms of international water trading and exchanges. Water has commonly been diverted between regions and uses via canals or pipelines, but growing demand and uncertainty of supply in the face of population growth, climate change, and other factors is motivating many states, provinces, and even individuals and corporations to examine new ways to transfer water from areas with water "surplus" to areas with unmet needs. These include longer and longer pipelines, the sale of various forms of bottled water, the physical transport of liquid water in tankers or large bags towed through the ocean, and even the capture and use of icebergs.

Such schemes have been considered for years, in various forms and designs. In the past, most large-scale transfers of water occurred within national and political borders. Now proposals for bulk water transfers are being made at international, and even global, levels. In recent years Alaskan, Canadian, Icelandic, Malaysian, Turkish, and other waters have been proposed as sources for international trade in bulk water. Besides the historically important environmental and socioeconomic implications of water transfers, the possibility of large-scale bulk trading of fresh water has now become an issue in international trade negotiations and disputes.

At the global scale, the possibility of bulk water transfers has caused concern in water-abundant regions that a global water-trading regime might lead to the requirement that their resources be tapped to provide fresh water for the rest of the world at the expense of their own environment and people. These issues turn the underlying question throughout this chapter—how water can be simultaneously managed as a social and economic good—into a significant trade concern.

The Current Trade in Water

Proponents of trade in water argue that natural resources, such as timber, finished lumber, minerals, fossil fuels, raw fish, and agricultural goods, are exported every day without generating nationalistic anti-export sentiment. Opponents of trade in water argue that water is different in important ways from other goods, including other natural resources, and that these differences require that water be treated differently in international markets. Is trade in water different in some important way from trade in other natural resource goods? If so, how should individuals, corporations, communities, countries, and even international trading agreements treat proposals to trade water?

Water Traded as a Raw or Value-Added Resource

Much of the international trade in resources involves raw natural resources that undergo some form of modification or finishing though human economic activity. Agricultural goods, livestock, fossil fuels, fish, and lumber typically involve some economic inputs, such as processing, refining, milling, or other time- and labor-intensive activities. In contrast, resources traded in highly raw form, such as crude oil, logs, or raw fish, involve much less investment in the country of origin. Investments are required to acquire the resource, but few or no additional inputs are needed. Some parties to the water globalization debate argue that activities that add significant value should be treated differently, for the purpose of trade agreements and other legal protections for investors, than activities that remove raw materials from the country of origin with minimal or "one-time" benefits for the local economy.

Water can be traded as either a raw (bulk) or value-added product. Indeed, a large and rapidly growing international market already exists for various forms of processed, value-added water—particularly bottled waters. Bottled-water sales worldwide in the mid-1990s exceeded 50 billion liters, and such sales have been increasing by nearly 10 percent a year since the 1970s (see Table 2.1). In 1999, the bottled-water industry in the United States alone generated $5 billion from the sale of more than 17 billion liters—up from less than 2 billion liters annually in the mid-1970s (see Figure 2.3). Most of this is domestically produced—about 8 percent was imported in 1999 (http://www.soc.duke.edu/~s142tm16/world.htm). Figure 2.4 shows the sources of bottled water imported into the

TABLE 2.1 Global Bottled Water Sales

Country/Region	1996 Sales (Million liters)	Projected 2006 Sales (Million liters)	Annual Percentage of Growth (%)
Australasia	500	1,000	11
Africa	500	800	4
CIS	600	1,500	13
Asia	1,000	5,000	12
East Europe	1,200	8,500	14
Middle East	1,500	3,000	3
South America	1,700	4,000	7
Pacific Rim	4,000	37,000	18
Central America	6,000	25,000	11
North America	13,000	25,000	4.5
Western Europe	27,000	33,000	2.5
Total	**57,000**	**143,800**	

Source: Modified from http://www.soc.duke.edu/~s142tm16/World%20Markets.htm

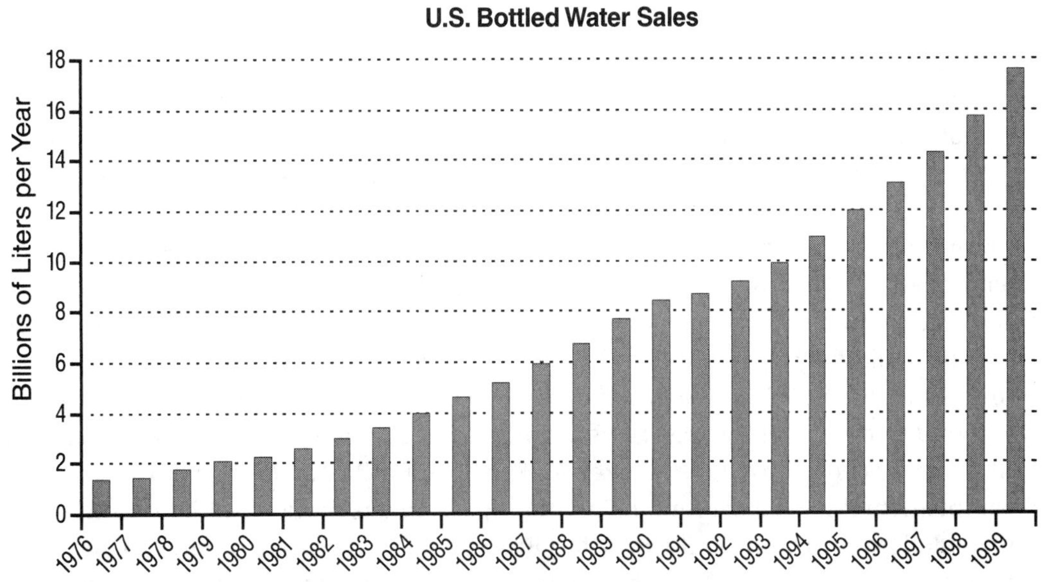

U.S. Bottled Water Sales

FIGURE 2.3 U.S. bottled-water sales. Sales of bottled water in the United States are increasing by 10 percent annually.

U.S. Bottled Water Imports, 1999

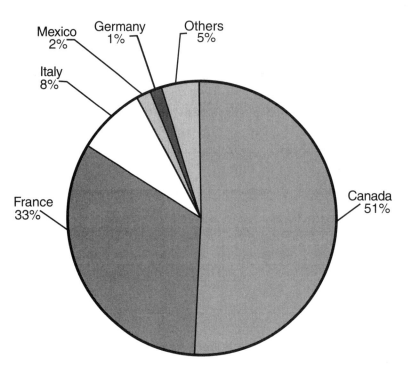

FIGURE 2.4   U.S. bottled water imports in 1999. These imports represent only about 8 percent of total U.S. bottled water sales.
Source: http://www.soc.duke.edu

United States, Canada, France, and Italy accounted for more than 90 percent of the U.S. imports in 1999.

Bottled-water sales are also increasingly prevalent and important in poorer countries (see Table 2.2). However, bottled-water sales must not be considered acceptable substitutes for an adequate municipal water supply. Bottled water rarely provides adequate volumes of water for domestic use, and the cost of such water is typically exorbitant. There may be circumstances when readily available (but nonpotable) water for domestic uses plus high quality and affordable bottled water for drinking are adequate, but we could not find any examples.

Water traded as bottled or value-added water is covered by international trade rules like any other good. Consequently, much of the debate and concern at present is focused on proposals to trade bulk, unprocessed water across international borders, either for later processing or for use for municipal or industrial purposes. We exclude from this in-basin trades or transfers of water by countries that share a watershed—such trades occur all the time, although typically through political agreements rather than market deals.

Interestingly, there have been relatively few long-term, international, out-of-basin water trades to date, although various proposals have been put forward. On occasions, bulk water has been brought by tankers to Pacific or Caribbean islands during drought to supplement limited local supplies. In the 1960s, tankers brought water to Hong Kong and loaded fresh water from the Houston River as backhaul cargo to Curaçao for use in refineries there (Meyer 2000). In the 1980s, there was a shuttle trade between the River Tees in Great Britain and Gibraltar, when Spain shut off its water supply for political purposes (Meyer 2000). Aruba imported water by tanker from Dominica. The small

TABLE 2.2   Bottled and Vended Water: Urban and Rural Use

Country	Year	Source of Water	Percentage of the Urban Population that Consumes Bottled or Vended Water	Percentage of the Rural Population that Consumes Bottled or Vended Water
Angola	1996	Tanker Truck	25.2	0.8
Cambodia	1998	Vendor	16	3.5
Chad	1997	Vendor	31.5	0.5
Dominican Republic	1996	Bottled Water	37	6.3
Ecuador	1990	Tanker Truck	16	7
Eritrea	1995	Tanker Truck	30.5	1.4
Guatemala	1999	Bottled Water	25.5	7.1
Haiti	1994	Bottled Water	26	0.3
Jordan	1997	Tanker Truck	1	10.6
Libyan Arab Jamahiriya	1995	Tanker Truck	6.8	13.9
Mauritania	1996	Vendor	53	0.9
Mongolia	1996	Vendor	16	1
Niger	1998	Vendor	26.4	1.9
Oman	1993	Bottled Water	39.5	42
Syrian Arab Republic	1997	Tanker Truck	4.1	11.3
Turkey	1998	Bottled Water	14.9	1
Yemen	1997	Bottled Water	14.6	0.1

Source: WHO 2000

island nation of Nauru imported water from Australia, New Zealand, and Fiji (see Box 2.2). In rare circumstances, such as the severe drought in mid-1994, Japan has imported limited quantities of water by tanker to maintain refinery and automobile production. This water has come from Alaska, Vietnam, South Korea, Hong Kong, and China (Sugimoto 1994, Jameson 1994, AFX News 1994, Brown et al. 1995). The provincial government of Mallorca contracted with a tanker company for shipments of water from western Spain until desalination plants could be built (Huttemeier 2000). In general, however, the very high cost of tankered water is a barrier to such transfers, especially if long-term supplies are needed. In such situations, the maximum amount a buyer will be willing to pay does not exceed the cost of alternatives, such as desalinated water. As a result, most of these transfers are phased out when other cheaper and more reliable solutions are found.

At present there are few major proposals pending for large-scale, long-term transfers of water across international borders (see Box 2.3). Turkey has offered water from the Manavgat River and has been negotiating with Israel (Ekstract 2000, Turner 2001). Tankers or giant bags would be used to transfer the water to an Israeli coastal port, where it would be treated and used. The alternative for Israel would be reallocation of existing water resources within Israel (for example, from agriculture to cities), improvements in water-use efficiency, or new supplies from elsewhere, such as desalination. Whether the proposed bulk transfer will be able to surmount the economic and political hurdles facing it remains to be seen, but as of this writing, solicitations are being considered and negotiations are continuing on price. Similarly, Spain is considering reviving an old proposal to import water from the Rhone River with a 320-kilometer aqueduct extending from Montpellier, France, to Barcelona (*Financial Times* 2000). If implemented, this would be the first trans-basin water deal in the European Union, but even if approved, it is unlikely to be completed within a decade.

Box 2.2 Transfers of Water Out of a Watershed

International, Out-of-Basin Transfers

Aruba (Netherlands Antilles) has received water from Dominica by tanker.

In the 1960s, Hong Kong received some water via tanker. Prior to the return of Hong Kong to China, the city received 75 percent of its potable water from China. Fifty percent was piped in from the mainland; the remaining 25 percent was piped from Lantau Island.

Nauru, an island nation located in the central Pacific, has received as much as a third of its water as return cargo from ships exporting phosphate. The water is from Australia, New Zealand, and Fiji.

In the 1990s Tonga regularly received water by tanker, and the Canary Islands imported practically all of its potable water as bottled water, before beginning to build desalination plants for local domestic and industrial use.

Unusual Domestic, Out-of-Basin Transfers

Hong Kong's islands, Lamma and Ma Wan, receive water by submarine pipeline.

Malaysia's Penang receives some of its water from the Malaysian peninsula via submarine pipes.

China's Xiamen Island receives 50 percent of its supply from the mainland.

St. Thomas and St. John (U.S. Virgin Islands) have received water from Puerto Rico transported by sea intermittently since 1955.

In the Bahamas, New Providence received an average of 21 percent of its total water supply from Andros Island from 1978 to 1987. The water was transported by barge. In 1987, 31 percent of the total supply was transported. The Bahamas now have about 54,000 m³/day of desalination capacity.

Mallorca receives water by tanker from the Spanish mainland but has drawn up plans to lay underwater pipes or build a desalination plant to avoid having to bring in drinking water by tanker each summer.

Turkey sends water to Turkish Cyprus by tanker.

The smaller Fijian Islands commonly received water from the larger islands starting in the early 1970s, especially during drought periods.

Sources: Coffin and Richardson, Inc. 1981, Water Supplies Department 1987, Zhang and Zhixin 1988, Fiji Country Paper 1984, Gattas 1998, Gleick 2000, Lee 1989, Swann and Peach 1989, Huttemeier 2000, Brewster and Buros 1985, Lerner 1986, UNESCO 1992, Meyer 2000.

Box 2.3 International, Out-of-Basin Transfers
under Consideration

Italy is considering importing water to its dry southern regions by building a pipeline under the Adriatic Sea to pump in supplies from Albania.

In October 2000, Austria claimed that it could supply all 370 million people in the European Union with well or surface water needing treatment, just ahead of EU plans to liberalize its water industry. Austrian ministers, including the agricultural and environmental minister, argue the economic advantages outweigh the political liabilities.

Spain is considering importing water from the Rhone River with a pipeline extending from Montpellier, France, to Barcelona.

Israel is negotiating to buy water from Turkey's Manavgat River. Negotiations will include an Israeli request for an annual amount of 15–25 million cubic meters of water with the possibility of doubling the amount for a period of 5–10 years. Israel sees Turkish water imports as a quick alternative to its plans to build a desalination plant. Water would be shipped from the Manavgat, in southwestern Turkey, to the Israeli port of Ashkelon, where it could be further distributed.

Sources: Rudge 2001, Demir 2001, *Financial Times* 2000

The Rules: International Trading Regimes

Rules governing international trade are complex and often contradictory. In recent years, efforts to implement standard rules have been developed in several international fora, and these rules have become increasingly sophisticated and important to the global economy. At the same time, they have become increasingly controversial, as their implications for the environment, civil society, and local economies become clearer. In this section we discuss two important agreements as they relate to the globalization of water resources: the General Agreement on Tariffs and Trade and the 1994 North American Free Trade Agreement (NAFTA). GATT is the overriding international trade agreement, and NAFTA is an excellent example of how the international water-trading debate has been influenced by a regional trade agreement. Other regional trade agreements have been signed or are being negotiated, and future assessments might consider these in more detail.

GATT provides the basic legal architecture that governs international trade for the member countries of the World Trade Organization. NAFTA governs trade in goods between the United States, Canada, and Mexico. While the two trade regimes have many similarities, there are also several provisions, described below, that distinguish the two. For the discussion about water, NAFTA is particularly important among the regional trade regimes because it includes the United States, the largest national economy in the world, and Canada, the nation that has expressed the greatest concern and taken the strongest actions with regard to international bulk water trading.

It is worth noting that there is little legal precedent pertaining directly to international trade in water, making it difficult to predict the outcomes of current and future trade disputes in this area with certainty. However, commercial pressures to export water are increasing, making resolution of these ambiguities an important goal. In addition, adverse, even virulent public sentiment over several proposed exports highlights the need to resolve and clarify issues (Barlow 1999; *FTGWR* 2001, 127).

The degree to which countries will be able to impose controls on the exportation of water will hinge upon the determination of whether bulk water in its natural state is considered a "product," and if so, whether exemptions in the trade agreements are applicable. Treating water as a product (a good) on the global market typically implies that transfers will be governed by international trading obligations. This has raised concerns among local communities and environmental groups that water resources may suffer the fate of other global common resources: overexploitation with significant restrictions on national environmental control and restraints. For example, some trade analysts interpret WTO and NAFTA rules as requiring that water must *continue* to be traded in bulk once it has *begun* to be traded. These issues are discussed below.

The General Agreement on Tariffs and Trade

GATT is a comprehensive international trade agreement and provides the basic legal structure that governs international trade for members of the WTO. All potential commodities are defined and described in the Harmonized Tariff Schedule (HTS). This schedule, used by the United States and all WTO countries, includes water of all kinds (other than seawater, which is described in a separate heading) under Section 2201 (Appleton 1994):

> 2201.90.0000: Other waters, including natural or artificial mineral waters and aerated waters, not containing added sugar or other sweetening matter nor flavored; ice and snow.

The existence of an HTS number means that there is a mechanism under which shipments of fresh water can be processed for import by U.S. Customs and comparable customs organizations of other nations.

Some communities and environmental groups are concerned that if bulk water is traded as a product anywhere in the international community, GATT Article XI, General Elimination of Quantitative Restrictions, would be interpreted to prohibit any government from prohibiting bulk water exports. Article XI, Section 1 states:

> No prohibitions or restrictions other than duties, taxes or other charges, whether made effective through quotas, import or export licenses or other measures, shall be instituted or maintained by any contracting party on the importation of any product of the territory of any other contracting party or on the exportation or sale for export of any product destined for the territory of any other contracting party.

Once out-of-basin bulk water transfers are initiated by domestic industry, Article XI plays a significant role in constraining WTO member governments' ability to establish policies, programs, or legislation that regulate, curtail, or eliminate such transfers.

However, the language of the agreement is unclear as to whether these restrictions apply only to specific, actual bulk water trades once they have begun or apply to all potential bulk water trading arrangements once trade in bulk water has begun between any WTO signatories.

Under either interpretation, however, other portions of GATT appear to be relevant to the question of whether some trade in bulk water may be limited or constrained by national laws, and hence exempt from the Article XI provisions. For bulk water, the two most relevant clauses upon which a government could base an exemption and adopt measures that restrict trade are found in Article XX:

> Article XX(b) necessary to protect human, animal or plant life or health;

> Article XX(g) relating to the conservation of exhaustible natural resources if such measures are made effective in conjunction with restrictions on domestic production or consumption; . . .

There is considerable debate among legal experts as to whether WTO member governments can control bulk water exports based on these resource conservation principles, and there are few legal precedents. The WTO has struck down two previous cases where national governments attempted to challenge unlimited resource trades using these exemptions (WCEL 1999). In the 1991 Tuna-Dolphin case and the 1998 Shrimp-Turtle case, the WTO expert panels ("arbitration committees" appointed, individually, for each trade dispute) did not accept domestic environmental protection legislation as a valid basis for imposing trade restrictions (Box 2.4). Some WTO member countries, such as South Africa, Angola, Ecuador, Chile, and Indonesia, have publicly stated during a recent WTO High-Level Symposium on Trade and Environment that environmental protection measures should not stand in the way of economic development (IISD 1999). The burden of proof that an environmentally based trade restriction is necessary rests on the country imposing the restriction (French 1999).

Because bulk water, unlike petroleum or minerals, can be a renewable resource depending upon the way it is extracted, some may argue that it falls outside the Article XX(g) exemption. The exemption appears to have been drafted with reference to mining of minerals and fossil fuels, but water also has non-renewable characteristics well understood by hydrologists:

> Freshwater resources typically are considered renewable: they can be used in a manner that does not affect the long-term availability of the same resource. *However, renewable freshwater resources can be made non-renewable by mismanagement of watersheds, overpumping, land subsidence, and aquifer contamination. Water policy should explicitly protect against these irreversible activities.* (Gleick 1998, 576; emphasis added)

Thus, this exemption could be interpreted to apply to stocks of water that were deposited long ago and are not being replenished at a significant rate compared with the rate of use. In such circumstances, a strong argument can be made to support an Article XX(g) exemption for bulk water resources where freshwater resources are "non-renewable" or exhaustible through overuse or abuse, assuming domestic production or consumption is also limited to prevent non-renewable uses.

There are many examples of groundwater overdraft, where human extraction exceeds natural replenishment, sometimes by a wide margin. In the mid-1990s, the

---

Box 2.4 The Turtle and Dolphin WTO Cases:
Relevant for Water?

In two separate cases related to protection of ocean resources, WTO expert
panels rejected protection of dolphins and turtles through import restrictions against tuna and shrimp caught with nets that were not dolphin- or
turtle-safe. The WTO argued that import restrictions on the basis of the
process of production of imported goods were in violation of GATT. Applying
this reasoning to possible bulk water trade restrictions raises all sorts of
unresolved questions. Moreover, the tuna/dolphin and shrimp/turtle cases
involved import rather than export restrictions. In both cases, the importing
country was trying to influence management of natural resources beyond
its borders. The concerns raised in the case of bulk water trading are associated with the power of the nation of origin to protect people or ecological
functions within their own borders. It is possible that the tuna/dolphin and
shrimp/turtle legal precedents for contests under the exemption provisions
would not greatly influence a case involving bulk water export controls
imposed by the nation of origin. This possibility cannot be understood any
further without an actual test case under GATT. GATT itself does not provide
sufficient guidance on this point: interpretation of GATT "case by case," or
amendment of GATT, will be required before the limits of and protections
offered for water by Article XX (b) and (g) are clearly defined. See the further
discussion in the text.

Ogallala Aquifer underlying seven states in the central United States was pumped at
rates three to four times faster than natural recharge. Groundwater in northern China,
India, Saudi Arabia, and many other places is also pumped unsustainably. Such usage
must be considered "non-renewable" just as stocks of oil are considered exhaustible. In
the case of exports of water from the Great Lakes of North America, some have argued
that only a tiny fraction of the lakes are "renewable" and that the vast bulk of the stored
water was laid down in geologic times (Barlow 1999).

In such circumstances, exports could, if large enough, lead to the irreversible
decline in lake levels. In 1999, the International Joint Commission between the United
States and Canada issued a report concluding that the Great Lakes are non-renewable,
with an eye to ensuring that they would be subject to a GATT Article XX exemption (IJC
2000). Cases where water stocks have been contaminated by human actions also
represent the conversion of renewable water resources into a non-renewable
resource—appropriate for an Article XX(g) exemption.

In some circumstances, bulk water exports could also be subject to an Article XX(b)
exemption if such exports threaten human or ecosystem health. Biologists and ecolo-
gists have long understood and demonstrated that some amounts of water are needed
in situ to protect animal and plant life and health, although the precise quantities
needed to maintain adequate instream flows and *in situ* resource values are subject to

study and debate. Many specific instream flow requirements have been set for particular watersheds to maintain ecosystem health. Although a specific Article XX(b) exemption has not been tested in the context of bulk water, we believe that it would also support a ban on bulk exports of water when such exports threaten ecosystem or human health.

The North American Free Trade Agreement

Canada, the United States, and Mexico have developed the North American Free Trade Agreement as a regional extension of GATT in many ways. For example, the Harmonized Tariff Schedule of the United States and the Canadian Customs Tariff, in which "ordinary natural water of all kinds" are classified under tariff heading 22.01, also implies that water can be traded internationally as a good. Article 201(1) of NAFTA defines a "good" as a "domestic product as these are understood in the [GATT] or such goods as the Parties may agree, and includes originating goods of that Party" (Yaron 1996). Hence the categorization of water provided above (HTS Number 2201.90.0000) is applicable under NAFTA as well as GATT.

Several other factors, however, complicate the interpretation of NAFTA rules for water. In 1993, the three NAFTA parties signed a joint declaration to provide explicit protection for *in situ* water resources and the rights of the country of origin under NAFTA and GATT:

> Unless water, in any form, has entered into commerce and becomes a good or product, it is not covered by the provisions of any trade agreement, including the NAFTA. And nothing in the NAFTA would obligate any NAFTA Party to either exploit its water for commercial use, or to begin exporting water in any form. Water in its natural state, in lakes, rivers, reservoirs, aquifers, water basins and the like is not a good or product, is not traded, and therefore is not and never has been subject to the terms of any trade agreement.

This joint declaration is the clearest exposition of the intent of the parties to NAFTA to protect natural waters from uncontrolled bulk withdrawals for international trade. But this form of agreement carries limited weight in international law. Some analysts argue that the protections offered by joint declarations are not legally binding and establish no legal obligations:

> It has long been recognized in international practice that governments may agree on joint statements of policy or intention that do not establish legal obligations. . . . These documents are sometimes referred to as non-binding agreements, gentlemen's agreements, joint statements or declarations. (Shrybman 1999)

Further, the 1993 joint declaration has received little subsequent formal support from the three governments (Shrybman 1999), and the sentiment expressed could be considered inconsistent with GATT and NAFTA tariff headings, "U.S. Law" and "International Law," both of which define water as a good (Shrybman 1999). Official U.S. policy in this area was further confused when U.S. Trade Representative Mickey Kantor wrote in 1993, "[W]hen water is traded as a good, all provisions of the agreements governing trade in goods apply."

On the one hand, Kantor's statement appears to be a simple reiteration that bulk water, once it has entered trade, must be subject to the existing trade agreements. But some are concerned that Kantor's statement could be interpreted as saying that the joint declaration would no longer apply once *any* NAFTA signatory government permits sale in bulk of any water for commercial purposes (Barlow 1999). This interpretation would put bulk water back into the realm of "goods to be traded," something environmentalists were hoping the joint declaration specifically and permanently excluded. A less extreme interpretation would be that the signers intended to exempt water in its natural state from trade agreement provisions, except for specific quantities of water that have been put into commerce with the approval of the country of origin. Under this interpretation not *all* waters become open to such trade provisions once *some* water has been traded. Under either interpretation, however, trade analysts agree that once international trade in bulk water has begun, at least the amount of water that has been authorized for trade cannot be withdrawn from trade by an action of the country of origin unless specific exemptions within GATT or NAFTA are satisfied.

The Canadian government took action to prevent bulk water trading in 1994 because it was concerned that its NAFTA obligations could impinge upon its ability to develop national water policy. On January 1 of that year, the North American Free Trade Implementation Act was proclaimed into force by the Canadian Parliament,[4] including a rider that specified that (1) nothing in either NAFTA or its implementing legislation, except the provision on tariff elimination, applied to water, and (2) "water" in this context meant natural surface and ground water in liquid, gaseous or solid state, not including water packaged as a beverage or in tanks. As appealing as this definition is for environmental and public interest groups looking to forestall bulk exports, it is a matter of Canadian domestic legislation and may not be binding on NAFTA dispute resolution panels (Appleton 1994). In February 1999, the government of Canada requested that each province implement a voluntary moratorium on bulk exports, but the federal government has yet to enact a national ban. Indeed, in spring 2001, the premier of Newfoundland called for lifting the ban on bulk water exports, reopening the contentious debate there (MacDonald 2001). Despite recent legislative actions in Canada, these issues remain unresolved.

As noted earlier, whether fresh water *in situ* falls under NAFTA's (or the GATT's) definition of a good has not been legally settled. If, however, bulk fresh water is considered a good under the NAFTA definition, there are three conditions of NAFTA that affect international trade. First, similar to GATT Article III, National Treatment, each signatory country must accord businesses and investors from the other signatory countries the same treatment that it accords its own businesses and investors for both goods and services. Article 1102, National Treatment, states:

> Each Party shall accord to investors of another Party treatment no less
> favorable than that it accords, in like circumstances, to its own investors
> with respect to the establishment, acquisition, expansion, management,
> conduct, operation and sale or other disposition of investments.

This means that any NAFTA country cannot treat other NAFTA bulk water exporters or importers any differently than it treats its own bulk water exporters or importers.

---

4. The implementing legislation of all signatory states came into force on January 1, 1994.

Second, NAFTA Chapter 11 also allows investors in any signatory country to sue the government of either of the other two signatories if that government takes some future action (usually legislation) to "expropriate" that company's profits (Barlow 1999). According to the provisions of Article 11.10, Expropriation and Compensation,

> No Party shall directly or indirectly nationalize or expropriate an investment of an investor of another Party in its territory or take a measure tantamount to nationalization or expropriation of such an investment ("expropriation"), except:
>
> (a) for a public purpose;
>
> (b) on a nondiscriminatory basis;
>
> (c) in accordance with due process of law and the general principles of treatment provided in Article 1105; and
>
> (d) upon payment of compensation in accordance with paragraphs 2 to 6.

Chapter 11 issues have already been raised in the context of bulk water exports. In the fall of 1998, a Santa Barbara, California, company called Sun Belt Water, Inc. sued the government of Canada under NAFTA Chapter 11. Sun Belt lost a contract to export water to California when the British Columbia (BC) provincial government banned bulk water exports in 1995 (Cordon 2000). While the Chapter 11 suit cannot overturn the BC law, it can make the government of Canada liable for the profits that Sun Belt would have made on this contract had BC not passed its export ban. This makes federal, state, and provincial governments reluctant to implement legislation regulating commerce in natural resources.

We note that while the government of Canada may be liable under these provisions of NAFTA, the profits Sun Belt might have actually received are highly uncertain—indeed, an argument can be made that any profits were unlikely. Water is very expensive to move from one place to another, and commands a high price only in the luxury form of bottled water—a form of water all participants in this debate agree is already covered by trading rules. Moreover, the amount that potential importers are willing to pay is capped at the cost of alternative sources, including desalination, making the size of possible profits highly speculative. Even assuming the sellers could command a price of $1.50 per cubic meter—more than double what most municipalities and industries currently pay for reliable urban supplies—optimistic tankering costs for water are between $2 and $4 per cubic meter or even higher (Bardelmeier 1995, Huttemeier 2000), depending on distance. This is why, historically, contracts to tanker water from one place to another have consistently given way to more local solutions, such as reallocation among end users, or desalination. Hence the size of the actual liability to Canada may be small or zero. This issue remains to be resolved.

Third, NAFTA Article 309 states that constraints on exports of any good must be shared proportionally across the signatory countries (Barlow 1999). This means that if Canada were to start exporting water in bulk and subsequently faced a drought or other shortage, it could not reduce the amount of water exported to the United States and Mexico in order to maintain *unreduced* deliveries to domestic customers. All customers must take proportional reductions; this is a de facto extension of the "national treatment" clause, wherein all customers are treated equally. These provisions reinforce the constraint on national sovereignty that arises under GATT: once bulk water is traded as a good under legally valid contracts, it must continue to be traded. This alarms many environmentalists who feel that the best method of protection for natural resources on a watershed scale is domestic legislation.

In sum, our analysis suggests that large-scale, long-term bulk exports of water across international borders are unlikely for many reasons, especially the high economic cost of moving water. Nevertheless, great uncertainty continues to revolve around the legal interpretation of international trade agreements in the context of globalizing water resources. Because of the risk of ecological damages and non-sustainable withdrawals of water for export, it will be important to clarify both national and international rules governing bulk exports of water. In this regard, it will be necessary for governments to institute national water policies that explicitly protect water necessary to support human and ecosystem health and prohibit the mining and export of non-renewable water resources.

REFERENCES

AFX News. 1994. "Mitsubishi Oil to import water from Alaska." August 24.

Appleton, B. 1994. *Navigating NAFTA: A Concise User's Guide to the North American Free Trade Agreement.* Carswell Publishing, Scarborough, Ontario, Canada. 214 pp.

Bardelmeier, W. 1995. "Water is too heavy a burden to carry." Lloyd's List, September 22.

Barlow, M. 1999. *Blue Gold: The Global Water Crisis and the Commodification of the World's Water Supply.* International Forum on Globalization. Sausalito, California.

Brewster, L., and N. Buros. 1985. "Non-conventional water resources: Economics and experiences in developing countries. (I)." *Natural Resources Forum,* Vol. 9, No. 2, pp. 133–142.

Briscoe, J. 1997. "Managing water as an economic good: Rules for reformers." Keynote Paper to the International Committee on Irrigation and Drainage Conference on Water as an Economic Good, September, Oxford, United Kingdom.

Brown, L.R., D. Denniston, C. Flavin, H. French, H. Kane, N. Lenssen, D. Roodman, M. Ryan, A. Sachs, L. Starke, P. Weber, and J. Young. 1995. *State of the World 1995.* Norton Press, New York. Via http://www.worldwatch.org/pubs/sow/sow95/ch01.html.

Coffin and Richardson, Inc. 1981. *Water Conservation under Conditions of Extreme Scarcity: The U.S. Virgin Islands.* Office of Water Research and Technology, National Technology Information Service, Washington, D.C.

Demir, M. 2001. "Turkey, Israel sign water deal." *Jerusalem Post,* January 29.

Ekstract, J. 2000. "Turkey and Israel set to complete water import deal." *Jerusalem Post,* July 5. Available at http://www.jpost.com/Editions/2000/07/05/News/News.9164.html.

Fiji Country Paper. 1984. *Technical Proceedings (Part 3) of the Regional Workshop on Water Resources of Small Islands, Suva, Fiji, 1984.* Commonwealth Science Council Technical Publication Series No. 182, Pp. 1–10.

Financial Times (London). 2000. "Thirsty markets turn water into valuable export: Drinking water is ever more scarce in the parched lands around the Mediterranean." November 7.

Financial Times. 2000. November 7. Available at http://www.internetional.se/toft/toft20011.htm#water.

Financial Times Global Water Report (*FTGWR*). 2001. "U.S. eyes Canadian water supplies." Vol. 127, p. 3 (August 6).

French, H. 1999. "Challenging the WTO." *Worldwatch,* November/December, pp. 22–27.

Garn, M. 1998. "Managing water as an economic good." From a conference on Community Water Supply and Sanitation, May 5–8, Washington, D.C. The World Bank, via http://www.wsp.org/english/focus/conference/managing.html.

Gattas, N. 1998. "Environment—Cyprus: Tourists get first sip of water shortages." *Inter Press Service,* November 3.

Gleick, P.H. 1996. "Basic water requirements for human activities: Meeting basic needs." *Water International,* Vol. 21, pp. 83–92.

Gleick, P.H. 1998. "Water in crisis: Paths to sustainable water use." *Ecological Applications,* Vol. 8, No. 3, pp. 571–579.

Gleick, P.H. 1999. "The human right to water." *Water Policy,* Vol. 1, pp. 487–503.

Gleick, P.H. 2000. *The World's Water 2000–2001: The Biennial Report on Freshwater Resources.* Island Press, Washington, D.C.

Graz, L. 1998. "Water source of life." *FORUM: War and Water.* International Committee of the Red Cross, Geneva, Switzerland, pp. 6–9.

Hudson, P. 1999. "Muddy waters; Argentina, the Latin American leader in private water works, struggles to create a transparent concession system." *Latin Trade,* March.

Huttemeier, J. 2000. Personal communication via email. Director, Maersk Tankers. Copenhagen, Denmark, December 7.

International Conference on Water and the Environment (ICWE). 1992. *The Dublin Principles.* Available in full at http://www.wmo.ch/web/homs/icwedece.html.

International Institute for Sustainable Development (IISD). 1999. *International Institute for Sustainable Development Report on the WTO's High-Level Symposium on Trade and Environment, 15–16 March 1999.* As found at http://www.wto.org/wto/hlms/sumhlenv.htm.

International Joint Commission (IJC). 2000. *Final Report to the Governments of Canada and the United States, Protection of the Waters of the Great Lakes.* Available at http://www.ijc.org.

International Union for the Conservation of Nature (IUCN). 2000. *Vision for Water and Nature: A World Strategy for Conservation and Sustainable Management of Water Resources in the 21st Century.* The International Union for the Conservation of Nature, World Water Forum, The Hague, Netherlands.

Jameson, S. 1994. "With Japan's relentless heat wave, it's boom or bust." *Los Angeles Times,* August.

Kemper, K.E. 1996. "The cost of free water: Water resources allocation and use in the Curu Valley, Ceara, northeast Brazil." Linkoping University, Linkiping, Sweden.

Lazaroff, C. 2001. "WTO upholds U.S. right to protect sea turtles." Environment News Service, June 19. As found at www.ens-news.com/ens/jun2001/2001l-06–19–07.html.

Lee, Yow Ching. 1989. *Development of Water Supply in Penang Island, Malaysia.* Proc. Seminar on Water Management in Small Island States. Cyprus Joint Tech. Council and Commonwealth Engineer's Council, Pp. 38–43.

Lerner. 1986. "Leaking pipes recharge ground water." *Groundwater,* Vol. 25, No. 5, pp. 654–662.

Linton, J. 1993. "NAFTA and water exports." Submission to the Cabinet Committee on NAFTA of the Government of the Province of Ontario, April. 15 pp.

MacDonald, M. 2001. "Brian Tobin has no sway in debate over water exports." Available at http://ca.news.yahoo.com/010329/6/3u8a.html.

Margat, J. 1996. "Comprehensive assessment of the freshwater resources of the world: Groundwater component." Contribution to Chapter 2 of the *Comprehensive Global Freshwater Assessment,* United Nations, New York.

McNeill, D. 1998. "Water as an economic good." *Natural Resources Forum,* Vol. 22, No. 4, November.

Meyer, T.A. 2000. Personal communication via email to Peter Gleick. December 9.

Morrison, J.I., S.L. Postel, P.H. Gleick. 1996. *The Sustainable Use of Water in the Lower Colorado River Basin.* Pacific Institute for Studies in Development, Environment, and Security, the Global Water Policy Project, and the United Nations Environment Programme. Oakland, California.

Rogers, P., R. Bhatia, and A. Huber. 1998. "Water as a social and economic good: How to put the principle into practice." Global Water Partnership/Swedish International Development Cooperation Agency, Stockholm.

Rudge, D. 2001. "Delegation going to Turkey to discuss importing water." *Jerusalem Post,* January 17.

Shrybman, S. 1999. "Legal opinion commissioned by the Council of Canadians re: water export controls and Canadian international trade obligations." Available at http://www.canadians.org.

Sugimoto, M. 1994. "Drought forces Japan to import Alaskan water." *South Bend Tribune,* August 28.

Swann and Peach. 1989. *Status of Groundwater Resources Development for New Providence [Bahamas].* Interregional Seminar on Water Resources. Management Techniques for Small Island Countries. UNDTCD, Suva, Fiji. ISWSI/SEM/4.

Turner, W. 2001. "Water export from Manavgat, Turkey." Available at http://www.waterbank.com/Newsletters/nws23.html.

UNESCO. 1992. "Small tropical islands—water resources of paradises lost." *IHP Humid Tropics Programme Series No. 2,* UNESCO, Paris, France.

United Nations (UN). 1997. *Comprehensive Assessment of the Freshwater Resources of the World.* United Nations, New York.

Water Supplies Department. 1987. *Hong Kong's Water.* Land and Work's Branch, Hong Kong Government, 32 pp.

West Coast Environmental Law (WCEL). 1999. Available at http://www.wcel.org/wcelpub/1999/12926.html

World Health Organization (WHO). 2000. *Global Water Supply and Sanitation Assessment: 2000 Report.* World Health Organization and United Nations Children's Fund. Report. Available in full at http://www.who.int/water_sanitation_health/Globassessment/GlobalTOC.htm.

Yaron, G. 1996. "Protecting British Columbian waters: The threat of bulk water exports under NAFTA." University of British Columbia, Canada, April, 61 pp.

Zhang, Z., and L. Zhixin. 1988. *Study of Water Supply to Xiamen Island.* Proc. South East Asian and the Pacific Regional Workshop on Hydrology and Water Balance of Small Islands, UNESCO/ROSTSEA, Nanjin, China, pp. 134–140.

CHAPTER 3

The Privatization of Water and Water Systems

Peter H. Gleick, Gary Wolff, Elizabeth L. Chalecki, and Rachel Reyes

Food and water are basic rights. But we pay for food. Why should we not pay for water?

ISMAIL SERAGELDIN AT THE SECOND
WORLD WATER FORUM, THE HAGUE

Water should not be privatized, commodified, traded or exported in bulk for commercial purposes.

MAUDE BARLOW

One of the most important—and controversial—trends in the global water arena is the accelerating transfer of the production, distribution, or management of water or water services from public entities into private hands—a process loosely called *privatization.* (See Chapter 2 for our definitions of this and other related terms.) Treating water as an economic good, and privatizing water systems, are not new ideas. Private entrepreneurs, investor-owned utilities, or other market tools have long provided water or water services in different parts of the world. What *is* new is the extent of privatization efforts underway today, and the growing public awareness of, and attention to, problems associated with these efforts.

The issue has resurfaced for several reasons. First, many public water agencies have been unable to satisfy the most basic needs for water for all humans. Second, major multinational corporations have greatly expanded their efforts to take over responsibility for a larger portion of the water service market than ever before. And third, several recent highly publicized privatization efforts have failed or generated great controversy. The privatization of water encompasses an enormous variety of possible water-management arrangements. Privatization can be partial, leading to so-called public/private partnerships, or complete, leading to the total elimination of government responsibility for water systems. At the largest scale, private water companies build, own, and operate water systems around the world with annual revenues of approximately $300 billion, excluding revenues for sales of bottled water (Gopinath 2000). At the smallest scale, private water vendors and sales of water at small kiosks and shops provide many more individuals and families with basic water supplies than they did 30 years ago. Taken all together, the growing roles and responsibilities of the private sector have important and poorly understood implications for water and human well-being.

As a measure of the new importance of privatization, the second World Water Forum in the Hague in March 2000 gave special emphasis to the need to mobilize new financial resources to solve water problems and called for greater involvement by the private sector. Indeed, the Framework for Action released at that meeting called for $105 billion per year in new investment—over and above the estimated $75 billion per year now spent—to meet drinking water, sanitation, waste treatment, and agricultural water needs between now and 2025. The Framework called for 95 percent of this new investment to come from private sources (GWP 2000). There was enormous controversy at this meeting about the appropriate role of governments and non-governmental organizations, and a planned public workshop and discussion on privatization and globalization of water was canceled.

Along with the growing efforts at water privatization, there is rapidly growing opposition among local community groups, unions, human rights organizations, and even public water providers. Protests—sometimes violent—have occurred in many places, including Bolivia, Paraguay, South Africa, the Philippines, and various globalization conferences around the world. Concerns are being expressed over the economic implications of privatizing water resources, the risks to ecosystems, inequities of access to water, and the exclusion of communities from decisions about their own resources. This chapter reviews why efforts are growing to turn over responsibility for public provision of water and water services to the private sector, the history of privatization efforts, and the risks of these efforts. It also reviews fundamental principles proposed to prevent inequitable, uneconomic, and environmentally damaging privatization agreements.

Drivers of Water Privatization

In 1992, the summary report from the water conference in Dublin set forth four "principles," including the concept that water should be treated "as an economic good" (ICWE 1992). This principle is, without doubt, the most important and controversial of the four. It was sufficiently vague to be accepted by the participants, and yet sufficiently radical to cause serious rethinking of water management, planning, and policy. In the years following Dublin, the concept of water as an economic good has been used to challenge traditional approaches to government provision of basic water services. Economists seized upon the idea to argue that water should be treated as a private good, subject to corporate control, financial rules, market forces, and competitive pricing.

Various pressures are driving governments to consider and adopt water privatization. These pressures fall loosely into five categories: *societal* (the belief that privatization can help satisfy unmet basic water needs), *commercial* (the belief that more business is better), *financial* (the belief that the private sector can mobilize capital faster and cheaper than the public sector), *ideological* (the belief that smaller government is better), and *pragmatic* (the belief that competent, efficient water-system operations require private participation) (Neal et al. 1996, Savas 1987). Privatization efforts in the United Kingdom and Europe were ideologically driven at first, but are increasingly characterized as commercial and pragmatic (Beecher 1997). Privatization efforts in the

United States were initially pragmatic but are now strongly ideological, as can be seen by the public policy push being given to water privatization by libertarian and free-market policy institutes. Privatization efforts in the developing world can primarily be described as financial and pragmatic, though some argue that the social benefits are significant (GWP 2000). Interestingly, several countries with strong ideological foundations have also chosen to explore water privatization for pragmatic reasons. China and Cuba, for example, have both recently awarded contracts to private companies to develop and operate municipal water-supply systems and build wastewater treatment plants.

Water-supply projects can be extremely capital intensive, though estimates of future needs vary widely. The World Bank estimates that new investment required for water infrastructure over the next decade will exceed $60 billion per year. In mid-2001, the American Water Works Association released a study suggesting that $250 billion may be needed over the next 30 years just to upgrade and maintain the existing drinking-water system in the United States (AWWA 2001). As noted above, the Framework for Action that emerged from Second World Water Forum in The Hague in March 2000 called for an additional $105 billion annually from private sources for the next 25 years to meet basic water needs (GWP 2000, World Commission 2000).

Whatever the actual investment required, emerging economies face significant hurdles finding the capital to expand coverage in rapidly growing urban areas, maintain existing infrastructure, and treat wastewater to even minimal quality standards. One option is for governments to turn to the private sector, with its greater access to private capital (Faulkner 1997). Because of this, private participation in the water sector is growing especially quickly in developing countries.

Governments must, of course, also spend limited public and international financial capital to meet other social needs (Yergin and Stanislaw 1999). By creating water systems that are self-supported through private investment and by implementing water pricing that pays back the investments, developing country governments can significantly reduce their fiscal and balance of payment problems (Shambaugh 1999). It is also difficult for government officials subject to political processes to raise water prices; privatization permits governments to give that problem to private entities.

The perception that companies are more competent and efficient than government also contributes to pressures to privatize water systems. The complexity of large water systems and their poor historical performance have encouraged the belief that the technical and managerial skills needed to improve water supply and management systems are only available, or at least more efficiently applied, through partial or complete privatization of water supply. Many developing country politicians also view introducing competition as desirable (Shambaugh 1999). Initially favorable results from a few privatization actions have supported these beliefs, although experience with government management of water systems in the industrialized countries demonstrates that government is not necessarily less efficient or competent than business.

History of Privatization

Private involvement in water supply has a long history. Indeed, in some places, private ownership and provision of water was the norm, until governments began to assume these responsibilities. In the United States, municipal services were often provided by private organizations in the early 1800s. Toward the latter half of that century, municipalities started to confront problems with access and service and began a transition toward public control and management. In particular, private companies were failing to provide access to all citizens in an equitable manner. Private water companies provided 94 percent of the U.S. market in the nineteenth century, dropping to around 15 percent by 2000 (Beecher, Dreese, and Stanford 1995). As Blake (1991) points out:

> Private companies supplied water to Boston from 1796 to 1848, and to Baltimore from 1807 to 1854. As late as 1860, 79 out of 156 water works in the United States were privately owned. But eventually most cities turned to municipal ownership. The profit motive was ill suited to the business of supplying water to city dwellers. Private companies were reluctant to invest enough capital; they preferred to lay their distributing pipes through the wealthier sections of the city and to hold back from carrying water into the poorer districts.

Anderson (1991) notes that the experience in Chicago and other cities in the United States was similar:

> Private companies were notorious for choosing a water source that would minimize the initial investment outlay, and for ignoring the concomitant shortcomings in water quantity and quality. Only municipal governments, so the argument goes, had the foresight and the latitude to invest large sums now in order to gain a future payoff in the form of years of excellent water.

In nineteenth-century France, the trend moved in the opposite direction: municipalities that previously had responsibility for providing water services began to contract services to private operators. Over the years, these operators expanded beyond the borders of France and as a result, they now have a dominant position in much of the world in providing private water services.

Major international efforts to privatize water systems and markets are still a relatively recent phenomenon, with major transfers taking place only over the past 10 to 15 years. By the end of 2000, at least 93 countries had partially privatized water or wastewater services (Brubaker 2001), including Argentina, Chile, China, Colombia, the Philippines, South Africa, Australia, the United Kingdom, and parts of Central Europe, but less than 10 percent of all water is currently managed by the private sector (LeClerc and Raes 2001).

In South America, public monopolies were the norm until the mid-1990s. As in many other regions, public water systems consistently failed to provide universal coverage, to treat most wastewater, and to find and reduce water losses that can be as high as 50 percent. Because of these failures, governments in South America increasingly seek private sector involvement. In some cases, such as Buenos Aires, governments have sold or leased water facilities, allowing private operators to sell services directly to the public, with government regulation. Mexico City took another approach to privatization, contracting the rights to operate parts of the city water system to multiple operators with the goal of stimulating competition among them (Waddell n.d.). By 2000, almost all countries in the region had begun to commit themselves to long-term private concessions. Chile has gone farther than most by combining the granting of concessions with private ownership of water resources.

Major cities in Asia also suffer from inadequate infrastructure, huge water losses, inadequate sewage treatment, and lack of service to large numbers of peri-urban residents. Australia, New Zealand, Malaysia, and the Philippines are all exploring various forms of privatization, and the water and wastewater utilities in almost every major city in Oceania have been taken over by private entities, or have contracted some important services. Recent efforts in Manila and Malaysia have run into political or economic controversy, causing private companies and governments to rethink the design of contracts and the conditions for concessions, but privatization efforts seem to be accelerating.

Nations in Europe have also explored a variety of different models recently, and the UK, Germany, France, and Italy all now encourage water privatization. In France and in the UK, the process is far advanced. The British under Prime Minister Margaret Thatcher, for example, sold state-owned water operations to private investors more than a decade ago. Those newly privatized companies have become multinational players in privatization markets. Service providers in all four countries initially tried to keep prices low, but they have recently imposed large price increases in order to upgrade their plants and distribution systems to accepted European standards. These price increases have led to growing consumer distrust, though some argue that governments would have similarly had to raise taxes or increase borrowing to make comparable improvements, or worse, would have failed to make them.

The United States and Canada have moved more slowly toward privatization. The United States has long had a mix of privately owned and publicly regulated water and wastewater utilities, though an estimated 85 percent of residences still receive water from public agencies (see Box 3.1 for the recent experience of Atlanta, Georgia).

The Players

There are a handful of major international private water companies, but two French multinational corporations dominate the sector: Vivendi SA and Suez Lyonnaise des Eaux (now called Ondeo). These two companies own or have interests in water projects in more than 120 countries, and each claims to provide water to around 100 million people (Barlow 1999; *FTGWR* 2000b, 94, http://www.suez.fr/metiers/english/index.htm) (Tables 3.1 and 3.2). Vivendi's water activities are, themselves, a part of the larger company Vivendi Universal, which was created in December 2000, when it merged with the Seagram Company to create a global media and telecommunications company. As an example of the diversity of Vivendi's activities, in spring 2001, Vivendi purchased MP3.com. Figure 3.1 describes many of the interlocking subsidiaries of Vivendi. Their total annual revenue in 2000 exceeded $37 billion, of which more than 25 percent came from the water business (Market Guide 2001).

Suez is active in more than 100 countries and claimed to provide 110 million people in 2000 with water and wastewater services. Of the 30 biggest cities to award contracts between 1995 and 2000, 20 chose Suez, including Manila, Jakarta, Casablanca, Santiago de Chile, and Atlanta. Suez also purchases stakes or full interests in other water companies: with its $1 billion purchase of United Water Resources, it became the second largest manager of municipal systems in the United States, just behind American Water Works. Suez also purchased Nalco and Calgon in the United States for $4.5 billion, making it the biggest provider of water treatment chemicals for both industry and cities. In 2000, Suez reported profits of 1.9 billion euros on sales of 35 billion euros: of this, 9.1 billion euros (or 44 percent) of revenues came from their water businesses (http://www.finance.suez-lyonnaise.com).

Box 3.1   Water Utility Privatization in Atlanta, Georgia

Throughout much of the 1980s and 1990s, Atlanta's wastewater system faced growing problems, including aging infrastructure and inadequate wastewater treatment. Federal, state, and private complaints resulted in millions of dollars in fines and a consent decree specifying expensive corrective action (Brubaker 2001). Faced with the need for almost $1 billion in capital for urgent improvements, the city government began to explore the possibility of privatization of some aspects of the local water system. The hope in Atlanta was that privatization would dramatically reduce annual operating costs, reduce the likelihood of rate increases, and free up money for new capital improvements.

In late 1998, the city signed a 20-year agreement to contract water services to United Water Services Atlanta (UWSA), a subsidiary of Suez Lyonnaise des Eaux. While the company made a number of innovative concessions, they also benefited from some significant tax breaks offered by the city. UWSA agreed to locate its regional headquarters in Atlanta, committed to hiring 20 percent of its work force from the area, and offered to provide $1 million in annual funding for water research at Clark Atlanta University. The firm is benefiting from tax incentives of as much as $8,000 per employee (see http://waterindustry.org/frame-8.htm, "Atlanta Project in Capsule").

The 20-year agreement between the city and UWSA went into effect on January 1, 1999, covering the operations and maintenance of two water-treatment plants serving 1.5 million people, 12 storage tanks, 7 pumping stations, fire hydrants, water mains, billing, collections, and customer service. The contract set UWSA's annual operations and maintenance fee at $21.4 million; thus UWSA can count on nearly half a billion dollars in service fees to be paid by the city of Atlanta over the term of the contract. This is substantially less than the city was expected to spend running the system itself over the same period. The city will continue to spend approximately $6 million annually on power, insurance, and monitoring the contract agreements. Atlanta retained responsibility for most capital investments. The agreement also stipulated that there would be no layoffs during the life of the contract, but staff reductions due to retirements and voluntary departures substantially reduced employment costs. At the time of turnover, many of the municipal employees objected to the privatization agreement. While it is too soon to know how well the goals of the effort to privatize the city's water system will be, other U.S. municipalities are watching closely (see Waddell n.d. and Brubaker 2001 for more details).

TABLE 3.1   Population Served by Vivendi Water and Wastewater Concessions

Population Supplied in 2000	Population (millions)
France	25.0
Western Europe	18.5
Central and Eastern Europe	6.3
Middle East and Africa	8.5
North America	16.8
Latin America	7.8
Asia	14.6
Total	97.5

Source: FTGWR 2000b:94, p.10.

TABLE 3.2   Population Served by Suez Lyonnaise des Eaux Water and Wastewater Concessions

Population Supplied in 2000	Population (millions)
Europe and Mediterranean	43
North America	14
South America	25
Asia Pacific	23
Africa	5
Total	110

Source: http://www.suez.fr/metiers/english/index.htm

Other companies also have major water interests, including Thames Water and United Utilities in Great Britain, Bechtel and Enron in the United States,[1] and Aguas de Barcelona in Spain. To add to the complexity, however, many of these companies have interlocking directorates or partial interests in each other. For example, in spring 1999, Vivendi purchased U.S. Filter Corporation. United Utilities of the UK has joint ventures with Bechtel. United Water Resources in the United States is partly owned by Suez Lyonnaise des Eaux. New joint ventures, consolidations, and reorganizations continue to develop.

Forms of Privatization

Despite the growing debate about privatization, there is considerable misunderstanding and misinformation circulating about what the term itself means. Privatization can take many forms. Only the most absolute form transfers full ownership and operation of water systems to the private sector. Much more common are forms that leave public ownership of water resources unaffected and include transferring some operational responsibilities for water supply or wastewater management from the public to the private sector. Privatization also does not, or should not, absolve public agencies of their responsibility for environmental protection, public health and safety, or monopoly oversight.

There are many different forms of privatization arrangements, agreements, and models. There is also a fundamental difference between public and private ownership

---

1. At least until the collapse of Enron in late 2001.

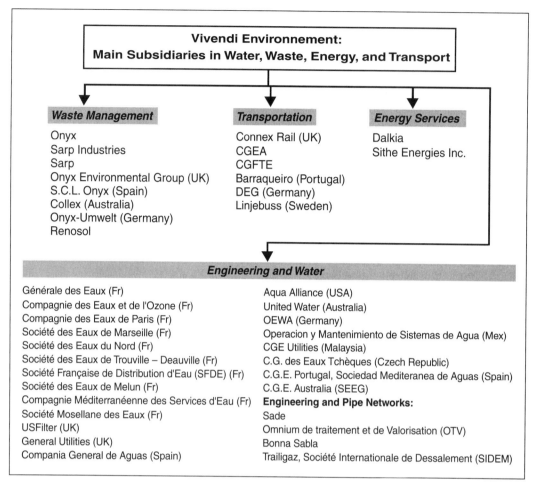

FIGURE 3.1 Vivendi is one of the world's largest providers of water and wastewater services. Vivendi Environnement is a subsidiary of a larger corporation and includes commercial activities in water, energy, engineering, transportation, and much more. The larger parent corporation includes vast holdings in the media and entertainment industry, including Vivendi Universal.

of water assets. Private ownership involves transferring assets to a private utility. Public ownership involves keeping the assets in the public domain, but integrating the private sector in various utility operations and activities through contract (Beecher 1997). Public- or private-sector employees can perform various functions. As an illustration, Table 3.3 lists several functions that could be assigned to private or public employees in thousands of combinations ranging from completely public to completely private operations. These different forms have very different implications worthy of more careful analysis than they have previously received.

The different functions in Table 3.3 can be combined or broken into even more sub-functions (e.g., design of reservoirs versus design of neighborhood-scale distribution piping). The functions can also be performed privately in one geographic area and publicly in another (e.g., northern and southern halves of a metropolitan area). In the remainder of this chapter we describe fully public water systems and compare them to four variations of models of privatization.

TABLE 3.3   Water System Functions That Can be Privatized

1. Capital improvement planning and budgeting (including water conservation and wastewater reclamation issues)
2. Finance of capital improvements
3. Design of capital improvements
4. Construction of capital improvements
5. Operation of facilities
6. Maintenance of facilities
7. Pricing decisions
8. Management of billing and revenue collection
9. Management of payments to employees or contractors
10. Financial and risk management
11. Establishment, monitoring, and enforcement of water quality and other service standards.

Fully Public Water Systems

Fully public management of water often takes place through national or municipal government agencies, districts, or departments dedicated to providing water services for a designated service area (in some cases, an entire country). Public managers make decisions, and public funds are used to finance construction, operation, and maintenance of facilities. Funds may be provided from general government revenues, in competition with other government investments or a water agency may be self-supporting via water charges. Governments are responsible for oversight, setting standards, and facilitating public communication and participation. Independent, special-purpose water agencies or districts can have technical and financial capacity equal to private corporations.

Another form of public management involves cooperatives and user associations in local water-system governance. Typically, local users join together to provide public management or oversight. An example is the public water cooperative in Santa Cruz, Bolivia, which serves nearly 100,000 customers. It arose out of a history of central government neglect and, consequently, a strong belief in decentralization. Customers are split into water districts, each covering approximately 10,000 people. All customers have decision-making powers through elections for different water authorities (Nickson 1998). In 1997, the cooperative compared well to other Bolivian utilities in terms of efficiency, equity, and effectiveness.

Public Water Corporations and Corporate Utilities

Private-sector participation in public water companies has a long history. In this model, ownership of water systems can be split among private and public shareholders in a corporate utility. Majority ownership, however, is usually maintained within the public sector, while private ownership is often legally restricted, for example, to 20 percent or less of total shares outstanding. Such organizations typically have a corporate structure, a Managing Director to guide operations, and a Board of Directors with overall responsibility. This model is found in the Netherlands, Poland, Chile, and the Philippines (Blokland, Braadbaart, Schwartz 1999).

A main benefit of the system is that it combines two potentially conflicting goals of water supply. Private owners seek to recover costs and maximize profits. Public owners may also seek to recover costs, but they are more likely to embrace concerns about affordability, water quality, equity of access, and expansion of service.

Empirical evidence suggests such models can attain a high level of operational efficiency and quality of service. Another strength is the stronger potential for public participation and protection of consumer rights. In the Philippines, the Board includes consumer rights associations and in the Netherlands consumer associations publish comparisons among similar companies. This type of customer representation encourages efficiency and discourages political exploitation of the water utility.

Service and Leasing Contracts: Mixed Management

In some cases, public water utilities may give private entities responsibility for operation and maintenance activities, general services contracts, or control over management of leased facilities. Ownership continues to reside in public hands. Such models do not usually address financing issues associated with new facilities, or create better access to private capital markets. They do, however, bring in managerial and operational expertise that may not be available locally.

Leasing contracts may include tariff (revenue) collection responsibilities as well as operation and maintenance (Rivera 1996). Such contracts may last for 10 to 15 years, and arrangements are sometimes made for the private company to share in the increases in revenues generated from better management and bill collection (Panos 1998). Service contracts range from smaller, one-time arrangements such as meter installation or pipeline construction to longer-term comprehensive arrangements. Areas in which service contracts have proven effective include maintenance and repair of equipment, water and sewerage networks, and pumping stations; meter installation and maintenance; collection of service payments; and data processing (Yepes 1992).

Concession Models

Much of the debate in recent years over privatization has revolved around more comprehensive concessions to the private sector. This is especially true in Latin America and Asia. The full-concession model transfers operation and management responsibility for the entire water-supply system along with most of the risk and financing responsibility to the private sector. Specifications for risk allocation and investment requirements are set by contract. To recoup heavy initial investments, concessions are usually long-term, as long as 25 to 50 years. Technical and managerial expertise may be transferred to the local municipality and community over time, as local employees gain experience.

Variations on full concessions include Build-Operate-Transfer (BOT), Build-Operate-Train-Transfer (BOTT), Build-Own-Operate-Transfer (BOOT), Rehabilitate-Operate-Transfer (ROT), and Build-Operate-Own (BOO). These arrangements can be thought of as "partial concessions" that give responsibilities to private companies, but only for a portion of the water-supply system. Ownership of capital facilities may be transferred to the government at the end of the contract. Training of local workers and managers over years prior to the transfer, with their jobs retained from some period after transfer, is a way of transferring skills along with the capital asset.

For both full and partial concessions, governments and companies are finding that responsibilities and risks must be defined in great detail in the concession contract, since such contracts are for a lengthy period, and ultimately govern how the concession will perform (Komives 2001). Case-by-case concession contract writing has led to vastly different outcomes for similar physical and cultural settings.

Fully Private Businesses and Small-Scale Entrepreneurs

The opposite extreme from government agency provision and management of water is supply and management by fully private actors, whether large corporations or small-scale entrepreneurs. In this model, water-quality regulation and other means of protecting society, such as basic water rights or protection of environmental resources, may be non-existent. Fully private businesses and entrepreneurs are already often found where the existing water utility has low coverage or poor service. They may obtain water directly from a water utility, indirectly from the utility through customers who have utility service, or from private water sources. In some cases, early settlers of an area have privately developed piped water systems, with later settlers becoming customers of, rather than partners in, the piped system. Private providers operate most often in poor urban and peri-urban areas, but they also serve higher income groups or businesses when water is scarce or inconvenient to obtain.

Private suppliers of water also coexist with public systems when the public system is unreliable, inconvenient, or rationed (e.g., the utility pressurizes pipes only a few hours each day). In Kathmandu, Nepal, water from privately controlled sources is sold by tanker truck to both low- and high-income areas of the city unserved by regular, reliable supply. Customers may turn to private vendors when they have more money than time for water collection.

Private businesses and small-scale entrepreneurs often operate free of regulation in less-developed countries. Private water companies are usually regulated to some extent (e.g., water quality) in more developed countries. Without regulation, high prices or low water quality can cause significant social problems. Numerous studies have shown that the poor often pay much more for water from private suppliers or small-scale vendors than they would pay if a regulated community water system, piped or otherwise, were put into place. For example, in El Alto, Bolivia, where a concession was granted in the mid-1990s, households with private connections spend around $2.20 per month for 10 cubic meters, while those relying on private vendors pay over $35.00 for the same amount of water (Komives 2001).

Risks of Privatization: Can and Will They Be Managed?

The move toward privatization of water services raises many concerns, and in some places, even violent opposition. In large part, opposition arises because of doubts about whether purely private markets can address the many different social good aspects of water, or whether some nonmarket mechanisms are necessary to serve social objectives. Other concerns relate to a fundamental distrust of corporate players and worries about the transfer of profits and assets outside of a community or even a country. The greatest need for water services often exists in those countries with the weakest public sectors; yet as we shall see, the greatest risks of failed privatization also exist where governments are weak. The rapid pace of privatization in recent years and the inappropriate ways several projects have been implemented have compounded the worries of local communities, non-governmental organizations, and policy makers. As a result, private water companies are increasingly seeing serious and sustained public opposition to privatization proposals. This section describes the major concerns and risks of privatization of water systems.

Usurping a Basic Responsibility of Governments

Governments have a fundamental duty to see that basic services, such as water, sewerage, and energy, are provided to their people. The failure to satisfy such basic needs, or at least enable the means for them, must be viewed as irresponsible. Efforts of international lending agencies and development organizations have, in the past, focused on helping governments to provide these services. More recently, these organizations have begun to shift their efforts, pushing privatization as a new solution. However, there are serious concerns about this transfer of responsibility and the loss of control it implies.

Bypassing Under-Represented and Under-Served Communities

One of the basic goals of any proposal to privatize water services should be the extent to which the needs of under-served communities are met through an expansion of access to water or wastewater services. Poor peri-urban populations have traditionally been underserved because they lack political power or representation, they come from unofficial "communities," or they may be unable to pay as much for water as residents in wealthier areas. Privatization can potentially worsen this neglect, or, if properly designed, can be used to expand the delivery of services.

In the past, private companies have been reluctant to make large investments in the water sector in poor economies. In some cases, however, reaching underserved populations has been an explicit part of privatization concessions or contracts. The concession granted to serve La Paz–El Alto in Bolivia was designed with performance requirements to expand service to the poor. These "expansion mandates" set obligations to achieve certain levels of coverage and water quality (Komives 2001).

Some multinational companies balk at provisions requiring expansion of coverage to marginal communities, stating that it is unrealistic to expect universal household connections in low-income areas in the immediate future, that lack of roads hinders expansion, and that rapid, uncontrolled peri-urban growth prevents proper water planning and service provision (Shambaugh 1999). When meeting these unmet needs is a top priority for governments, tools for inducing concessionaires to invest in coverage in low-income areas should be part of any agreement, with provisions for mandates, quantitative performance indicators, and economic incentives. One benefit of such mandates is that they provide companies with an incentive to develop innovative, lower-cost options for residents. Many private companies also request exclusivity over certain service areas, which may help increase the attractiveness of a concession, but exclusivity may suppress competition and the provision of equitable service.

Worsening Economic Inequities and Water Affordability

One of the greatest concerns of communities and individuals is that privatization will lead to increases in the cost of water to consumers. Water pricing is a complicated issue; indeed, we note three major different types of pricing and affordability questions associated with privatization.

Are price increases necessary? The problem of water rates is perhaps the most controversial issue around privatization efforts. One of the leading arguments offered by proponents of privatization is that private management or ownership of water systems can reduce the water prices paid by consumers. Ironically, one of the greatest concerns of local communities is that privatization will lead to *higher* costs for water and water services. The actual record is mixed—both results have occurred. Significant price increases for some groups of water users may also take place even when overall prices do not rise.

Savings from privatization can result from reduction in system inefficiencies, overhead, labor costs, and management expenses. Economies of scale may also exist for developing new infrastructure. Conversely, water supply is a costly business, and substantial improvements in water systems can lead to increased need for revenue from rates. In addition, the requirement that private companies make a profit may drive up rates in systems where government subsidies were the norm. Private utilities may have better access to capital than some public systems, but they may also have to pay a higher cost for that capital, as well as pay taxes. Finally, a move toward full-cost pricing may improve overall economic efficiency but contribute to rate increases as subsidies are removed or reduced.

There is abundant evidence that people, even those with low incomes, are willing to pay for water and sanitation when the services are reliable and the cost of delivering services is reasonably transparent and understandable to customers (Wright 1997). Experience also suggests that people and businesses will pay more for water without significant resistance when they receive new or improved services that they desire. In the context of privatization, this suggests that dissemination of detailed information about the improvement in services, and the capital investments needed to create those improvements, is essential to public acceptance of increases in overall water prices. The new or improved services should be clearly described and rate changes should be phased in together with strong education and information programs describing the changes and their reason.

Rapid and large increases in water rates cause strong social and political reactions. Public protests and political demonstrations over price increases have taken place in such diverse settings as Cochabamba, Bolivia, Tucuman, Argentina, Puerto Rico, the United States, and Johannesburg, South Africa. In Argentina and Bolivia, rate concerns along with other factors led to privatization efforts being canceled (see Box 3.2). Across Southeastern Asia, disputes over water tariffs are raging. In Indonesia, Jakarta's city council recently approved a rate increase. In Malaysia, rate increases just prior to privatization led to protests. In Manila, Maynilad Water Services has been lobbying for a rate adjustment to cover losses caused by currency fluctuations and threatening to return the concession if its petition is refused by the Philippine government (*FTGWR* 2001b, 117). These experiences fuel public skepticism over arguments that water prices will decline as a result of privatization. In the La Paz–El Alto, Bolivia privatization agreement, the

Box 3.2 Failed Privatizations: Tucuman, Argentina and Cochabamba, Bolivia

Aguas del Aconquija, a subsidiary of Vivendi and local Argentinean companies, won a 30-year concession contract in July 1995 to run the water-supply system for the 1.1 million people of Tucuman, Argentina. Aguas del Aconquija doubled water tariffs within a few months time in order to meet aggressive investment requirements specified in the concession. A new governor, Antonio Bussi, took office around the same time and objected to the privatization. He and his supporters encouraged residents to stop paying bills. Soon afterward, delivered water turned brown, an incident Aguas del Aconquija assured was not harmful and was attributable to the naturally high sediment content of the city's water. The explanation did not convince residents: approximately 80 percent stopped paying their bills. In October 1998 the concession was terminated, but on the condition that the company continue to operate the water system for 18 months. Vivendi agreed, but quickly filed a US $100 million suit against the government, and joined several other companies who had filed complaints against Argentina with the World Bank Arbitration Panel.

An outbreak of violence resulting from a proposal to privatize the public water system in Cochabamba, Bolivia's third largest city, exemplifies the severe problems that can result from rushing toward management of water as an economic good while disregarding its social good aspects. In 1999, the Bolivian government privatized the water system of Cochabamba, partly in response to pressures from the World Bank to make structural adjustments to its economy. The government granted a 40-year concession to run the water system to a consortium led by Italian-owned International Water Ltd. and U.S.-based Bechtel Enterprise Holdings. The consortium also included minority investment from Bolivia. The newly privatized water company immediately modified the rate structure, putting in place a tiered rate and rolling in previously accumulated (but not recovered) debt. As a result, many

General Manager of Aguas del Illimani publicly stated that he would not raise rates in the first five years of the contract, even if the company's costs rise (Komives 2001).

Are water subsidies appropriate and desirable? Subsidies, especially water subsidies, have been a controversial topic for many years (Myers and Kent 1998). On the one hand, economic theory acknowledges that they can be socially desirable and economically efficient in some circumstances. On the other hand, they are often applied as policy favors or social gifts far more widely than necessary to meet critical social goals. Many groups claim they deserve subsidies. Businesses threaten bankruptcy or job cuts if water prices increase. Other users argue that their products or water uses are socially critical or particularly beneficial.

Water-pricing systems often already include some subsidies. Government policies often keep water tariffs low to benefit public welfare. "Lifeline" rates for basic water needs local residents received (or anticipated receiving) increases in their water bills. Aguas de Tunari maintained that the rate hikes would have a large impact only on industrial customers; however, the poor peasants of the town claimed that some residents saw increases as high as 100 percent. Water collection also required the purchase of permits, which threatened access to water for the poorest citizens.

Local farmers had already expressed concerns about privatization. In October 1998, 3,000 farmers organized a march protesting a draft law to charge for water that they believe they own, and for the lack of government attention to a drought that was ruining farmers and livestock owners. After the contract was signed, local groups, including rural farmers, community organizers, and even wealthier groups with an interest in maintaining the parallel private water tanker/trucking market held several protests to demand that the water system stay under local public control. During the protests, the Bolivian army killed as many as nine (reports range from one to nine killed), injured hundreds, and arrested several local leaders. The government also reportedly cut off drinking water to Villa Tunari during this time. Martial law was declared on April 8, 2000, but in late April the government gave in and canceled its contract with Aguas de Tunari. While the cancellation of the contract and the violence have helped put a spotlight on problems with privatization elsewhere, nearly 60 percent of the population still are not served with any water other than expensive water from private tankers. Moreover, these segments of the population are likely to remain the long-term losers from the continued failure to provide adequate clean water.

Sources: Brook Cowen 1999a; Mandell-Campbell 1998; Hudson 1999; Pilling 1996, *FTGWR* 2000a, 2000c, 93, 94; Goldman Foundation 2001; *Business Wire,* October 11, 1999; International Press Service, October 9, 1998; *Earth Island Journal,* September 22, 2000.

are sometimes available for the lowest income groups in a community. Most governments offer substantial subsidies for agricultural water use by farmers and the poorest urban users. A 1997 rate study from the World Bank notes that agricultural water users may pay as little as 20 percent of the total costs of providing irrigation water and may never fully repay capital costs of projects that benefit them (Dinar and Subramanian 1997).

When properly designed, subsidies can satisfy social goals without causing serious problems for the overall market. In practice, however, subsidy design is often unsatisfactory. The quality of water services and coverage are inadequate in many countries. Subsidies directed at the poor often end up benefiting the wealthier populations, while many poor remain unconnected to the system.

One of the potential benefits of privatization is elimination of inappropriate subsidies, a point not lost on those who argue for increased private control. The public sector is often sensitive—some might say too sensitive—to calls for subsidies from various interest groups. Shifting responsibility to the private sector can lead to prices that better reflect costs and allow governments to discontinue subsidies, while letting private providers take most of the heat for price increases. This can be a clear advantage to privatization.

We also note, however, that *lack* of water subsidies in some cases can have disastrous results, especially when combined with pressures to recover costs. In South Africa in 2000, a massive outbreak of cholera occurred in the KwaZulu-Natal region when the local water agencies began requiring repayment of fees for water services. This led some of the poorest communities to abandon clean utility services and switch to free but contaminated water from other sources (http://www.saep.org/forDB/forDBOct00/HEALTHcholeraBEELD001025.htm).

Inadequate attention has been given in privatization negotiations and debates to identifying the difference between appropriate and inappropriate subsidies. When water systems or operations are privatized, it may be desirable to protect some groups of citizens or businesses from paying the full cost of service, perhaps permanently. For example, affordable supply of water sufficient to meet basic needs may be a fundamental human right or may be socially desirable for other reasons, and hence worthy of a subsidy (Gleick 1996, 1999). Water-dependent industries that are critical to local employment patterns or long-term economic growth may be worth subsidizing, either with revenue from other water users or with general tax revenues.

How should rates be designed? Public acceptance of efforts to privatize water services often hinges on decisions about the design and size of rates. One of the problems with privatization is that the incentives for companies to put in place innovative rate structures often conflict with the incentives for companies to generate revenues. When income is a function of how much water a company sells, rate structures that encourage efficient use and conservation may simply reduce overall income. Similarly, inequitable rate structures that favor one class of user over another may be economically beneficial to a company, but socially undesirable.

Rates depend on a wide range of factors, including the balance between one-time system connection fees, fixed fees, and volumetric water prices. Water can be expensive on a volumetric basis, but affordable for low-income families if connection charges and periodic fixed fees are set at zero for these families. Conservation-oriented rate structures, however, are still not well understood or consistently applied even in more developed parts of the world. For example, in a recent survey of California water utilities, more than half of all rate structures used either flat or declining block rates, which are usually less effective at encouraging efficient use of water than increasing block rates (Black and Veatch 2001) (see Figure 3.2).

In the La Paz–El Alto concession, a progressive rate structure was developed that subsidized low-volume residential users and imposed an increasing four-block rate—the more water used, the higher the tariff. Industrial customers pay a single rate, equal to the long-run marginal cost. Two tiers were set for commercial users (see Table 3.4).

Significant customer concerns about changes in water prices are to be expected. As a result, transparent and reasonable explanations for proposed changes are essential for public acceptance. When subsidies from general tax revenues are eliminated, causing rates to increase in general and rate structures to be changed, detailed information about the alternative use of general tax revenues may also be essential to public acceptance.

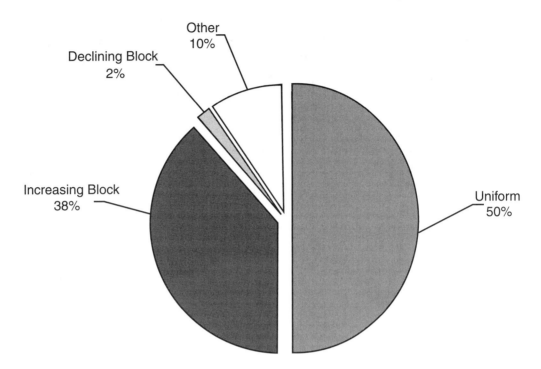

FIGURE 3.2 Various rate structures are used. For California, the diversity of approaches can be seen in the figure, where only 38 percent of water utilities reported using increasing block rates in 2000.

Source: Black and Veatch 2001

TABLE 3.4 Tariff Structure for Aguas del Illimani, Bolivia

Tariff (US$/m³)	Residential	Commercial	Industrial
0.2214	1 to 30 m³		
0.4428	31 to 150 m³		
0.6642	151 to 300 m³	1 to 20 m³	
1.1862	Above 300 m³	Above 20 m³	All water

Notes:
1. 99 percent of all residential customers use less than 150 m³ per month.
2. The long-run marginal cost is estimated at $1.18 per m³.

Source: Komives (2001).

Privatization efforts have also been opposed when rate changes have occurred rapidly and without public education. Rapid rate increases, for any reason, tend to engender opposition and protests, even in publicly operated water systems. Phasing in such increases allows people and businesses to adjust to price changes if the schedule of change is communicated in advance and people believe that it will actually be implemented. In many instances, measures to reduce water use can be adopted before price changes take place, which reduces the financial burden of the changes to consumers. In fact, phased changes in water prices and rates are not only less burdensome for customers, they create greater revenue stability for the water supplier and make financial projections less difficult and burdensome.

Failing to Protect Public Ownership of Water and Water Rights

Privatization of water management can, under some circumstances, lead to the loss of local ownership of water systems, which in turn can lead to neglect of the public interest. Many of the concerns expressed about privatization relate to the control of water rights and changes in water allocations rather than explicit financial or economic problems. In part, this is the result of the deep feelings people have for water. It is also the result, however, of serious neglect of these issues by some who promote privatization.

Water rights and control. Control of water has enormous implications for any society or culture. Many different forms of water "rights" exist—this is not the place to review them (for more information see Bruns and Meinzen-Dick 2000). But each of these cultural, social, or legal controls has developed over time to address some aspect of public ownership, control, and participation over water and water policy. Among other things, they may ensure equitable access to water service, minimize impacts on downstream water users, protect water quality, or resolve disputes.

While some privatization contracts and proposals do not lead to any formal change in water rights, a growing number either intentionally or unintentionally change the status quo. Some even explicitly transfer ownership of water resources from public to

---

Box 3.3 Establishing Public Property Rights for *In-situ* Water

The Edwards Aquifer of South Central Texas is the sole source of drinking water for 1.5 million people in parts of eight counties, including all of San Antonio, the ninth largest city in the nation (according to the 2000 U.S. Census; www.census.gov). The aquifer provides 300 million cubic meters of irrigation water annually for about 34,000 hectares of agricultural land. It also supports an extremely diverse wildlife population in surface springs and underground. At least nine endangered species rely on springflows for their survival, baseflow in the Guadalupe and San Antonio Rivers depends in part on the aquifer, and its subterranean aquatic ecosystem is believed to be the most diverse in the world.

Historically, Texas law granted complete ownership of groundwater to the landowner above it. This common-law rule was replaced long ago in most other U.S. states. Several serious droughts (1984 and 1996), legal decisions to enforce the Endangered Species Act (between 1990 and 1996), and citizen action that raised public understanding of the importance of the aquifer led the Texas legislature to gradually impose public control over (and hence partial public ownership of) water in this and other aquifers in Texas. In 1993 the legislature created an Edwards Aquifer Authority to limit water pumping, penalize violators, issue permits, control the transfer of water rights, and institute water quality programs.

private entities. For example, the Edwards Aquifer Authority in the central United States has considered selling water rights for either a limited period of time (e.g., one year) or in perpetuity (EAA 2001) (Box 3.3). Granting perpetual withdrawal rights would reduce the public's ability to ensure that the aquifer is managed as a social good.

Despite numerous legal challenges, this and other actions to establish public ownership of underground water in Texas have been upheld. Most strikingly, the Supreme Court of Texas rejected a claim that action creating the Edwards Aquifer Authority deprived landowners of a property right vested to them by the Texas Constitution. The Edwards Aquifer Authority is an excellent example of the type of changes in property rights and rules that are necessary if water is to be managed effectively as both a social and an economic good (EAA 2001). Conversely, Box 3.4 shows how easy it is for public ownership in water to be reversed.

Changes in access and water rights may also occur without explicit agreement. One of the causes of tensions in Bolivia over the proposal to privatize the water systems in Cochabamba were efforts to restrict unmonitored groundwater pumping by rural water users and to bring them into the private system. While this may make sense from a purely economic and efficiency perspective, it imposed a fundamental change in the historical use rights in the region.

Another challenge associated with privatization is the degree to which the process of privatization leads to the transfer of government or public assets into the hands of those who are friends of government, or already wealthy. When privatization results in a redistribution of wealth in an inequitable way, there will be strong pressure to oppose or cancel reforms. Confidence in the fairness of the process, in turn, depends on both the design and the transparency of the rules and legal system (Yergin and Stanislaw 1999).

Lack of Public Participation and Contract Monitoring

Oversight and monitoring of public-private agreements are key public responsibilities. Far more effort has been spent trying to ease financial constraints and government oversight, and to promote private-sector involvement, than to define broad guidelines

---

Box 3.4   Backsliding in Texas

Despite success in Texas in establishing some degree of public ownership of groundwater, the Edwards Aquifer Authority has moved in the opposite direction with the interim water trading system it has established. The Authority has sold some water rights for a limited period of time (e.g., one year) but has sold others in perpetuity. Granting perpetual withdrawal rights reduces the public's ability to ensure that future water from the aquifer is managed as a social good. Full implementation of public ownership of water at the source requires that ownership cannot be permanently transferred to private hands.

Source: EAA 2001

for public access and oversight, monitor the public interest, and ensure public partici-pation and transparency.

Weaknesses in monitoring progress can lead to ineffective service provision, dis-criminatory behavior, or violations of water-quality protections. In the late 1980s, Guinea had one of the least developed urban water-supply systems in West Africa. Fewer than 40 percent of urban residents had access to piped water, services were irregular, and water quality was unreliable. In 1989, the government of Guinea entered into a lease arrangement for the capital and sixteen other cities and towns. Consider-able improvements have resulted (Brook Cowen 1999b), but problems with weak mon-itoring and enforcement have led to fewer gains to consumers than expected.

One option is to have regulators set and monitor explicit indicators of service per-formance—"benchmarking." Benchmarking can focus attention on service quality and provide incentives for long-term performance. Performance benchmarking has become standard practice in the water-sector reforms in England and Wales. Ofwat, the public regulator, collects and publishes sets of indicators on an annual basis from water and sewerage companies. These scorecards help pressure the worst providers to improve service and boost the reputations of the best providers (Kingdom and Jagan-nathan 2001). Figure 3.3 shows the "scorecard" Ofwat produced in 2000 based on a variety of performance criteria, including customer service, water pressure, billing factors, public complaints, supply interruptions, water quality, and more.

In São Paulo, Brazil, the introduction of pollution tests and public reporting has led 95 percent of polluting industries to install waste-treatment units to avoid paying fines and seeing their names published. Procuraduría Federal de Protección al Ambiente, Mexico's environmental enforcement agency, will shortly publish information on the

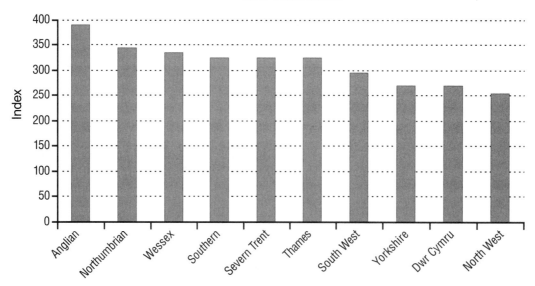

FIGURE 3.3   Ofwat is responsible for oversight and monitoring of private water agencies in the United Kingdom. They produce an annual report addressing the performance of water companies. The higher the score, the better the performance.

Source: http://www.ofwat.gov.uk/pdffiles/los2000.pdf

environmental performance of industries in an effort to encourage improvements in environmental quality (Kingdom and Jagannathan 2001).

There are many barriers to public reporting, including inadequate data, vested interests that block exposure of poor practices, conflicts of interest among agencies that both provide and regulate services, and costs. Nevertheless, the clear advantage of performance monitoring is a strong argument for more universal programs to collect and disseminate benchmarks as a basic part of privatization efforts.

Ignoring Impacts on Ecosystems and Downstream Water Users

Many privatization contracts include provisions to encourage the development of new water supplies, often over a long time period. If privatization contracts do not also guarantee ecosystem water requirements, development of new supply options will undermine ecosystem health and well-being (for both public and private developments). Famous examples of this problem include the Aral Sea in central Asia and the San Francisco Bay–delta ecosystem in central California. Similarly, the largest lake in Mexico, Lake Chapala, is shrinking due to overextraction of groundwater, strong expansion of irrigated areas, reduced flows to the Lerma River, and unchecked urban water demand in the watershed. Once *in situ* flows fall below minimum levels, significant and costly ecosystem damage occurs or society in general is required to purchase water rights from those who have obtained them for free. Authorities in Mexico are now trying to buy back water from agricultural producers (Muñoz 2001).

Decisions about water supply and system operations affect natural flows of water and ecosystem health. Water withdrawals and use come at the expense of riparian and riverine ecosystems. Timing and magnitude of flows may change. Private operators have little incentive to operate reservoirs to maintain minimum downstream flows required for ecosystem health, fishing or recreational interests, and so forth. Balancing ecological needs with water supply, hydroelectric power, and downstream uses of water is a complex task involving many stakeholders. In addition to our growing understanding of the ecological impacts of water development, there has been new attention given in recent years to the economic impacts of these environmental changes as well. We will not review here the growing literature on quantifying the ecological benefits of water systems in economic terms (see, for example, Postel and Carpenter 1997, Daily 1997), but we point out the growing economic costs—typically billions of dollars—being spent to restore previously degraded systems such as the Everglades in Florida and California's Sacramento–San Joaquin delta.

Neglecting the Potential for Water-Use Efficiency and Conservation Improvements

Selling water itself is much easier than selling water conservation and efficiency improvements. One of the greatest concerns of privatization watchdogs is that efficiency programs are typically ignored or even canceled after authority for managing public systems is turned over to private entities. Improvements in efficiency reduce water sales, and hence may lower revenues. As a result, utilities or companies that provide utility services may have little or no financial incentive to encourage conservation. In addition, conservation is often less capital intensive and therefore creates fewer opportunities for investors. Consequently, it may be neglected in comparison with traditional, centralized water-supply projects such as new reservoirs.

Where water scarcity is an important issue, or when new sources of supply are expensive, water-use efficiency improvements may be particularly cost effective. Many of the benefits of such improvements, however, may not be easily or directly measured, including improvements to ecosystem health, energy savings, and reduction in wastewater treatment costs. Capturing those improvements may also be a challenge, requiring policies ranging from proper rate design and pricing to rebates to education and information transfers. Water prices are important tools to encourage improvements in water conservation and use efficiency.

Lessening Protection of Water Quality

Private suppliers of water have few economic incentives to address long-term (chronic) health problems associated with low levels of some pollutants. In addition, private water suppliers have an incentive to understate or misrepresent to customers the size and potential impacts of problems that do occur. As a result, there is widespread agreement that maintaining strong regulatory oversight is a necessary component of protecting water quality. Concerns about the ability of private water providers to protect water quality led the National Council of Women of Canada, a nonpartisan federation of organizations, to adopt a policy in 1997 of opposition to the privatization of water purification and distribution systems (NCWC 1997). The Water Environment Federation in the United States supports "national policy to encourage public/private partnerships (privatization)," but with appropriate public oversight (WEF 2000).

When strong regulatory oversight exists, privatization can lead to improvements in water quality. Standard and Poor's notes that with the water privatization in the UK over the past 11 years, water and wastewater quality have improved (S&P 2000). Indeed, prior to privatization, there was a distinct reluctance of government agencies to monitor and fine other government water providers who were violating water-quality standards—a classic conflict of interest. Governments that own, operate, and finance water and wastewater utilities have shown that they cannot always properly regulate them, too. Privatization has the potential to reduce those conflicts and permit governments to regulate. In the UK, government regulators have greatly increased their successful prosecutions for violations (Orwin 1999).

In Buenos Aires, privatization in the early 1990s led to rapid improvements in wastewater treatment. Aguas Argentinas increased the capacity of water-treatment plants and brought on-line wastewater plants that were previously inoperable (Idelovitch and Ringskog 1995). In Chile, municipal water companies have been run by concessions for many years, with different regional companies granting concessions for water supply and treatment, water distribution, operation of sewers, and sewage treatment. Starting in 1997, the Chilean government began to privatize wastewater treatment as well. All operators are kept under close scrutiny by the Superintendencia de Servicios Sanitarios, an autonomous government agency (Orwin 1999).

Lack of Dispute-Resolution Procedures

Public water companies are usually subject to political dispute-resolution processes involving local stakeholders. Privatized water systems are subject to legal processes that involve nonlocal stakeholders and perhaps nonlocal levels of the legal system. This change in *who* resolves disputes, and the rules for dispute resolution, is accompanied by increased potential for political conflicts over privatization agreements. Public-private partnerships have not often developed clear mechanisms for open participation in dispute resolution, and contracts are often ambiguous in this area. Carefully worded contracts can avoid some such problems. But the water market is relatively new, and some problems are likely to occur even with carefully developed contract language.

It is becoming clear that governments in some developing countries or even local governments in developed countries may not be experienced in negotiating complex contracts that specify level and quality of service, monitoring and success indicators, water-quality protection, and so forth. Contracts also have cultural contexts that differ widely and should be accounted for in specific contract language, such as that related to dispute resolution (Calaguas 1999).

Privatization May Be Irreversible

When governments transfer control over their water system to private companies, the loss of internal skills and expertise may be irreversible, or nearly so. Many contracts are long-term—10 to 20 years or more. Management expertise, engineering knowledge, and other assets in the public domain may be lost for good. Indeed, while there is growing experience with the transfer of such assets to private hands, there is little or no recent experience with the public sector reacquiring such assets from the private sector.

Transferring Assets Out of Local Communities

In the past, revenues generated from local sales of water and services went to local agencies for reinvestment in the community. Because of the multinational character of most water privatization companies, some opponents of privatization fear the loss of a wide range of assets that could be transferred out of local communities. These assets include jobs that may go to outside parties and the profits from operations that go to corporate entities in other countries.

Principles and Standards for Privatization

Despite the vociferous, and often justified, opposition to water privatization, proposals for public-private partnerships in water supply and management are likely to become more numerous in the future. There are many forms of water privatization, or public-private partnerships, making unilateral support or opposition to privatization illogical. However, privatization agreements should meet certain standards and accept specific principles. Consequently, we conclude this chapter with guiding principles and standards for privatization of water-supply systems and infrastructure.

The responsibility for providing water and water services should still rest with local communities and governments, and efforts should be made to strengthen the ability of governments to meet water needs. The potential advantages of privatization are often greatest where governments have been weakest and failed to meet basic water needs. Where strong governments are able to provide water services effectively and equitably, the attractions of privatization decrease substantially. Unfortunately, the greatest risks of privatization are also where governments are weakest, where they are unable to provide the oversight and management functions necessary to protect public interests. This contradiction poses the greatest challenge for those who hope to make privatization work successfully.

1. Continue to Manage Water as a Social Good

a. Meet basic human needs for water. All residents in a service area should be guaranteed a basic water quantity under any privatization agreement. Contract agreements to provide water services in any region must ensure that unmet basic human water needs are met first, before more water is provided to existing customers. Basic water requirements should be clearly defined (Gleick 1996, 1999).

b. Meet basic ecosystem needs for water. Natural ecosystems should be guaranteed a basic water requirement under any privatization agreement. Basic water-supply protections for natural ecosystems must be put in place in every region of the world. Such protections should be written into every privatization agreement, enforced by government oversight.

c. The basic water requirement for users should be provided for less than cost when necessary for reasons of poverty. Subsidies should not be encouraged blindly, but some subsidies for specific groups of people or industries are occasionally justified. One example is subsidies for meeting basic water requirements when that minimum amount of water cannot be paid for due to poverty.

2. Use Sound Economics in Water Management

a. Water and water services should be provided at fair and reasonable rates. Provision of water and water services should not be free. Appropriate subsidies should be evaluated and discussed in public. Rates should be designed to encourage efficient and effective use of water.

b. Whenever possible, link proposed rate increases with agreed-upon improvements in service. Experience has shown that water users are often willing to pay for improvements in service when such improvements are designed with their participation and when improvements are actually delivered. Even when rate increases are primarily motivated by cost increases, linking the rate increase to improvements in service creates a performance incentive for the water supplier and increases the value of water and water services to users.

c. Subsidies, if necessary, should be economically and socially sound. Subsidies are not all equal from an economic point of view. For example, subsidies to low-income users that do not reduce the price of water are more appropriate than those that do because lower water prices encourage inefficient water use. Similarly, mechanisms should be instituted to regularly review and eliminate subsidies that no longer serve an appropriate social purpose.

d. Private companies should be required to demonstrate that new water-supply projects are less expensive than projects to improve water conservation and water-use efficiency before they are permitted to invest and raise water rates to repay the investment. Privatization agreements should not permit new supply projects unless such projects can be proven to be less costly than improving the efficiency of existing water distribution and use. When considered seriously, water-efficiency investments can earn an equal or higher rate of return to that earned by new water-supply investments. Rate structures should permit companies to earn a return on efficiency and conservation investments.

3. Maintain Strong Government Regulation and Oversight

a. Governments should retain or establish public ownership or control of water sources. The "social good" dimensions of water cannot be fully protected if ownership of water sources is entirely private. Permanent and unequivocal public ownership of water sources gives the public the strongest single point of leverage in ensuring that an acceptable balance between social and economic concerns is achieved.

b. Public agencies and water-service providers should monitor water quality. Water suppliers cannot effectively regulate water quality. Although this point has been recognized in many privatization decisions, government water-quality regulators are often under-informed and under-funded, leaving public decisions about water quality in private hands. Governments should define and enforce laws and regulations. Government agencies or independent watchdogs should monitor, and publish information on, water quality. Where governments are weak, formal and explicit mechanisms to protect water quality must be even stronger.

c. Contracts that lay out the responsibilities of each partner are a prerequisite for the success of any privatization. Contracts must protect the public interest; this requires provisions ensuring the quality of service and a regulatory regime that is transparent, accessible, and accountable to the public. Good contracts will include explicit performance criteria and standards, with oversight by government regulatory agencies and non-governmental organizations.

d. Clear dispute-resolution procedures should be developed prior to privatization. Dispute resolution procedures should be specified clearly in contracts. It is necessary to develop practical procedures that build upon local institutions and practices, are free of corruption, and difficult to circumvent.

e. Independent technical assistance and contract review should be standard. Weaker governments are most vulnerable to the risk of being forced into accepting weak contracts. Many of the problems associated with privatization have resulted from inadequate contract review or ambiguous contract language. In principle, many of these problems can be avoided by requiring advance independent technical and contract review.

f. Negotiations over privatization contracts should be open, transparent, and include all affected stakeholders. Numerous political and financial problems for water customers and private companies have resulted from arrangements that were perceived as corrupt or not in the best interests of the public. Stakeholder participation is widely recognized as the best way of avoiding these problems. Broad participation by affected parties ensures that diverse values and varying viewpoints are articulated and incorporated into the process. It also provides a sense of ownership and stewardship over the process and resulting decisions. We recommend the creation of public advisory committees with broad community representation to advise governments proposing privatization, formal public review of contracts in advance of signing agreements, and public education efforts in advance of any transfer of public responsibilities to private companies. International agency or charitable foundation funding of technical support to these committees should be provided.

Conclusions

Private provision of water is not a new idea, nor is it a radical one—as examples from more than a century ago demonstrate. But in the past decade, the trend toward privatization of water has greatly accelerated, with both successes and spectacular failures. Insufficient effort has been made to understand the risks of water privatization, and to put in place guiding principles and standards to govern privatization efforts.

Vast expenditures of funds over the past several decades have permitted urban water utilities to develop systems that adequately serve middle- and upper-class residents. But these government agencies may be bureaucratic, inefficient, and overstaffed and many in less-developed countries are increasingly failing to provide water services to poorer populations. Indeed, the urban poor in developing countries rarely have adequate access to either water or sanitation, and as a result, they often pay far more than middle-class customers for water of uncertain quality from private vendors. Efficient management of the world's limited water resources can be improved by the wise application of economic and financial tools. For this reason, proponents of water privatization argue that privatizing water offers a chance for fixing the deficiencies of public water systems.

Unfortunately, the move toward private markets has failed to address some of the most important issues and concerns about water. In particular, water has vital social, cultural, and ecological roles to play that cannot be protected by purely market forces. In addition, certain management goals and social values require direct and strong government support and protection, yet privatization efforts are increasing rapidly in regions where strong governments do not exist. Therefore, any efforts to privatize or commodify water must be accompanied by formal guarantees to respect certain principles and support specific social objectives. Among these are the need to provide for basic human and ecosystem water requirements as a top priority, independent monitoring and enforcement of water-quality standards, equitable access to water for poor populations, inclusion of all affected parties in decision making, and increased reliance on water-use efficiency and productivity improvements. Openness, transparency, and strong public regulatory oversight are fundamental requirements in any efforts to share the public responsibility for providing clean water with private entities.

Water is both an economic and social good. As a result, completely "free" market forces can never completely and equitably satisfy social objectives. Given the legitimate concerns about the risks of privatization, efforts to capture the positive characteristics of the private sector must be balanced with efforts to address privatization's flaws, gaps, and controversial aspects. Thus, efforts to privatize water must be accompanied by efforts to clearly define these public objectives and to determine how best to satisfy them. Water is far too important for human and ecological well-being to be left to narrow economic interests. This balance requires careful thought and the intelligent application of policy. Whether that balance will be achieved remains to be seen.

REFERENCES

American Water Works Association (AWWA). 2001. *Reinvesting in Drinking Water Infrastructure.* American Water Works Association, Denver, Colorado (May). Available at http://www.awwa.org/govtaff/infrastructure.pdf.

Anderson, L. 1991. "Water and the Canadian city." In *Water and the City.* Public Works Historical Society, Chicago.

Barlow, M. 1999. *Blue Gold: The Global Water Crisis and the Commodification of the World's Water Supply.* International Forum on Globalization, Sausalito, California.

Beecher, J.A. 1997. "Water utility privatization and regulation: Lessons from the global experiment." *Water International,* Vol. 22, No. 1, pp. 54–63.

Beecher, J.A., G.R. Dreese, and J.D. Stanford. 1995. *Regulatory Implications of Water and Wastewater Utility Privatization.* National Regulatory Research Institute, Columbus, Ohio (July).

Black and Veatch. 2001. *California Water Charge Survey 2001.* Black and Veatch Corporation, Irvine, California.

Blake, N.P. 1991. "Water and the city: Lessons from history." In *Water and the City.* Public Works Historical Society, Chicago.

Blokland, M., O. Braadbaart, and K. Schwartz (eds.) 1999. *Private Business, Public Owners: Government Shareholdings in Water Enterprises.* Ministry of Housing, Spatial Planning and the Environment, The Hague, Netherlands.

Brook Cowen, P.J. 1999a. "Bail out: The global privatization of water supply." *Urban Age Magazine,* Winter. Urban Development. The World Bank, Washington, D.C.

Brook Cowen, P.J. 1999b. *Lessons from the Guinea Water Lease.* Public Policy for the Private Sector, Note No. 78. The World Bank Group, Washington, D.C., April.

Brubaker, E. 2001. *The Promise of Privatization.* Energy Probe Research Foundation, Toronto, Ontario (April). Available in full at http://www.environmentprobe.org/enviroprobe/pubs/Ev548.htm.

Bruns, B.R., and R.S. Meinzen-Dick (eds.). 2000. *Negotiating Water Rights.* International Food Policy Research Institute. Vistaar Publications, New Delhi, India.

Calaguas, B. 1999. "Private sector participation." Thematic Paper prepared for Vision 21 process, Export Group Meeting, April, Wageningen, The Netherlands.

Daily, G.C. (ed.). 1997. *Nature's Services: Societal Dependence on Natural Ecosystems.* Island Press, Washington, D.C.

Dinar, A., and A. Subramanian. 1997. *Water Pricing Experiences: An International Perspective.* World Bank Technical Paper No. 386. The World Bank, Washington, D.C.

EAA. 2001. Web site of the Edwards Aquifer Authority: www.e-aquifer.com.

The Economist. 2000. "Nor any drop to drink." March 25, p. 69.

Faulkner, J. 1997. "Engaging the private sector through public-private partnerships." In *Bridges to Sustainability, Yale Bulletin Series,* No. 101. Yale University Press, New Haven, Connecticut.

Financial Times Global Water Report (FTGWR). 2000a. "Bolivia: The Cochabamba crisis." April 14, Issue 93. pp. 1–3.

Financial Times Global Water Report. 2000b. "French giants slug it out." April 28, Issue 94. pp. 10–12.

Financial Times Global Water Report. 2000c. "International water's rebuttal: Bolivia." April 28, Issue 94, pp. 4–6.

Financial Times Global Water Report. 2001a. "Azurix writes down its activities in Argentina." January 26, Issue 114. p. 8.

Financial Times Global Water Report. 2001b. "Maynilad pull-out threat in Manila." January 26, Issue 114. pp. 1–3.

Financial Times Global Water Report. 2001c. "Maynilad fights on in face of government opposition." March 12, Issue 117. pp. 6–7.

Gleick, P.H. 1996. "Basic water requirements for human activities: Meeting basic needs." *Water International,* Vol. 21, pp. 83–92.

Gleick, P.H. 1999. "The human right to water." *Water Policy,* Vol. 1, pp. 487–503.

Global Water Partnership (GWP). 2000. *Toward Water Security: A Framework for Action to Achieve the Vision for Water in the 21st Century.* Global Water Partnership, Stockholm, Sweden.

Goldman Foundation. 2001. Goldman Prize winners 2001, Oscar Olivera. Available at http://www.goldmanprize.org.

Gopinath, D. 2000. "Blue gold." In *Institutional Investor International Edition.* February.

Hudson, P. 1999. "Muddy waters; Argentina, the Latin American leader in private water works, struggles to create a transparent concession system." *Latin Trade.* March.

ICWE. 1992. *The Dublin Principles.* Available in full at http://www.wmo.ch/web/homs/icwedece.html.

Idelovitch, E., and K. Ringskog. 1995. *Private Sector Participation in Water Supply and Sanitation in Latin America.* The International Bank for Reconstruction and Development/The World Bank, Washington, D.C. (May). Available at http://www.worldbank.org/html/lat/english/papers/ewsu/ps_water.txt.

Kingdom, B., and V. Jagannathan. 2001. *Utility Benchmarking: Public Reporting of Service Performance.* The World Bank Group, Note No. 229. World Bank, Washington, D.C. (March).

Komives, K. 2001. "Designing pro-poor water and sewer concessions: Early lessons from Bolivia." *Water Policy,* Vol. 3, No. 1, pp. 61–80.

LeClerc, G., and T. Raes. 2001. *Water: A World Financial Issue.* PriceWaterhouseCoopers, Sustainable Development Series, Paris, France.

Mandell-Campbell, A. 1998. "Argentina sell-off sparks litigation." *National Law Journal,* July 20.

Market Guide. 2001. Vivendi Universal. Available at http://yahoo.marketguide.com/mgi/MG.asp?nss=yahoo&rt=ageosegm&rn=A26C7.

National Council of Women of Canada (NCWC). 1997. *Privatization of Water Purification and Distribution Systems.* Policy 97.14EM. Available at http://www.ncwc.ca/policies/water_waterquality.html.

Neal, K., P.J. Maloney, J.A. Marson, and T.E. Francis. 1996. *Restructuring America's Water Industry: Comparing Investor-Owned and Government Water Systems.* Reason Public Policy Institute, Policy Study No. 200. January.

Nickson, A. 1998. *Organizational Structure and Performance in Urban Water Supply: The Case of the SAGUAPAC Co-operative in Santa Cruz, Bolivia.* International Development Department, University of Birmingham, United Kingdom.

Orwin, A. 1999. *The Privatization of Water and Wastewater Utilities: An International Survey.* Environment Probe, Canada. Available at http://www.environmentprobe.org/enviro-probe/pubs/ev542.html.

Panos. 1998. *Liquid Assets: Is Water Privatization the Answer to Access?* Panos Briefing, London, United Kingdom.

Pilling, D. 1996. "Generale des Eaux in row over Argentine water." *Financial Times* (London), February 13.

Postel, S., and S. Carpenter. 1997. "Freshwater ecosystem services." In G.C. Daily (ed.), *Nature's Services: Societal Dependence on Natural Ecosystems.* Island Press, Washington, D.C., pp. 195–214.

Rivera, D. 1996. *Private Sector Participation in the Water Supply and Wastewater Sector: Lessons from Six Developing Countries.* The World Bank, Washington, D.C.

Savas, E.F. 1987. *Privatization: The Key to Better Government.* Chatham House, Chatham, New Jersey.

Shambaugh, J. 1999. *Role of the Private Sector in Providing Urban Water Supply Services in Developing Countries: Areas of Controversy.* Research Paper, United Nations Development Program/Yale University Research Clinic, New Haven, Connecticut.

Standard and Poor's (S&P). 2000. "European water: Slow progress to increase private sector involvement." September 12. Available at http://www.sandp.com/Forum/Ratings Commentaries/CorporateFinance/Articles/eurowater.html.

Trémolet, S. 2001. "Putting the poor at the heart of privatization." *Financial Times Global Water Report,* Issue 114.

Tully, S. 2000. "Water, water everywhere." *Fortune,* 15 May 2000.

Waddell, J.A. N.d. *Public Water Suppliers Look to Privatization.* PriceWaterhouseCoopers. Available at http://Waddell.UtilitiesProject.com.

Water Environment Federation (WEF). 2000. "Financing water quality improvements." Available at http://www.wef.org/govtaffairs/policy/finance.jhtml.

World Commission on Water for the 21st Century. 2000. *A Water Secure World: Vision for Water, Life, and the Environment.* World Water Vision Commission Report. The Hague, Netherlands.

World Health Organization (WHO). 2000. *Global Water Supply and Sanitation Assessment 2000.* World Health Organization and United Nations Children's Fund. Report. Available at http://www.un.org/Depts/unsd/social/watsan.htm#srce.

Wright, A.M. 1997. *Toward a Strategic Sanitation Approach.* UNDP-World Bank Water and Sanitation Program, Washington, D.C.

Yepes, G. 1992. "Alternativas en el manejo de empresas del sector agua potable y saneamiento." Departamento Nacional de Planeación. Seminario Internacional Eficiencia en la Prestación de los Servicios Públicos de Agua Potable y Saneamiento Básico. Bogotá, Colombia, p. 111–16.

Yergin, D., and J. Stanislaw. 1999. *The Commanding Heights: The Battle between Government and the Marketplace That Is Remaking the Modern World.* Simon and Schuster, New York.

CHAPTER 4

Measuring Water Well-Being: Water Indicators and Indices

Peter H. Gleick, Elizabeth L. Chalecki, and Arlene Wong

The connections between freshwater availability and use and human well-being have been the focus of extensive writings and analysis in the past decade. It is well understood that adequate clean water is a fundamental necessity for human and ecological health, the production of food, goods, and services, the generation of energy, and much more. As our understanding of the links among these variables has grown, so too has the desire to measure, track, and evaluate various aspects of these connections. Accordingly, the use of water-related "indicators" or "indices" to measure the vulnerability of water systems, the quality of human or ecosystem well-being, or the level of development, has greatly accelerated in recent years. In part, this is the result of a growing interest in human development. It also results from a growing understanding of the limitations of the traditional measures used during the twentieth century. These measures were often simple estimates of economic wealth such as income or gross domestic product.

Many possible indicators can be developed to measure aspects of well-being related to water. These can vary in complexity and content. Among the simplest are straightforward estimates of water supply or demand, measured as total quantity, or quantity per unit time, or quantity per person ("per capita"). More complex indices have also been discussed or developed, such as the Human Poverty Index (HPI-1) prepared by the United Nations. The HPI-1 was not developed for the purposes of evaluating water conditions but as a broader measure of human well-being. It works from the premise that well-being includes factors other than simple measures of income or expenditures or economic growth in monetary terms. Instead, the HPI-1 is a composite, incorporating measures of health, economics, and more.

Other indicators have been developed specifically to address water-related issues. These can include elements such as water availability, access to clean water and safe sanitation, the time and effort required to collect domestic water, cost and price, quality, vulnerability of water systems to climate change, and more. This chapter discusses the design and development of water indices to measure various aspects of human and ecosystem well-being. There is a long history of attempts to create such indices and measures—several of which are reviewed below.

Quality-of-Life Indicators and Why We Develop Them

What Are Indicators?

Access to information is essential in planning, designing, monitoring, and implementing policies and actions. As a result, there is a growing need for both actual data and more sophisticated ways of combining and interpreting those data. In the last few decades, the demand for data on environmental and natural resource issues has exploded, as interest in the protection and management of resources has become more sophisticated and urgent. Since the 1972 Stockholm environmental conference, there has been increasing emphasis on the creation of, and access to, good data for planning and decision making at all levels of government—local, regional, national, and international. At a global level, we find initiatives related to data assessment, collection, and dissemination. For example, the *Critical Trends* report prepared for the five-year review of the United Nations Conference on Environment and Development, the United Nations Environment Programme's *Global Environmental Outlook* (GEO) reports, and the UNEP/World Resources Institute/United Nations/World Bank *World Resources Reports* all provide environmental, resource, consumption, and waste data for many different countries as well as related information on poverty, food security, and other environmentally derived conditions.

Such data are valuable by themselves. More and more, however, the environmental policy community is interpreting and packaging data into policy and planning tools. This has given rise to the use of "indices"—combinations of data in different forms. Definitions of the terms *indices* and *indicators* are provided in Box 4.1. While data are simply facts or individual measurements, indices are qualitative or quantitative measures, typically tracked over time, that provide information about the conditions of a system or phenomenon. Sometimes indicators are combinations of separate data that provide integrated assessments of complex relationships; sometimes indicators are more simple proxies for something that is not easily measurable. For example, a proxy indicator for the health of an aquatic system may be a fish count for a single species in a particular location. Thus, simple measures may be used to provide at least

---

Box 4.1 Definitions

Indicator/measure: a single number that is derived from a variable, such as literacy rate or carbon dioxide concentration in the air (can be measured at a single time or tracked over time).

Index: a single number that is a mathematical aggregation of two or more indicators.

Indicator profile: a number of indicators presented together but not aggregated.

Source: Ott 1978.

some information on more complex, but difficult to quantify, conditions of ecological health and well-being.

When measured over time, an indicator can denote the direction and magnitude of change. It must be recognized that indicators represent a value system as well. Indicators are selected with an end or a goal in mind, and their trends indicate the bettering or worsening of the condition over time. This information can then be used to develop appropriate actions.

The Development and Use of Indicators and Indices

Indicators inform decision making. Policy and decision makers have long relied on economic indicators (such as employment, gross national product, productivity, inflation) to measure and report on economic health. Similarly, they have used social indicators (such as infant mortality rates, poverty rates, and life expectancy) to evaluate quality of life and human conditions. Indices can be used as a research and planning tool, for education, or for advocacy promoting environmental protection and sustainable development goals. In the last several decades, the environmental community has used indicators to measure environmental conditions and to lobby for policies to improve those conditions.

The development and use of indicators and indices at all levels of government has become increasingly common in the last decade for a variety of reasons and goals. Different measures have been developed to evaluate the vulnerability of water systems to human activities, such as climate change. By reporting on conditions and simplifying processing of complex information, they have become tools to measure progress and evaluate the impacts or effectiveness of policies, programs, or actions.

A prime characteristic of indicators is their utility in providing a composite measure of development as a way of evaluating the change in a condition over time. Indicators are also a valuable way of communicating information, educating and informing audiences, and creating accountability. The public educational value of indicators can be seen in the use of a water scarcity index in recent major articles in general audience publications such as *Scientific American* and *National Geographic* (see the February 2001 issue of *Scientific American* and EarthPulse in the April 2001 issue of *National Geographic*).

Indicators will also define areas of concern and actions that should be taken. For some countries or international donors, a measure like the Human Development Index (HDI) can help rank regions in need of assistance and plan actions and funding to address focused problems. Indicators reflect values and assumptions about strategies to address the conditions they report on, and in doing so, they can help define priorities and build consensus around long-term goals. They can therefore be tools for advocates to galvanize public opinion around a course of action.

The emergence of a more comprehensive view of "sustainability" as a goal has led to the desire for a more comprehensive set of measures on economic, social, and environmental interrelationships. Thus, sustainability and environmental indicators at the international, national, and local levels are now used as tools to communicate existing conditions, to demonstrate the concept and value of sustainability, and to evaluate long-term trends. By identifying measures that show the relationships between economic, human, and environmental well-being, indices can be a powerful communication tool for decision makers and the public.

Current Water Indicators

To date, there have been several assessments of world water conditions (such as the 1997 *Comprehensive Assessment of the Freshwater Resources of the World* requested by the UN Commission for Sustainable Development and the series of works done in preparation for the World Water Forum in the Hague in March 2000). These assessments tend to use aggregated country information: water indicators specific to a narrow aspect of water management or planning, such as scarcity or availability, water quality, health, water services, or resource management. This holds true whether the indicator/index is intended to be a measure about water resources or conditions or whether the water measures are embedded in a larger indicator or assessment project.

At an international level, there are development organizations that implement specific programs influencing water management or planning. Nations or international aid organizations invest in specific water infrastructure projects such as irrigation systems, ecosystem restoration, or water-system conservation programs that affect water management and planning. Health organizations such as the WHO may develop programs specific to water-borne diseases or other health-related issues tied to water treatment, water availability, or water quality. International agricultural organizations such as the Food Agricultural Association (FAO) or Consultative Group on International Agricultural Research (CGIAR) focus on issues related to crop productivity, irrigation requirements, and plant water use. Many of these organizations are developing indicators to evaluate impacts and help direct planning. The World Bank, for example, has established numerous indicator programs to monitor and evaluate the positive and negative environmental impacts of Bank-supported activities.

Still other international bodies, such as the United Nations, the Organization for Economic Co-operation and Development (OECD), and non-governmental environmental organizations, support broad programs related to sustainable development and environmental protection, which are influenced by freshwater issues. The United Nations supports an Indicators of Sustainable Development project that includes 30 agencies, other international organizations, and governments (see http://www.un.org/esa/sustdev/indi6.htm). In addition, 22 countries are currently testing indicators to establish a common overall framework and methodologies.

Agenda 21 is a comprehensive plan of action to be taken globally, nationally, and locally in every area in which there are human impacts on the environment. Agenda 21, adopted by more than 178 governments at the United Nations Conference on Environment and Development (UNCED) held in Rio de Janeiro, Brazil, in 1992, calls for recommendations for sustainable development indicators at the national, regional, and global levels. The goal is for a set of measures and indicators that can be regularly updated in accessible reports and databases for use at the international level.

In response to the calls for a comprehensive index of progress toward sustainable development, the United Nations Development Programme (UNDP) has put considerable resources into the *Human Development Report* and the Human Development Index. In an acknowledgment of the limitation of traditional economic measures of well-being, UNDP proposed that a new index, the HDI, be created that "focuses not just on poverty of income but on poverty from a human development perspective—on poverty as a denial of choices and opportunities for living a tolerable life" (UNDP 1997). The first version of the HDI was published in 1990 and is described later in this chapter.

Numerous other organizations have developed and publish various kinds of indicators and indices to track and compare environmental conditions, including the World Health Organization (see Box 4.2 for a composite health index presented in 2001), Worldwatch Institute, World Wildlife Fund, and World Resources Institute. The OECD has also developed a core set of environmental indicators, published every two years, to keep track of environmental progress, ensure integration of environmental concerns into sectoral and economic policies, measure environmental performance, and help determine whether countries are on track toward sustainable development.

Starting in 1996, the OECD launched its Shaping the 21st Century project to formulate and implement strategies for sustainability for every member country. As part of the effort, the OECD/DAC (Development Assistance Committee), the United Nations, the World Resources Institute, and national representatives are working to develop a new set of indicators to measure progress toward sustainable development, including measurements of water resources, impacts on water resources, and water-related impacts on human populations. The purpose for these indicator projects is largely to promote goals for sustainable use of resources, evaluate and report on progress, and identify priorities.

Box 4.2   The World Health Organization Health Index

In 2001 the World Health Organization developed a composite index measuring the performance of health systems in 191 countries. According to *World Health Report 2000,* advances in the quality of human health can be achieved by improving the way available health interventions are organized and delivered. Differences in the quality of "health" in different countries often reflect differences in the performance of their health systems. And differences in outcomes among groups within countries can be attributed to disparities in access to health services. The composite index summarizes performance in terms of both the overall level of goal achievement and the distribution of that achievement, giving equal weight to these two aspects.

Five components make up the index: overall good health, distribution of good health, overall responsiveness of the health system, distribution of responsiveness, and fairness in financial contributions. "Good health" is measured by disability-adjusted life expectancy. The distribution of good health is measured by an equality of child survival index. The overall responsiveness of the health system and the distribution of responsiveness are measured on the basis of survey responses looking at respect for patients and client orientation. Fairness in financial contributions is estimated using the ratio of total household spending on health to their permanent income above subsistence.

Source: Based on WHO 2000b.

At a national level, many governments are developing indicators as tools to support national decision-making processes and to meet reporting obligations for international conventions, agreements, and other forms of cooperation. Academics and non-governmental organizations have also been key in helping to collect and disseminate national and international data. Many of these indicator sets or indices are targeted toward selected national policies or sectoral issues—sustainable development, energy, transportation, environment/natural resources, and so on. Many of these national indices on sustainability, environment, or health may contain some water-related measures. However, for many countries, much of the specific water planning and management takes place at the subnational to local level.

In the United States, there are numerous efforts to measure sustainability, community improvement, and quality of life, organized at the regional, county, city, and even neighborhood levels. Many of these incorporate indicators to both communicate priorities and measure progress of these efforts (see, for example, http://www.neip.org).

Limitations to Indicators and Indices

There is considerable interest in water-related indices, as shown by the continued attempts to develop and apply them. While we recognize their potential value, we are also wary of their considerable limitations, and we offer some cautionary thoughts for potential users of such information. No index can completely describe the characteristics of water problems or needs, or the quality of life related to access to water. Below we describe a number of important problems with indices, many of which are common to general efforts to develop indices of well-being. Some specific points about these indices include the following:

- Data problems and constraints are probably the most important limitation to the development of comprehensive and useful indices. The availability, quality, and dissemination of water data all have serious problems.
- Water issues are, by their nature, interdisciplinary and multifaceted. No single index can provide all appropriate information about water availability, use, quality, and equity.
- Aggregating different measures into a single index is attractive but dangerous. Single aggregate measures can be misleading and uninformative. Composite indices are more valuable and flexible, allowing different kinds of comparisons to be made at different regional scales.
- Definitions of terms are often uncertain and change with time. Care must be taken to clearly define different measures.
- Scales and locations are critical factors for determining the nature of a water-related problem.
- Overinterpretation and overuse of indices is possible. The debates concerning water indices must recognize the complex nature of water policy and planning and the fact that there are often widely conflicting interests involved. An index that measures one particular aspect of water may say nothing about another aspect of concern to a group of people or a region.

Data Problems and Constraints

The greatest problem in producing useful indices of well-being, for water or any other aspect of development, comes not from lack of computer capability or from our ability to conceive of appropriate measures but from limitations on the quality, availability, and regional resolution of data. Because each possible index varies in the kind and scope of data required, data problems will also vary. Some of the most typical and important problems include gaps in data availability, lack of time-series, and inaccuracies:

- Gaps in data on water availability still exist and are unlikely to be filled soon. While precipitation, temperature, and runoff are relatively well measured in most developed countries, many regions of the world suffer from gaps in both present-day instrumental coverage and from lack of any long-term records. Even in richer countries, pressures to cut funds for observations and monitoring stations threaten the continuity of time-series data.

- Water-use data are especially hard to find. Many indices need data on how much water is used, by individuals or sectors. Yet far fewer data are collected on water use than on water supply and availability (see Chapter 1 and the Introduction to the Data Section). Domestic water use is often not measured directly and details on how that water is used are rarely collected. Industrial and commercial water use is inventoried infrequently or not at all. Agricultural water-use data, including both irrigation water and rainfall, are uneven and unreliable. Groundwater withdrawals are rarely measured or regulated. While the problems are worse in developing countries, even rich countries have a poor record in compiling water-use data. In Canada, water-use data for the nation as a whole have not been compiled since 1991. Even when water-use data are collected, information on changing water-use patterns over time is often not available, making analysis of trends difficult.

- Some countries or regions still restrict access to water data. Even in this era of easy Internet access, some countries refuse to share water-related data with neighbors or even their own scientists. In regions where water is shared internationally, nations are tempted to restrict information when there is a perceived political advantage in doing so. We have previously commented on this problem (see Gleick 2000) and believe that open sharing of water data is critical for proper and effective water planning and management. Further, the development of informative, comprehensive indices is not possible unless data are collected and shared.

- Some water uses or needs are unquantified or unquantifiable. In all likelihood, some uses and needs are unlikely ever to be accurately determined or included in scenario projections. For example, ecological needs, recreational uses, water for hydropower production or navigation, and reservoir losses to seepage or evaporation are often difficult to calculate with any accuracy. Nevertheless, these water uses and activities will eventually need to be quantified if they are to be included in indices. Excluding them would mean excluding critical factors related to well-being.

Interdisciplinary Nature of Water Problems

Many water indices or sets of water indicators fail to recognize the interconnection of water well-being with social, economic, and environmental aspects of human development, instead measuring a single part of a problem: scarcity, use, quality, and so on. At subnational and local levels, data may be collected on water use or water quality, but again, they are often not integrated into indicator development around freshwater resources as a whole, or even as components of broader planning. Without such integration, it is difficult to see how such measures can contribute to more comprehensive decision making.

Aggregating Different Measures

Building an "index" involves aggregating several indicators together to calculate a single index. Indices have the benefits of further simplification and aggregation of numerous measures and the improved comparability across units. Some disadvantages of aggregating data include the difficulty of weighting diverse parameters, which may often include measures that cannot utilize the same units; the difficulty of interpreting the movement over time of a single number, given the many aspects that may influence it; the difficulty in discerning these influences in a single number (i.e., a few indicators may be driving the results but that influence will be hidden in the single number); and the difficulty in comparing results across units when influential indicators may be different for different countries.

The aggregation of a variety of water-related data sets into a single index is more of an art than a science. In aggregating individual indicators, choices must be made about which indicators to include; how to weight the different components; and the specific form of the aggregation (e.g., additive, multiplicative, etc.). Not all the variables are independent, which can lead to hidden imbalances and weights. There are often many correlations, dependencies, and relationships among the various indicators.

Aggregation of multiple indicators requires explicit or implicit weighting systems. These always have values, which may come to dominate the analysis if the indicators do not move in the same direction over time. In aggregating five indicators of vulnerability to climate change, Gleick (1990) used a linear counting approach that summed the number of individual indicators classified as critical. Sullivan (2001) offers a more complex approach to weighting, permitting the user to define and apply different nonlinear weights.

Changing Definitions

Among the most important and commonly cited measures of water well-being are the fractions of the global population with access to clean water and sanitation services. Data on these measures are available going back to 1970, yet the actual definitions of *access, clean water,* and *sanitation services* have changed over the years, making it very difficult to draw robust conclusions about trends and the effectiveness of efforts to meet basic needs. In the 1990 World Health Organization assessment, they noted in a tiny footnote, "There is no standard regional or global definition for coverage" (WHO 1992). In 1996, WHO published information on how some countries defined several related terms (see Box 4.3). For example, 54 countries (out of 84 reporting in 1994) provided definitions of *convenient distance,* and 38 countries offered definitions of *adequate amount.* Table 4.1 shows the number of countries defining *convenient distance* in different ways. Figure 4.1 shows how different countries (reporting to WHO) define *minimum* acceptable water quantities for rural areas.

In 2000, these definitions changed again. The World Health Organization replaced the terms *safe* and *adequate* with *improved* and defined a set of technologies as *improved.* Populations with access to those technologies were assumed to be covered (http://www.who.int/water_sanitation_health/Globassessment/Global1.2.htm).

As definitions to various terms and measures change, getting a clear picture of how particular measures may have improved or deteriorated is extremely difficult. Users should be aware of definitions and changes in those definitions before claiming victory (or failure) in efforts to improve human well-being.

TABLE 4.1   Definitions of "Access to Safe Drinking Water Source"

Number of countries defining access as "Water source at a distance of less than…"								
	50 m	100 m	250 m	500 m	1000 m	2000 m	5 minutes	15 minutes	30 minutes
Urban	20	6	3	8	1	–	1	–	1
Rural	10	1	6	17	4	4	–	1	1

Source: WHO 1996.

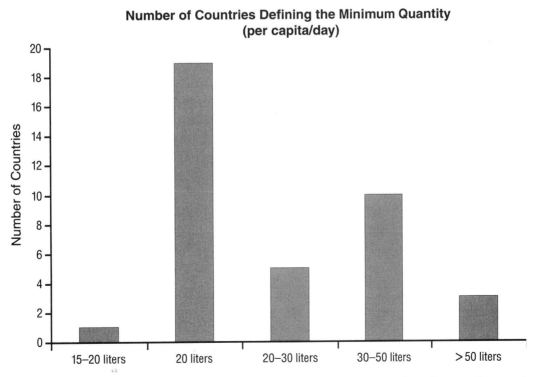

Number of Countries Defining the Minimum Quantity (per capita/day)

FIGURE 4.1   The number of countries defining a minimum quantity of water for rural inhabitants, in per capita use per day.

Source: WHO 1992

Scale and Location Mismatches

An indicator reported at one scale may not be informative for policy making at another. For example, tracking the percentage of population with access to sanitation services or safe drinking water may show improvement at a country level, but fail to identify those local areas that have no services. Some water indicators are necessarily limited in their ability to depict differences in important regional contexts that could affect levels of stress and vulnerability. What might be indicative of a critically threatened system in one region may be less critical in another. China, for example, is not considered "water stressed" in terms of average per capita water availability. Yet some regions within China have considerably less water than the average per capita availability for the country as a whole. This kind of problem is extremely common.

Over-Interpretation of the Results

The results of different indices and indicators should not be "over-interpreted." Because of the limitations of indices, any one result should not be interpreted as an absolute identification of a particular vulnerability or issue but, rather, as indicative of relative problems worth more detailed attention. An index that measures one particular aspect of water may say nothing about another aspect of concern to a group of people or a region. Hence, an index developed for one region or issue should be applied elsewhere with particular caution.

Examples of Single-Factor or Weighted Water Measures

The simplest water indicators use a single data type or measure such as water availability or use. Sometimes these factors are weighted, most typically by population. Below we present some of the most commonly used and well-developed single-factor indicators used in the water community.

Access to Drinking Water and Sanitation Services

The two most commonly used single-factor measures in the water area are the fraction of the population in a country with (or without) access to clean drinking water and sanitation services. Data of varying qualities are available for both going back to the 1970s, permitting some perspective on changes over time. The actual definitions of "access" were first developed by the United Nations and international aid organizations, but these definitions have varied over time, making it difficult to track time-series trends for specific countries or regions in a consistent way.

Despite the fact that these measures are the two most commonly used in the water area, the available information is not comprehensive. For the 1990 baseline (WHO 1992), only 70 countries out of 134 even responded to the United Nations survey. In Africa only 58 percent of the countries reported data, representing 68 percent of the continent's population. In West Asia, 33 percent of the countries reported data, representing a mere 2 percent of the population. In 1992, a total of 82 countries out of 130 provided data—in all representing 55 percent of developing countries and 76 percent of developing country populations (WHO 1993).

For the 1994 status report (WHO 1996), 84 countries out of 130 responded. In addition, this report included more detailed definitions of *access*. In 2000, a new assessment was published (the *Global Water Supply and Sanitation Assessment 2000 Report*), with data on 89 percent of the global population (WHO 2000a).

As already noted, the differences in definition make it very difficult to make direct comparisons. Box 4.3 presents some of the definitions of *access* and how they have changed over time. Nevertheless, because of the long period over which these two measures have been collected, and because data are available for a large number of countries, they remain the most commonly cited indicators of water "well-being."

Box 4.3   Definitions of Access to Clean Water and Sanitation Services

Definitions of Access to Drinking Water and Sanitation, 1990

In the mid-1990s the following definitions were used for "access." Note that all words in **boldface** were defined individually at the country level, leading to inconsistent measures.

Safe drinking water coverage: proportion of population with **access** to an **adequate amount** of **safe** drinking water located within a **convenient distance** from the user's dwelling.

Sanitary means of excreta disposal coverage: Proportion of population with **access** to a **sanitary facility** for human excreta disposal in the dwelling or located within a **convenient distance** from the user's dwelling.

Definitions of Access to Drinking Water and Sanitation, 2000

In the 2000 *Global Water Supply and Sanitation Assessment 2000 Report*, this assessment replaced the terms *safe* and *adequate* with *improved*. This assessment assumed that certain types of technology were safer or more adequate than others. The new data, therefore, show the population with access to *improved* water supply and sanitation, with no assessment of safety. Tables 3 and 4 in the Data Section list these technologies and definitions. Essentially, this new definition assumes that technology is an indicator of improved water and sanitation. The coverage figures produced by technology indicators do not provide information about the quality of the water provided or about its use.

Gleick (1998) presents maps of both measures using data from the mid-1990s. In 2000, new data were presented by the World Health Organization (WHO 2000a), updating the access estimates for many countries, changing definitions, and adding data from some countries that had previously not been included. Tables 3 and 4 (in the Data Section) present the current WHO data. According to these estimates, there are 2.4 billion people without access to adequate sanitation services and nearly 1.2 billion people without access to "improved" water and sanitation in the countries for which data are presented.

Falkenmark Water Stress Index/Water Competition Index (1980s and 1990s)

Another early measurement of water scarcity is the Falkenmark Water Stress or Competition Index. After access to clean drinking water and sanitation services, this measure has been the most influential and powerful water measure for the past two decades. Proposed by the Swedish water expert Malin Falkenmark, the Water Stress Index or Water Competition Index measures the amount of water available in a country as a function of population. Early versions appeared in 1974 (Falkenmark and Lindh 1974). In the original index, Falkenmark applied these two factors to develop a measure of how many people could be supported by a region's natural endowment of water. Over time, many other researchers have inverted this to measure how much water is available per capita.

Using Israel as a benchmark for a society's ability to develop in arid regions, Falkenmark identifies 2,000 people per flow unit [1 million cubic meters of water] per year as "the maximum number [of people] that an advanced society is able to support and manage" (Falkenmark 1990). This means that for every flow unit of water that a country has, it can support a maximum of 2,000 people at a high level of development. The more people trying to survive off each flow unit, the greater the "water scarcity."

Falkenmark tries to distinguish between genuine water scarcity and human-induced water scarcity. Genuine water scarcity is controlled by local and regional climatic conditions and can include both general climate dryness and large interannual fluctuations in limited rainfall. Human-induced water scarcity, on the other hand, is exacerbated by human behavior. Activities such as overgrazing can leave the soil desiccated and unable to hold moisture, resulting in excessive runoff. Population growth can also result in per capita progressive scarcity. Human behavior of this type can be minimized or controlled (Falkenmark 1991). Using a per capita minimum of 100 liters per day for basic health and household needs (cooking, drinking, washing), she defined water stress thresholds (see Table 4.2), including all water for food production, industrial and commercial use, and domestic needs.

Falkenmark, Lundqvist, and Widstrand (1989) defined (and others have adopted [Engleman et al. 2000]) 1,700 m^3/capita/yr as the level above which water shortages are rare and localized; below 1,000 m^3/capita/yr, water supply begins to hamper health, economic development, and well-being; below 500 m^3/capita/yr, water availability is a primary constraint to life. However, these levels of water stress occur at different amounts for various countries depending on circumstances such as level of development, conservation measures, groundwater use, and so forth. These levels can also be delayed or ameliorated by technological advances, efficient water use, and conservation.

TABLE 4.2 Water Stress Definitions (modified from the works of Falkenmark)

Annual Renewable Fresh Water (cubic meters per capita per year)	Level of water stress
> 1,700	Occasional or local water stress
1,000 – 1,700	Regular water stress
500 – 1,000	Chronic water scarcity (lack of water begins to hamper economic development and human health and well-being)
< 500	Absolute water scarcity

Many people and organizations have relied upon the Falkenmark index as the only measure of water-related quality of life. As an example, in 1993 and again in 1997, Population Action International (PAI) used the Falkenmark Water Stress Index to rank the countries of the world in order of scarcity (PAI 1993, 1997). Using data on water availability by country and population, PAI estimated that in the mid-1990s 166 million people in 18 countries suffered from water scarcity, while almost 270 million more in 11 other countries were water stressed (PAI 1997). Figure 4.2 identifies the water scarcity status of the most stressed countries.

Over time, as populations grow, countries move into different levels of scarcity. India, for example, was projected to experience water stress as early as 2015, but due to new information about fertility and population rates, now they are projected to cross the water stress threshold in 2035 and returning to a state of relative water sufficiency by 2045 (PAI 1997). In all of these measures, water availability is assumed to be constant; the only thing that changes over time is population.

While the Falkenmark Water Stress Index has been the benchmark for water adaptive capacity, it has a number of shortcomings. Some of these are well known; others are subtle. For example, at the most fundamental level, the index assumes that water availability, measured as the total average annual renewable water resources of a country, is a proxy for well-being. In fact, water availability is simply a measure of the natural endowment of a region and says little about how that region or country mobilizes its natural endowment, or uses the water that is available. In recent years, some attempts have been made to recalculate this measure with actual water use, rather than availability, as a more explicit estimate of well-being.

The index also assumes, as noted above, that water availability is constant over time. In fact, water availability fluctuates dramatically from year to year, and even on a seasonal basis. Some countries receive all of their annual renewable water within a short rainy season; others have more even seasonal distributions. The index also assumes that water availability is evenly distributed within a country, when the regional disparities are often enormous. Estimates of average annual water availability within the United States or China, for example, hide tremendous inequalities of regional availability.

Perhaps the greatest limitation to the Falkenmark index is its exclusion of water used to support vegetation, both natural and agricultural. Precipitation that falls within a country and goes directly to ecosystems is not included in estimates of national water availability, which usually only include renewable surface and groundwater flows. In recent years, Falkenmark has worked to modify her index to address this problem. For example, she now also classifies water according to its different uses, such as precipitation used by natural or human-modified systems. Unfortunately, most data sets still exclude such water; and most users of the Falkenmark index are unaware of (or ignore) this limitation.

Freshwater Availability by Population 2000

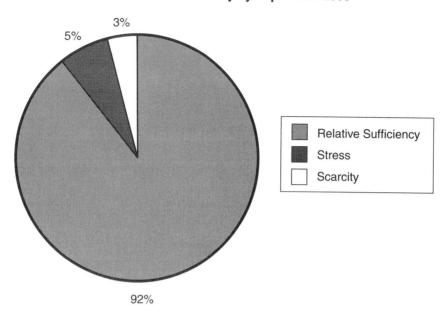

Freshwater Availability by Population 2025

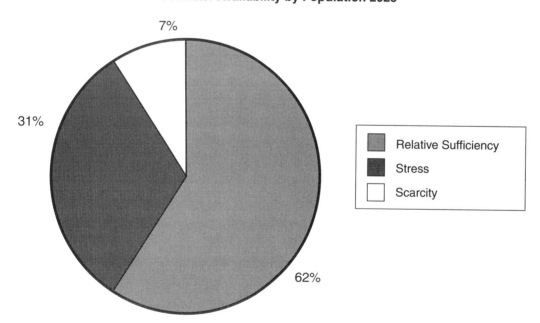

FIGURE 4.2   Water scarcity is defined by Falkenmark (and adopted by Engleman et al. 2000) as availability less than 1,000 cubic meters per person per year. The less-severe category of "stress" is defined as availability between 1,000 and 1,700 cubic meters per person per year. These two pie charts show the world population falling into these categories for 2000 and 2025, using the medium UN projection. The population in countries classified as under stress or scarcity grows from 480 million to nearly 3 billion.
Source: Engleman et al. 2000

Gleick: Basic Human Needs (1996)

Major international environmental meetings, including the historically important Dublin water conference and the Rio Earth Summit in 1992, have emphasized the importance of meeting basic human needs for water as a top priority in water policy decisions. In an effort to address the question of meeting these basic needs, Gleick (1996) developed an indicator that shifted the focus from measuring water availability to measuring some aspects of water use. In this assessment, he quantifies a "basic water requirement (BWR)" for drinking, cooking, bathing, and sanitation and hygiene at 50 liters per person per day (lpcd), and then presents estimates of the population, by country, without access to this BWR. Table 4.3 presents those countries for which reported (1990) domestic water use is less than 50 lpcd. According to this measure, nearly a billion people are not using even the most basic water quantities necessary to meet fundamental needs. Some specific data limitations are discussed in the Table and in Gleick (1996).

This indicator was developed to help set specific policy—in this case to encourage communities, governments, and private water providers to meet basic human needs for water as a top priority. Over time, regular estimates of the populations with access to this water can measure progress in meeting this goal. Limitations of this indicator include highly inadequate country data on domestic water use and an inability to distinguish regional water problems hidden by country-level aggregation. This measure, like most single-factor indicators, includes no information on water-quality issues.

Multifactor Indicators

Simpler indices like the ones presented above often leave out important factors and may present a misleading, or at least incomplete, picture of a water problem. Conversely, more complicated indices that include more types of specific data run the risk of including so many variables that finding adequate data sets becomes difficult and the relative importance of each variable is diminished. Increased complexity also makes interpretation more difficult, especially when it comes to policy actions. Looking at the subcomponents may offer insights in some cases. Despite these potential difficulties, efforts have begun to develop more sophisticated water indicators. These efforts take advantage of improving data sets and addressing some of the limitations of single-factor indicators. Several examples are offered below.

Gleick: Vulnerability of Water Systems (1990)

As part of a comprehensive assessment of the potential impacts of climate change for water resources and water systems, Gleick developed a series of measures for watersheds in the United States. A more comprehensive set of indices to look at the same kinds of problems was developed by Hurd, Smith, and Jones (1999). Gleick developed

TABLE 4.3   1990 Populations Using Less than 50 Liters per Person per Day

Country	1990 Population (million people)	Total Reported Domestic Water Use in liters/person/day
Gambia	0.86	4.5
Mali	9.21	8.0
Somalia	7.50	8.9
Mozambique	15.66	9.3
Uganda	18.79	9.3
Cambodia	8.25	9.5
Tanzania	27.32	10.1
Central African Republic	3.04	13.2
Ethiopia	49.24	13.3
Rwanda	7.24	13.6
Chad	5.68	13.9
Bhutan	1.52	14.8
Albania	3.25	15.5
Zaire	35.57	16.7
Nepal	19.14	17.0
Lesotho	1.77	17.0
Sierra Leone	4.15	17.1
Bangladesh	115.59	17.3
Burundi	5.47	18.0
Angola	10.02	18.3
Djibouti	0.41	18.7
Ghana	15.03	19.1
Benin	4.63	19.5
Solomon Islands	0.32	19.7
Myanmar	41.68	19.8
Papua New Guinea	3.87	19.9
Cape Verde	0.37	20.0

continues

and evaluated five different quantitative indicators to look at different aspects of regional vulnerabilities. These indicators offer insights into the sensitivity of U.S. water resources to water demand, floods and droughts, groundwater overdraft, reliance on hydroelectricity, and variability. These indicators were then evaluated for the 21 official U.S. water-resources regions (shown in Figure 4.3). Gleick notes that a wide range of measures can be selected and that many water problems are local. As a result, these estimates of vulnerability may hide smaller-scale problems or may miss important issues entirely—a limitation of any measure or indicator aggregated to a given geographic scale (Gleick 1990).

Regions with low storage volume relative to renewable supply (S/Q). This measure is an indicator of the ability of a region to withstand prolonged drought or severe flooding. With large storage compared to supply, short-term droughts are less likely to cause major water shortages. When this ratio is small, changes in the intensity of floods and droughts may be more strongly felt.

TABLE 4.3   *Continued*

Country	1990 Population (million people)	Total Reported Domestic Water Use in liters/person/day
Fiji	0.76	20.3
Burkina Faso	9.00	22.2
Senegal	7.33	25.4
Oman	1.50	26.7
Sri Lanka	17.22	27.6
Niger	7.73	28.4
Nigeria	108.54	28.4
Guinea-Bissau	0.96	28.5
Vietnam	66.69	28.8
Malawi	8.75	29.7
Congo	2.27	29.9
Jamaica	2.46	30.1
Haiti	6.51	30.2
Indonesia	184.28	34.2
Guatemala	9.20	34.3
Guinea	5.76	35.2
Cote D'Ivoire	12.00	35.6
Swaziland	0.79	36.4
Madagascar	12.00	37.2
Liberia	2.58	37.3
Afghanistan	16.56	39.3
Uruguay	3.09	39.6
Cameroon	11.83	42.6
Togo	3.53	43.5
Paraguay	4.28	45.6
Kenya	24.03	46.0
El Salvador	5.25	46.2
Zimbabwe	9.71	48.2
Total Population	**960.19 million**	

Source: Gleick (1996)
Notes: Many other people may receive less than 50 liters per person per day, but live in countries where the average use is higher than that. At the same time, many people in these countries may receive more than 50 lpcd. Better regional data are needed for both situations.

Regions with high consumptive use relative to available supply (D/Q). The ratio of use to supply is a combination of some of the single-factor measures described above. When this ratio is high, water shortages may be a problem and pressures on ecosystems and limited water resources will be high. In some parts of the western United States, water use is often a large fraction of total renewable supply.

Regions with a high proportion of hydroelectricity relative to total electricity (H/E). In regions that depend on hydroelectricity, changes in water availability (such as during droughts) affect total electricity availability and its cost. When this ratio is high, other kinds of environmental impacts associated with fossil fuel or nuclear energy generation are reduced.

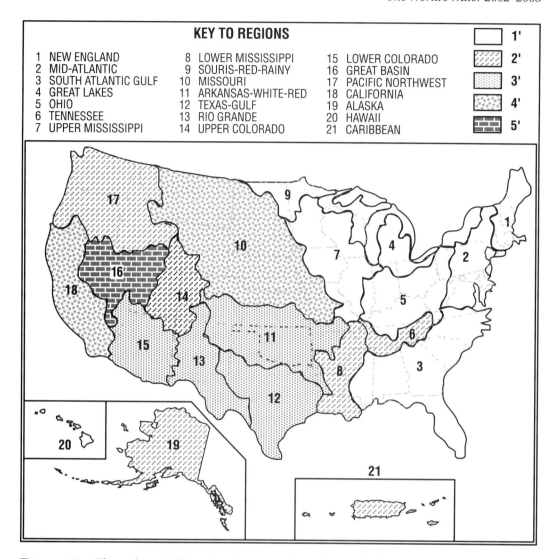

KEY TO REGIONS

1 NEW ENGLAND	8 LOWER MISSISSIPPI	15 LOWER COLORADO
2 MID-ATLANTIC	9 SOURIS-RED-RAINY	16 GREAT BASIN
3 SOUTH ATLANTIC GULF	10 MISSOURI	17 PACIFIC NORTHWEST
4 GREAT LAKES	11 ARKANSAS-WHITE-RED	18 CALIFORNIA
5 OHIO	12 TEXAS-GULF	19 ALASKA
6 TENNESSEE	13 RIO GRANDE	20 HAWAII
7 UPPER MISSISSIPPI	14 UPPER COLORADO	21 CARIBBEAN

FIGURE 4.3 The vulnerability of major watersheds in the United States to various indicators is shown here. Five indicators, described in the text, are evaluated. All U.S. watersheds are vulnerable to at least one; one region—the Great Basin— is vulnerable to all five.

Source: Gleick 1990

Regions with large groundwater overdraft relative to total groundwater withdrawals (G_o/G_w). Many regions depend on groundwater for part of their freshwater needs. Some of these regions already pump groundwater faster than the natural recharge rates. Where the ratio of groundwater overdraft (G_o) to total groundwater withdrawal (G_w) is high, water availability is already a problem.

Regions with variable streamflow (Q_{05}/Q_{95}). When a region has a high variability in streamflow, the risks of both floods and droughts are high. Q_{05} is the flow quantity exceeded only 5 percent of the time; Q_{95} is the quantity exceeded 95 percent of the time. Thus, when the ratio of the two is low, the variability in a basin is low. This particular indicator is fixed unless climate changes alter one or both of the factors.

Stockholm Environment Institute: Water Resources Vulnerability Index (1997)

Many factors influence the adequacy of a nation's water system: withdrawals, ecosystem conditions, supply infrastructure, and variability of resources. One effort to expand the complexity of water indices is the Water Resources Vulnerability Index (WRVI), calculated in 1997 by researchers at the Stockholm Environment Institute. The WRVI is made up of three sub-indices, which in turn may be made up of other indicators.

Many water scarcity indicators and indices face the same problem: how to include enough data to account for regional and temporal variability while still maintaining simplicity and robustness of function. For the WRVI, three overarching issues were included: a measure of use to availability, a measure of "coping" ability, and the question of "reliability of supply"—itself composed of three factors (Figure 4.4).

The Use-to-Resource Ratio Sub-index measures the average water-related stress that both ecological and socioeconomic systems place on a country's usable resources. The Coping Capacity Sub-index measures the economic and institutional ability of countries to endure water-related stresses. The three indicators that make up the Reliability Sub-index all examine different aspects of uncertainty of water supply. Each of these indicators and sub-indices is divided into four classes, wherein any given value can denote *no stress, low stress, stress,* and *high stress.* Indicator scores are then averaged to produce the Reliability Sub-index, and the sub-index scores are then averaged to produce the WRVI. A variant of the WRVI relies not on averages, but the highest value of any of the three sub-indices. This produces a stronger signal of vulnerability, reasoning that if a country is vulnerable in any one of these areas, it is considered "vulnerable." Figure 4.5 shows the WRVI with the three components averaged.

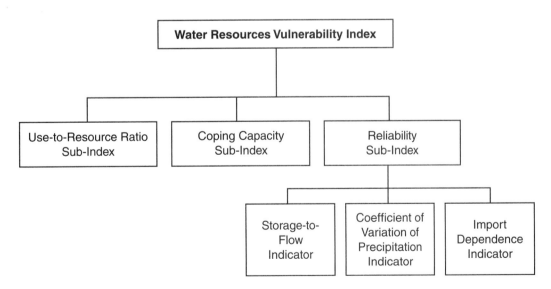

FIGURE 4.4 The SEI Water Resources Vulnerability Index.
Source: Raskin 1997

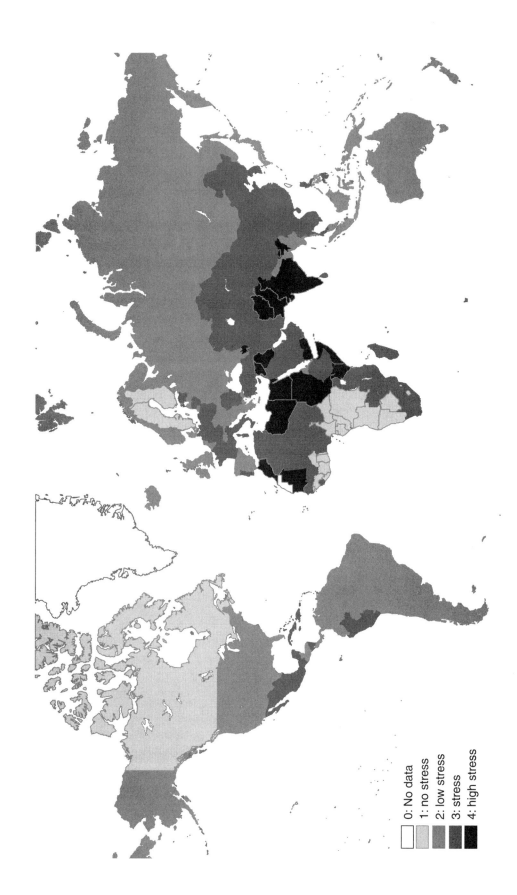

FIGURE 4.5 Water Resources Vulnerability Index 1 for 2025.
Source: Raskin 1997

Legend:
0: No data
1: no stress
2: low stress
3: stress
4: high stress

International Water Management Institute: Indicator of Relative Water Scarcity (1998)

David Seckler and his colleagues at the International Water Management Institute (IWMI) use two basic criteria to calculate an Index of Relative Water Scarcity (IRWS): the percentage increase in water "withdrawals" over the 1990–2025 period and water withdrawals in 2025 as a percentage of the Annual Water Resources (AWR) of the country. Thus, the IWMI indicator measures both how fast a country's water use is growing and how close it is to its total available water limit (see Figure 4.6).

Group 1 countries, with 8 percent of the world's population, are water-scarce by both criteria listed above. Their projected 2025 withdrawals are 191 percent of 1990 withdrawals and 91 percent of AWR. According to IMWI, unless they begin serious desalination efforts, they will reach an absolute limit on development of their water resources, including drawing down groundwater supplies. Water scarcity will become a major constraint on food production, human health, and environmental quality, and as these countries divert water from irrigation to domestic and industrial needs, they may need to import more food.

Group 2 countries, with 7 percent of the world's population, must develop more than twice the amount of water they currently use to meet their projected 2025 requirements. Group 3 countries, with 16 percent of the world's population, need to increase water withdrawals on average 48 percent to meet future needs. Group 4 countries, with 16 percent of the world's population, need to increase water withdrawals by less than 25 percent, and Group 5 countries, with 12 percent of the world's population, will require no additional withdrawals. Because the extreme variations between wet and dry areas in China and India would have rendered inclusion in any particular group meaningless, they were considered separately in the IWMI study. Russia and several other countries indicated in white were also not included.

It is interesting to compare this map with the one prepared by the Stockholm Environment Institute (Figure 4.5). While the main conclusions are the same (e.g., North America is facing little water stress, while the Middle East is facing extreme water stress), conclusions for individual countries vary greatly. In southeast Africa, the WRVI places Angola, Gabon, Congo, and the Democratic Republic of the Congo at low water stress, but the IRWS places those same countries in Group 2, high water stress. Alternatively, the IRWS indicates that Germany, Belgium, and the Netherlands, in Group 5, face no water stress, whereas the WRVI indicates that they will face considerable water stress. Likewise, in Southeast Asia, Thailand, North Korea, South Korea, and Sri Lanka are ranked as Group 5 (no stress) under the IWRS but are categorized as water-stressed countries under the WRVI. There are several possible reasons for these discrepancies, including the consistency of the input data across regions, how the individual indicators are aggregated, or the focus of the index itself (general use, in the case of the WRVI, irrigation in the case of the IRWS).

University of Victoria: Index of Human Insecurity (2000)

The Index of Human Insecurity (IHI), developed by the Global Environmental Change and Human Security project at the University of Victoria, folds water well-being into the environmental component of a larger index measuring nontraditional security

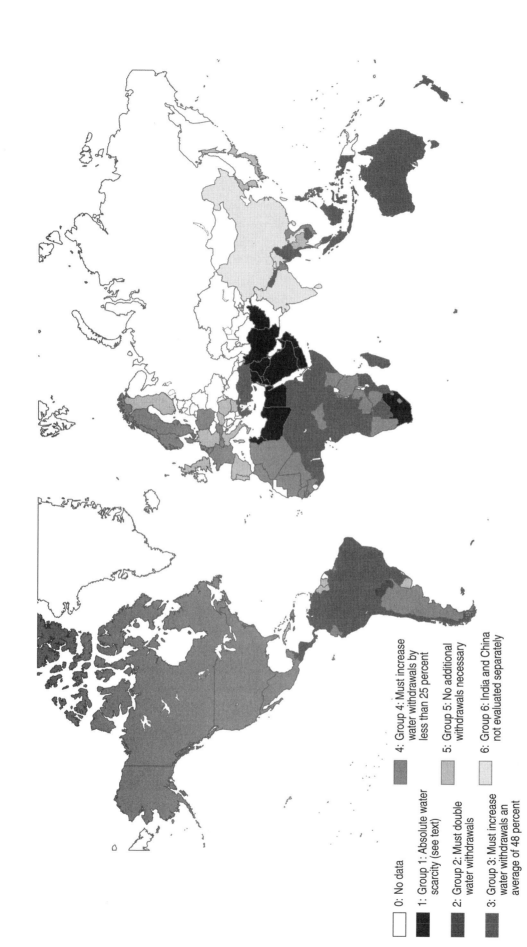

FIGURE 4.6 IWMI Index of Relative Water Scarcity.

Source: Seckler et al. 1998

0: No data

1: Group 1: Absolute water scarcity (see text)

2: Group 2: Must double water withdrawals

3: Group 3: Must increase water withdrawals an average of 48 percent

4: Group 4: Must increase water withdrawals by less than 25 percent

5: Group 5: No additional withdrawals necessary

6: Group 6: India and China not evaluated separately

threats: resource scarcity, human rights abuses, rapid population growth, and environmental degradation. The IHI was developed as a mechanism to help identify vulnerable or insecure regions, and to help inform policy makers as to development assistance. It incorporates data on water as well as other stressed environmental resources.

Like the WRVI, the IHI consists of several sub-indices that address different facets of human insecurity: environment, economy, society, and institutions on a country-by-country basis. The environmental indicators are listed in Figure 4.7.

As with other complex indices, data issues arise: the IHI should be meaningful over time, not just relatively between countries in any single year. Observing how the IHI changes from year to year will provide an indication of countries that could be facing medium- and long-term security crises. The IHI should be able to be broken down into specific resource-related sub-indices, which can be used to measure the condition of different resources within a country. The Environment Sub-index contains data from the energy and land use sectors, as well as water. Observing how these sub-indices change from one year to the next can help national governments identify where declining resources may become security problems (IHI 2000).

Because water is only a small factor in this index, it is not well suited for measuring water "well-being" per se. Indeed, because the water measure used is the same as "access to clean drinking water," the overall index actually provides less information about water well-being than less comprehensive water indices.

United Nations Development Programme: Human Development Index (1990s)

One of the most interesting indices of well-being is the Human Development Index, calculated annually by the United Nations Development Programme, starting in 1990. The HDI is a summary index designed to provide a measure of aggregate national well-being. It was developed because of the growing uneasiness felt by many about the use of simple economic indices, such as gross domestic product, to measure progress. The

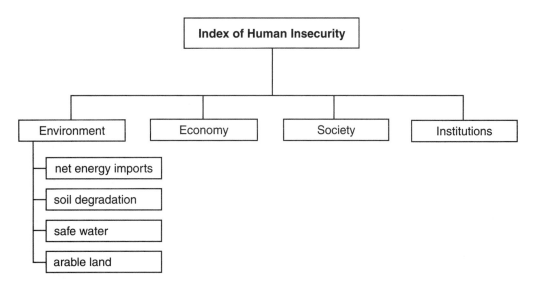

FIGURE 4.7 University of Victoria Index of Human Insecurity
Source: IHI 2000

HDI and its related indices consist of indicators that measure more diverse aspects of human society.

The HDI measures the average achievements in a country in three basic dimensions: longevity, knowledge, and a "decent standard of living." These dimensions are actually measured by a combination of life expectancy, adult literacy, enrollment in school, and real gross domestic product (GDP) per capita. The HDI uses national-level data because of data limitations on the regional level. As a result, it also masks regional or social disparities within a country. The actual HDI has no water component at all (indeed no environmental component of human well-being), although some of the indicators within the HDI, such as life expectancy, are functions of ecological health.

The *Human Development Report* does include several other indices, however. A second index called the Human Poverty Index (HPI-1) is calculated for a subset of nations and includes a somewhat different set of measures, including one related to access to water. The water component of the HPI-1 derives from the use of data on the "Percentage of Population Not Using Improved Water Sources." It is worth noting that the UNDP lacks data on this factor for about one-fifth of the applicable countries. Moreover, this water indicator is combined with another indicator ("Percentage of Children Under 5 Who Are Underweight") into a sub-index called Deprivation in a Decent Standard of Living, which is then combined with two other indicators to produce the HPI-1. As a result, water data comprises approximately one-sixth of the overall index. The water indicator would have to show significant movement to result in any noticeable change in the HPI-1.

In an acknowledgment that many other factors are relevant to quality of life, the *Human Development Report,* issued annually, includes extensive tables on a wide variety of other factors, permitting the public, researchers, and policy makers to greatly expand their assessment of trends and critical issues for any single issue. For more information on the HDR or HPI-1, see the 2001 *Human Development Report* at www.undp.org, or the data themselves at http://www.undp.org/hdr2001/back.pdf.

Centre for Ecology and Hydrology: Water Poverty Index (2000)

In an effort to develop a comprehensive and sophisticated index of water well-being, a new measure of water well-being, the Water Poverty Index (WPI), is being explored by researchers at the Centre for Ecology and Hydrology in the United Kingdom. The WPI aims to reflect both the physical availability of water and the degree to which human populations are served by that water, subject to the constraints imposed by the maintenance of ecological integrity. It can then be used to develop more effective procedures and guidelines for sustainable water management at the national level. The researchers at the Centre hope that if the WPI is widely accepted, it could be used to facilitate local, regional, and international comparisons of water resources and allocation. In particular, it may be useful in determining where the most serious water resource problems are likely to occur (Sullivan 2001).

The WPI can be constructed in a number of ways. The two most comprehensive are either a conventional aggregate index, or as a gap-measure index. The former is the conventional method, made up of relevant components that are weighted by importance. This is a relative method that would allow the comparison of the index from one year to the next, thereby measuring progress over time. The latter is an absolute method that would consider how much water provision and use in a given country or region deviate from pre-determined standards of ecosystem and human health, economic welfare, and community well-being. One obstacle to the formulation of such an absolute index is expert disagreement over where and how such standards should be set. As Falkenmark points out, more water can support a higher level of development, and many countries used to a higher standard of living would argue that the "minimum" amount of water needed is larger than the BWR of 50 liters per person per day (Gleick 1996). These data could be useful as a separate sub-index, measuring how well basic human needs are met.

The WPI can also be constructed as either a time-analysis function or as a part of a larger measure of general environmental sustainability. The advantage of forgoing a conventional aggregated index in favor of a time function is its simplicity: a certain amount of time is necessary to gain access to a certain per capita quantity of water, and the larger the amount of time required, the more water-stressed the country. However, its simplicity has drawbacks, in that it leaves out a number of water allocation issues among users, as well as failing to account for supply fluctuations. Finally, either the WPI itself or the data that support it can be included in a larger, "green-inclusive" alternative to gross national product (GNP). While any kind of environmental data would be a welcome inclusion to traditional economic measures of well-being, water indicators are already aggregated to form the WPI, and further aggregation may dilute their meaning.

Although much more research must be done on both the construction and the weighting, the originators of the WPI point out that domestic water poverty is more small-scale (and hence more rectifiable) than agricultural water poverty. This suggests that changes in agricultural water use, via conservation or greater efficiency, have the potential to bring about great measurable improvements in "aggregate" water poverty measurements.

Conclusions

The growing importance of water worldwide is leading to new and better indicators and indices that look at the connections between water and human well-being. Over the past decade a variety of new ways of evaluating the condition or scarcity of water resources have been proposed and developed, providing more and better information to policy makers. These indicators serve many purposes, from raising public awareness to pinpointing specific vulnerabilities and threats. In part, the increased awareness of water problems in recent years is the direct result of better information from these indicators.

Indicators have a number of drawbacks and limitations. Lack of adequate or accurate data hinders the development of comprehensive indicators. The complexities of water problems are hard to describe with only a few measures, yet the more measures used, the less likely that clear policy recommendations and options can be developed. Many water problems are local, yet most water data are large-scale or national. Definitions often change over time.

We expect that development of water-related indicators will continue. Their potential value is such that policy makers and the public have come to expect the water community to identify problems and to provide ways of measuring progress in solving those problems. If appropriate indicators are to be found, caution must be taken to ensure that they are clearly defined, their limitations understood, and their assumptions made transparent.

REFERENCES

Engleman, R., R.P. Cincotta, B. Dye, T. Gardner-Outlaw, and J. Wisnewski. 2000. *People in the Balance*. Population Action International, Washington, D.C., 28 pp.

Falkenmark, M. 1990. "Global water issues confronting humanity." *Journal of Peace Research*, Vol. 27, No. 2, May, pp. 177–190.

Falkenmark, M. 1991. "Living at the mercy of the water cycle." Water Resources in the Next Century. Stockholm Water Symposium Proceedings, August 12–15, Stockholm, Sweden, pp. 11–29.

Falkenmark, M., and G. Lindh. 1974. "Impact of water resources on population." Paper submitted by the Swedish Delegation to the UN World Population Conference, August 19–30, Bucharest, Hungary.

Falkenmark, M., J. Lundqvist, and C. Widstrand. 1989. "Macro-scale water scarcity requires micro-scale approaches: Aspects of vulnerability in semi-arid development." *Natural Resources Forum*, Vol. 13, No. 4, pp. 258–267.

Gleick, P.H. 1990. "Vulnerability of water systems." In P.E. Waggoner (ed.), *Climate Change and U.S. Water Resources*. John Wiley and Sons, Inc., New York.

Gleick, P.H. 1996. "Basic water requirements for human activities: Meeting basic needs." *Water International*, Vol. 21, pp. 83–92.

Gleick, P.H. 1998. *The World's Water 1998–1999*. Island Press, Washington, D.C.

Gleick, P.H. 2000. *The World's Water 2000–2001: The Biennial Report on Freshwater Resources*. Island Press, Washington, D.C.

Hurd, B., J. Smith, and R. Jones. 1999. *Water and Climate Change: A National Assessment of Regional Vulnerability*. Stratus Consulting, Boulder, Colorado.

Index of Human Insecurity (IHI). 2000. *AVISO*, Issue No. 6, January.

National Geographic. 2001. "EarthPulse: World of water—enough for all?" Vol. 199, No. 4. (April).

Ott, W.R. 1978. *Environmental Indices: Theory and Practice*. Science Publishers, Inc., Ann Arbor, Michigan.

Population Action International (PAI). 1993. *Sustaining Water: Population and the Future of Renewable Water Supplies*. Population Action International, Washington, D.C., 56 pp.

Population Action International. 1997. *Sustaining Water, Easing Scarcity: A Second Update*. PAI, Washington, D.C., 20 pp.

Raskin, P. 1997. *Comprehensive Assessment of the Freshwater Resources of the World Water Futures: Assessment of Long-Range Patterns and Problems*. Stockholm Environmental Institute, Stockholm, Sweden, 77 pp.

Seckler, D., U. Amarasinghe, D. Molden, R. de Silva, and R. Barker. 1998. *World Water Demand and Supply, 1990 to 2025: Scenarios and Issues*. Research Report 19. International Water Management Institute, Colombo, Sri Lanka, 40 pp.

Sullivan, C. (ed.). 2001. *The Development of a Water Poverty Index: A Feasibility Study*. Centre for Ecology & Hydrology, Wallingford, United Kingdom.

United Nations Development Programme (UNDP). 1997. *Human Development Report, 1997*. Oxford University Press, New York.

World Health Organization (WHO). 1992. *Water Supply and Sanitation Sector Monitoring Report 1990: Baseline Year*. WHO/UNICEF Joint Monitoring Program, New York and Geneva.

World Health Organization. 1993. *Water Supply and Sanitation Sector Monitoring Report 1993: Sector Status as of 31 December 1991*. WHO/UNICEF Joint Monitoring Program, New York and Geneva.

World Health Organization. 1996. *Water Supply and Sanitation Sector Monitoring Report 1996: Sector Status as of 31 December 1994*. WHO/UNICEF Joint Monitoring Program WHO/EOS/96.15, New York and Geneva.

World Health Organization. 2000a. *Global Water Supply and Sanitation Assessment 2000 Report*. Available at http://www.who.int/water_sanitation_health/Globassessment/GlobalTOC.htm.

World Health Organization (WHO). 2000b. *World Health Report 2000: Health Systems Improving Performance*. The World Health Organization, Geneva, Switzerland.

Pacific Island Developing Country Water Resources and Climate Change

William C.G. Burns

As one commentator has observed, "[O]ne ironic and tragic aspect of this environmental crisis of greenhouse emissions is the fact that those parts of the world least responsible for creating the global warming problem will be the first to suffer its horrifying consequences" (Panjabi 1993, Burns 2000). Pacific Island Developing Countries (PIDCs) are responsible for only 0.03 percent of the world's carbon dioxide emissions, and the average island resident produces only one-quarter of the emissions of the average person worldwide (Hay 1999). Yet it is anticipated that these nations will experience some of the earliest and most severe consequences of climate change over the next two centuries (IPCC 2001a, Burns 2001). While the popular press has focused on the threat of inundation of island coastal areas by rising sea levels, perhaps the most critical near- and long-term threat to these nations is the possible impacts of climate change on freshwater quality and availability (Meehl 1996, East-West Center 2001). This chapter focuses on the potential impacts of climate change on the already severely strained freshwater resources of PIDCs.

PIDCs and Freshwater Resources

There are nearly 30,000 islands in the Pacific Ocean, 1,000 of which are populated (UNEP 1998) (see Figure 5.1). Polynesian peoples populated Tonga, the Cook Islands, and French Polynesia during the present interglacial period (Wilkinson 1996). Melanesian peoples began to colonize the high islands of the western Pacific (Papua New Guinea, Solomon Islands, Vanuatu, and New Caledonia) as far back as 40,000 years ago (Tutangata 1996), millennia before settlement began in Europe (Tutangata 1996), and Micronesian peoples settled on many low islands, such as Micronesia and the Marshall Islands, in the last 2,000 years (Wilkinson 1996).

　　PIDCs consist of 22 political entities, 15 of which are politically independent (Campbell 1996), spread out over 28 million square kilometers of ocean (Tutangata

FIGURE 5.1 Pacific Island Developing Countries.

2000) (see Table 5.1). Almost all PIDCs lie between 30° north and 30° south latitude, between the Tropic of Cancer and the Tropic of Capricorn in the southwestern portion of the Pacific (U.S. Naval Meteorology and Oceanography Command 2001). With the exception of Papua New Guinea and Fiji, all PIDCs fall within the United Nations' definition of "small island states," which are islands with land areas smaller than 10,000 square kilometers and with fewer than 500,000 inhabitants (Pernetta 1992, Bequette 1994). The majority of countries in the region fall well below these thresholds, with populations below 200,000 and land areas well below 1,000 square kilometres, and many fall into the category of "very small islands," with land areas less than 100 square kilometers or a maximum width of 3 kilometers (Dijon 1994). The combined population of PIDCs in the region is slightly over 6 million, the lion's share being in Papua New Guinea (Kaluwin and Smith 1997). PIDCs comprise only 10 percent of the land area in the Pacific (Asian Development Bank 2001).

Pacific islands are traditionally classified as "high" or "low," with a further subdivision into continental and volcanic islands in the former category and atolls and raised limestone islands on the other (Campbell 1996). In most cases, PIDCs are a combination of these island types (see Table 5.2). High islands primarily consist of rugged volcanic mountains surrounded by fringing or barrier reefs. They may also have a fringe of low-lying coastal plains surrounding the mountainous interior. Atolls consist of limestone reef deposits laid down on an underlying volcanic cone. In most cases, the portion of atolls above sea level is usually not more than a few square kilometers

TABLE 5.1 Profile of Pacific Island Developing Countries

Country	Political Status	Population (2000 Est.)	Land Area (Sq. Km.)
America Samoa	Territory of the United States	65,446	199
Commonwealth of the Northern Marianas	Commonwealth in political union with the United States	71,912	477
Cook Islands	Self-governing, in free association with New Zealand	20,407	240
Federated States of Micronesia	Independent Nation	133,144	702
Fiji	Independent Nation	832,494	18,270
French Polynesia	Overseas territory of France	249,110	4,167
Guam	Territory of the United States	154,623	541.3
Kiribati	Independent Nation	91,985	717
Nauru	Independent Nation	11,845	21
New Caledonia	Overseas territory of France	201,816	19,060
Niue	Self-governing, in free association with New Zealand	2,113	260
Palau	Independent Nation	18,766	458
Papua New Guinea	Independent Nation	4,926,984	462,840
Pitcairn Islands	Overseas territory of the United Kingdom	54	47
Republic of the Marshall Islands	Independent Nation	68,126	181.3
Samoa	Independent Nation	179,466	2,860
Solomon Islands	Independent Nation	466,194	28,450
Tokelau	Territory of New Zealand	1,458	10
Tonga	Independent Nation	102,321	748
Tuvalu	Independent Nation	10,838	26
Vanuatu	Independent Nation	189,618	14,760
Wallis and Futuna	Overseas territory of France	15,283	274

Source: All data derived from: U.S. Central Intelligence Agency, *The World Factbook 2000*, http://www.odci.gov/cia/publications/factbook/, site visited on Sept. 20, 2001.

(Granger 1996). Raised atolls are uplifted coral atolls that consist almost entirely of limestone and dolomite, some rising 60–70 meters above sea level (Solomon and Forbes 1999, Granger 1996).

Description and Status of Freshwater Resources

The freshwater resources of small island states can be classified as either "conventional" or "nonconventional." Conventional resources include rainwater collected from artificial or natural surfaces, groundwater, and surface water. Nonconventional resources include seawater or brackish groundwater desalination, water importation by barge or submarine pipeline, treated wastewater, and substitution (such as the use of coconuts during droughts) (Falkland 1999a).

Rainwater. Rainwater collections systems, such as rainwater catchments on the roofs of individual houses or paved runways, are commonly used on many PIDCs. On some very small low-lying PIDCs, such as Tuvalu, the northern atolls of the Cook Islands, and some of the raised coral islands of Tonga, rainwater collection on roofs or

TABLE 5.2   Terrain of Pacific Island Developing Countries

Country	Terrain
America Samoa	5 volcanic islands and 2 coral atolls
Commonwealth of the Northern Marianas	Limestone (southern islands); volcanic (northern islands)
Cook Islands	Coral atolls (north); volcanic (south)
Federated States of Micronesia	High volcanic islands; low coral atolls
Fiji	Primarily mountains of volcanic origins
French Polynesia	Rugged high islands and low atolls
Guam	Volcanic origin, limestone plateau
Kiribati	Primarily low-lying atolls
Nauru	Raised atolls
New Caledonia	Coastal plains with interior mountains
Niue	Raised atolls
Palau	High mountainous islands to low coral islands
Papua New Guinea	Primarily mountainous with coastal lowlands
Pitcairn Islands	Rugged volcanic formation
Republic of the Marshall Islands	Raised atolls
Samoa	Narrow coastal plain with volcanic, rocky, rugged mountains in interior
Solomon Islands	Primarily rugged mountains with low coral atolls
Tokelau	Low-lying coral atolls enclosing large lagoons
Tonga	Limestone bases with uplifted coral formation; limestone overlying volcanic bases
Tuvalu	Very low-lying and narrow coral atolls
Vanuatu	Primarily mountains of volcanic origin
Wallis and Futuna	Volcanic origin, low hills

Source: All data derived from: U.S. Central Intelligence Agency, *The World Factbook 2000*, http://www.odci.gov/cia/publications/factbook/, National Communications, Non-Annex I parties, United Nations Framework Convention on Climate Change, http://www.unfcc.de/resource/natc, sites visited on Sept. 18, 2001.

community buildings is the sole source of fresh water. On other islands in the region, rainwater is a supplementary source of water that is used for more essential needs, such as cooking and drinking (Falkland 1999b).

Groundwater. Groundwater occurs on small island states in two main types of natural reservoirs (aquifers), perched and basal. Perched aquifers occur where an impermeable layer exists in the zone of aeration,[1] creating a groundwater formation above the water table, or when groundwater is retained in compartments by a series of vertical volcanic dikes (Falkland and Custodio 1991). They are similar to the aquifers found on large islands or continents. Basal aquifers may occur on either high or low islands in the form of coastal aquifers or rainwater that percolates through an island

---

1. The zone of aeration is "a subsurface zone containing water under pressure less than that of the atmosphere, including water held by capillarity; and containing air or gases generally at atmospheric pressure. [I]t extends from the ground surface to the water table." Society of Professional Well Log Analysts, Glossary of Water Terms, http://www.spwla.org/gloss/reference/glossary/glossz/glossz.htm>, site visited on September 25, 2001.

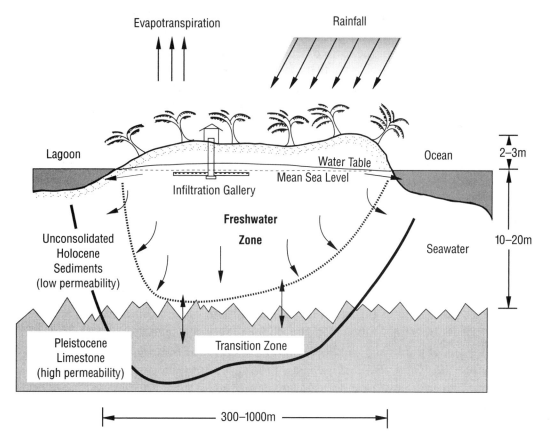

FIGURE 5.2   Cross-section through a small coral island showing main features of a freshwater lens (exaggerated vertical scale).
Source: Falkland 1999

and floats on the denser salt or brackish water in what is termed a Ghyben-Herzberg lens (Whittaker 1998) (see Figure 5.2).[2]

The volume of freshwater lenses is roughly proportional to the width and surface area of an island and is also influenced by factors such as rainfall levels, the permeability of the rock beneath islands, and salt mixing due to storm- or tide-induced pressure (Roy and Connell 1991). While such lenses may be as thick as 20 meters on some islands, on raised coral atolls, such as Nauru and many of the islands of Tonga, they may be no more than 10–20 centimeters thick (Falkland and Custodio 1991).

Most PIDCs rely on groundwater resources for at least some of their water needs, and it is the primary source on several of these islands, including Kiribati (UNESCO 2001). Many low-lying atoll islands supplement rainwater supplies with storage capacity in the freshwater lens (Small Island Developing States Net 2001).

Surface water. Surface water, in the form of ephemeral and perennial streams, springs, lakes, and swamps, is found mainly on high islands in PIDCs (Falkland 1999a). Surface water is rarely found on low islands such as coral atolls and limestone islands

---

2. The Ghyben-Herzberg principle provides that "freshwater recharge to an island will tend, when restrained and protected from wave and tidal mixing by the porous medium, to form a lens of freshwater which floats upon the underlying seawater in the formation." M.E. Herman, R.W. Buddemeier, and S.W. Wheatcraft, "A layered aquifer model of atoll island hydrology: Validation of a computer simulation," *J. Hydrology,* vol. 84, 1986, pp. 303, 304.

because of the high infiltration capacity of the soils and rocks found on such islands (Falkland and Custodio 1991). In those rare cases where surface water is found on low islands, it is likely to be in the form of shallow, often brackish lakes (UNEP 1998).

On some high islands in the Pacific, surface water is the predominant source of fresh water because gravity-fed water systems are more cost-effective than developing and pumping from groundwater sources (Falkland and Custodio 1991). For example, surface water provides more than 95 percent of the water requirements in French Polynesia (Falkland and Custodio 1991). Surface water also contributes to the freshwater supply in Nauru, Palau, and on the high islands of the Cook Islands (Falkland and Custodio 1991).

Nonconventional sources. Given the costs or nonfeasibility of nonconventional alternatives for many PIDCs, especially small coral islands, most fresh water on PIDCs is derived from conventional sources (UNEP 1998). However, approximately 60 percent of the water requirements on the island of Nauru are produced from desalination (Ghasemmi 1997), and some other islands have established desalination plants for specific uses, such as tourist resorts (Falkland 1999a).

Nauru received most of its water in ships until it installed a desalination plant (see Chapter 2), and some of the small islands of Fiji and Tonga also receive water from nearby islands by barge or boat (Falkland 1999a). During severe droughts or natural disasters some of the small islands of Fiji, Kiribati, and the Marshall Islands have relied on coconuts for water (Falkland 1999a). Finally, nonpotable sources, including sea water, brackish groundwater and wastewater, are used for toilet flushing and fire fighting on a number of PIDCs, including Kiribati and the Marshall Islands (Falkland 1999a).

Threats to Freshwater Resources

Fresh water is an imperiled and extremely limited resource on most PIDCs. As the United Nations Economic and Social Commission for Asia and the Pacific recently concluded, "Freshwater is an essential, and threatened, resource. Throughout the Pacific, smaller islands, and the leeward side of large high islands, experience difficult, and sometimes life threatening, deficiencies of unpolluted water supplies" (Ministerial Conference 2001).

A recent survey by the Asian Development Bank (ADB) found that only 50–75 percent of the residents of Samoa and only 44 percent of the residents in Kiribati had access to safe water (Hoegh-Guldberg et al. 2000). In Papua New Guinea, only 10 percent of the rural population has access to safe drinking water (Lauerman 1997).

Rapid population increases in PIDCs, largely attributable to advances in health care, are placing increasing strain on the limited water resources of many PIDCs (Falkland and Custodio 1991). Over-pumping from fragile groundwater resources to meet increasing water needs can lead to saltwater intrusion and the ultimate loss of supply (South Pacific Applied Geoscience Commission 2001a). These pressures will be exacerbated in the future. Population in the region may double in the next quarter century, compared to an estimated global population doubling over the next 50 years or so (East-West Center 2001, Larson 2000). Water supplies are also threatened by aging and poorly installed infrastructure that results in water leaks of up to 70 percent from the reticulation system (South Pacific Applied Geoscience Commission 2001b).

Water shortages on PIDCs are also exacerbated by other natural and anthropogenic factors. Rapid urbanization in many PIDCs has resulted in a substantial increase of pollutants over the past few decades, including nutrients, biochemical oxygen demand, solids, and microbial pollution (UNEP 2000). Moreover, many PIDCs have inadequate sewage disposal infrastructure (Brodie, Prasad, and Morrison 2001). This has resulted in the contamination of many water systems, producing high levels of diarrheal and other infectious diseases (Falkland 1999b). The groundwater resources of small coral islands are particularly threatened by pollution because these islands are characterized by thin, highly permeable soil zones (Falkland and Custodio 1991).

Many PIDCs rely on a single source of water, rendering them extremely vulnerable to natural variability in precipitation patterns or changes in storm tracks. This is particularly true for the atoll states in the region (Salinger et al. 1995). El Niño/Southern Oscillation (ENSO) episodes in the past two decades have reduced precipitation by as much as 87 percent in the western Pacific while resulting in unusually high rainfall in the central Pacific. ENSO is the most important mode of the Earth's seasonal and year-to-year climate variability. El Niño, the warm phase of ENSO, refers to climate conditions in the tropical Pacific characterized by unusually warm sea-surface temperatures and weak trade winds. La Niña refers to the climate state characterized by anomalously cool sea-surface temperatures and stronger than average trade winds (CIG 2001; Kripalani, Kulkarni, and Sabade 2001).

Drought associated with ENSO events have depleted rainfall collection supplies and the freshwater lenses and perched aquifers on many PIDCs. For example, in 1998, 40 atolls of Micronesia ran out of water during an ENSO event (Tutangata 1996), resulting in the declaration of a national emergency. In the same year, rainwater tanks in substantial parts of Kiribati dried up and shallow groundwater reserves became brackish (World Bank 2000). The main island of the Marshall Islands also had access to drinking water for only seven hours every fourteen days, and rationing occurred on all islands in the North Pacific (East-West Center 2001). Tropical storms may also imperil water supplies. Ocean waves may overtop small low-lying islands, contaminating freshwater lenses with saltwater for months and damaging rainfall collection systems (Falkland and Custodio 1991, Falkland 1999b). As I will outline in the next section of this chapter, the water problems of PIDCs are likely to be severely increased by projected human-induced climate change over the next century and beyond.

Climate Change and PIDC Freshwater Resources
A Science Overview

The surface of the earth is heated by solar radiation emanating from the sun at short wavelengths between 0.15 and 5 μm. Each square meter of the earth receives an average of 342 watts of solar radiation throughout the year (IPCC 2001b). Approximately one-third of the incoming solar radiation is reflected back to space in the form of thermal infrared, or longer-wave radiation, at wavelengths of 3–50 μm (IPCC 1994). Of the remainder, a portion is partly absorbed by the atmosphere, but most (168 watts per square meter) is absorbed by land, ocean, and ice surfaces (IPCC 2001b).

Some of the outgoing infrared radiation is absorbed by naturally occurring atmospheric gases—principally water vapor (H_2O)—as well as carbon dioxide (CO_2), ozone (O_3), methane (CH_4), nitrous oxide (N_2O), and clouds (Benarde 1992). This absorption is termed the *natural greenhouse effect* because these gases, which are termed *greenhouse gases,* operate much like a greenhouse: they are "transparent" to incoming short-wave radiation, but "opaque" to outgoing infrared radiation, trapping a substantial portion of such radiation and re-radiating much of this energy back to the earth's surface (University of Capetown 1999). This natural process is critical to the sustenance of life on earth, elevating surface temperatures by about 33° Celsius (C) (University of Capetown 1999).

In the past, the net incoming solar radiation at the top of the atmosphere was balanced by net outgoing infrared radiation, contributing to climatic stability. For the past 8,000 years, the world's climate has been very stable, varying only within a range of + or − 1 degree C (Anonymous 1995). However, with the advent of fossil-fuel burning plants to support industry, mass use of automobiles, and the energy demands of modern consumers, "humans began to interfere seriously in the composition of the atmosphere" (Pearce 1995).

The burning of fossil fuels, mainly coal, oil and gas, has soared since the beginning of the Industrial Revolution. We now produce approximately 5.5 gigatons of carbon annually, nearly all of which enters the atmosphere as CO_2. An additional 1.5 gigatons is released into the atmosphere from land-use changes, such as deforestation. Cement production contributes a small additional amount (Hadley Centre 1999a). As a consequence, concentrations of carbon dioxide in the atmosphere have increased approximately 25 percent since 1850, from 270 to 280 parts per million (ppm) by volume in pre-industrial times to over 370 ppm today (Jardine 1994, IPCC 2001c),[3] most of the increase occurring in the past fifty years (Wigley 2001). These concentrations have not been exceeded in the past 420,000 years, and the increase this century is "unprecedented" in the last 20,000 years (IPCC 2001d). Anthropogenic activities have also resulted in substantially increased atmospheric concentrations of other greenhouse gases, including methane (up 145 percent since 1750) and nitrous oxides (up 15 percent since 1750) (Warbrick and McGoldrick 1998), as well as new sources, such as chlorofluorocarbons and halons (Brasseur 1994). Tropical land use changes from the clearing of natural vegetation and secondary forests adds approximately another 2.4 gigatons annually to the atmosphere, representing 29 percent of the combined total of anthropogenic emissions (Fearnside 2001).

Increases in the concentration of greenhouse gases reduce the efficiency with which the earth's surface radiates to space. It results in an increased absorption of the outgoing infrared radiation by the atmosphere, with this radiation re-emitted at higher altitudes and lower temperatures (IPCC 2001d). This resulting change in net radiative energy, which is termed *radiative forcing*, tends to warm the lower atmosphere and the earth's surface (Wigley 2001). The amount of radiative forcing that occurs is dependent on the magnitude of increases in the concentrations of greenhouse gases, the radiative properties of the gases, and the concentrations of existing greenhouse gases in the atmosphere. Overall, CO_2 accounts for 65 percent of the total radiative forcing resulting from anthropogenically released greenhouse gases, methane contributes an additional 19 percent, chlorofluorocarbons, 10 percent, and nitrous oxide about 6 percent (Aplin 1999).

The latest assessment by the Intergovernmental Panel on Climate Change (IPCC)[4] concluded that rising concentrations of greenhouse gases are the primary cause (IPCC

---

3. Parts per million is the ratio of the number of molecules of a gas (in this case carbon dioxide) to the total number of molecules of dry air.

4. The IPCC, comprised of 2,500 climate scientists from throughout the world, was established by the United Nations in 1988 to gather information and coordinate research related to climate change, to evaluate

2001c) for the increase in average global temperatures of about 0.6° C in the past century (United Kingdom Department 1999). "20th Century global mean temperature is at least as warm as any other century since at least 1400 AD," notes the IPCC (1995). In the Northern Hemisphere, "the increase in temperature in the 20th century is likely to have been the largest of any century during the past 1000 years" (IPCC 2001c). Warming has accelerated in the last 25 years, more than doubling that of the twentieth-century average (Stevens 1999).

In the South Pacific, surface air temperatures have increased by 0.3–0.8° C during the twentieth century, with the greatest increase in the zone southwest of the Southern Pacific Convergence Zone (IPCC 2001a). Temperature increases in this region are well in excess of global rates over the past century (IPCC 2001a).

Unfortunately, as we indicate in the next section, the unprecedented increases in temperatures over the last century are likely to be ratcheted up dramatically in this century, including in PIDCs.

Projections for the Twenty-first Century

Global projections. Predicting future climate is an extremely imposing task because it requires an assessment of the future state of a wide array of complex climatic components, including the atmosphere, the ocean, the cryosphere, land surfaces, the stratosphere, and the sun, as well as the interactions among these components (Wigley 2001). The only practical method to make such projections is through the use of mathematical models, derived from weather forecasting, to represent the earth's energy and water cycles (Burns 2001a).

The most sophisticated of these models, general circulation models (GCMs), use a three-dimensional grid overlaying the surface of the earth with grid points 300–500 kilometers per side, within which cells are stacked about 20 layers deep (Hadley Centre 1999). Coupled atmosphere-ocean general circulation models seek to integrate atmosphere and ocean GCMs with sea-ice models and those of land-surface processes in an effort to obtain a realistic simulation of the earth's climatic system, including feedbacks among these components (Hadley Centre 1999, Heilman 1999).

The vertical layers of a GCM represent levels in the atmosphere and depths in the ocean, dividing the surface of the planet into a series of horizontal boxes separated by lines similar to latitudes and longitudes (Hennessy 2001). Within each grid point, a series of equations are run on a supercomputer, producing simulations of key climatic components, including wind, air pressure, temperature, humidity, ice coverage, and land-surface processes (Barron 1995).

Climate models are usually run for several simulated decades, with the derived results compared to actual statistics on climatic conditions over this period, such as mean temperatures and precipitation. The models are then run with changes in external forcing, such as projected increases in atmospheric greenhouse gas concentrations, over a series of decades or centuries. The differences between these two "climates" provide an estimate of the consequent climate change (Schlesinger 1993).

In its *Third Assessment Report* on climate change, the IPCC concluded that by 2100 carbon dioxide concentrations will have risen 90–250 percent above pre-industrial

---

proposals for reducing greenhouse gas emissions, and to assess the viability of response mechanisms. Their most recent of three reports was released in mid-2001.

levels, to between 540 and 970 parts per million (IPCC 2001c). This will result in an increase of global average temperatures of 1.4 to 5.8° C by the end of this century, "very likely to be without precedent during at least the last 10,000 years" (IPCC 2001c). In a subsequent study, Wigley and Raper (2001) sought to assess the IPCC's projected temperature range in probabilistic terms, concluding that the likeliest increase, with a 50 percent probability, will lie between 2.4–3.8° C, with further warming through the twenty-second century "virtually certain."

The IPCC assessment also concluded that global mean sea level would rise by 9 to 88 centimeters by 2100 (with a mid-range estimate of approximate 50 centimeters) due to thermal expansion and the loss of mass from glaciers and ice caps (Wigley and Raper 2001). This is equivalent to a 200–500 percent acceleration over sea-level rise in the twentieth century (Nicholls and Mimura 1998).

The United Kingdom's Hadley Centre's most recent assessment on climate change concluded that temperatures could rise by 3° C and sea levels by 50 centimeters by 2100 if projected increases in greenhouse gas emissions are not reduced (Hadley Centre 1999).

Climate change in PIDCs. It must be emphasized that general circulation models remain a crude instrument for assessing regional climate trends because their resolution is often too coarse at this level:

> [General circulation models] have difficulty in reproducing regional climate patterns, and large discrepancies are found among models. In many regions of the world, the distribution of significant surface variables, such as temperature and rainfall, are often influenced by the local effects of topography and other thermal contrasts, and the coarse spatial resolution of the GCMs can not resolve these effects. (Solman and Nunez 1999, Burns 2001b)

Resolution problems with GCMs are particularly acute for small landmasses in oceanic regions (SPREP 1999). In the case of PIDCs, many of the circulation features that dominate the climate of the islands in the region are inadequately simulated by the current generation of GCMs (SPREP 1999).

However, modeling of regional climate conditions has improved substantially in the past few years. In its most recent assessment of climate change impacts on small island states, the IPCC conducted a model validation exercise to assess the ability of leading GCMs to simulate present-day climate in the four major regions where the majority of small islands are located, including the Pacific. It concluded that several models "have reasonable capability in simulating the broad features of present-day climate and its variability over these regions" (IPCC 2001a). Moreover, Australia's Commonwealth Scientific and Industrial Research Organization (CSIRO) has recently "nested" a higher resolution regional climate model within a GCM. The nested model is able to broadly simulate climatic conditions in the Pacific, although it diverges substantially from observed patterns of rainfall (Kondratyev and Cracknell 1998).

With the caveat that regional climate assessments remain speculative, the following section summarizes current projections of climate change in PIDCs by the IPCC and CSIRO.

The twenty-first century. The IPCC in its *Third Assessment Report* projects that temperatures in the Pacific will rise by approximately 2.0° C by 2050 and 3.0° C by 2080

(IPCC 2001a). While it is anticipated that temperatures in the region will rise less than global mean averages over the next century (IPCC 2001a), this range of change will constitute a dramatic increase over the substantial rise witnessed in the twentieth century. CSIRO's projections are consistent with this assessment, though its analysis reveals differential temperature increases in the region, with the greatest increases in north Polynesia (0.7–0.9° C) and the least warming in south Polynesia (0.7° C) (SPREP 1999).

The IPCC's *Third Assessment* projects that sea levels may rise in the region by as much as 5 millimeters per year over the next century (IPCC 2001a) and continue to rise in the twenty-second century due to lags in the climate effect (Klein and Nichols 1999). Moreover, sea-level rise will raise the baseline for storm surges (Hay and Kaluwin 1993), significantly increasing the vulnerability of coastal areas to inundation (Leatherman 1997). "Sea-level rise and storm surge effects are linearly additive; in other words, whatever storm surge occurs at a particular location under the current climate can simply be added to the sea-level rise at that location" (Hay and Kaluwin 1993, South Pacific Regional Environmental Programme 1999).

Several GCMs also predict more frequent ENSO-like patterns (IPCC 2001a). This could result in a 26–200 percent increase in rainfall over the central and east-central Pacific, with possible decreases in the Melanesian and Polynesian regions (IPCC 2001a, Jones 2000). Additionally, warming could also lead to increased extreme rainfall intensity and frequency. One model projects a doubling of the frequency of 100 millimeter per day rainfall events and a 15–18 percent increase in rainfall intensity over large areas of the Pacific (IPCC 2001a).

Projected buildups in greenhouse gas emissions will likely raise ocean temperatures and ocean surface water temperatures to above 26° C in the next century (NASA 2001; Karl, Nicholls, and Gregory 2001). This could result in a greater exchange of energy and add momentum to the vertical exchange processes critical to the development of tropical typhoons and cyclones. Some researchers estimate, therefore, that the occurrence of tropical typhoons and cyclones could increase by as much as 50–60 percent (NASA 2001, Schlesinger 1993, Haarsman 1993), and their intensity by 10–20 percent (IPCC 2001a; Knutson, Tuleya, and Kurihara 1998; Druyan 1999).

However, there is by no means universal agreement that climate change will visit an increase in violent weather events on PIDCs. Some researchers believe that the purported linkage between increased ocean temperatures and violent weather events is overly simplistic, citing other factors that influence storm development, including atmospheric buoyancy, instabilities in the wind flow, and vertical wind shear (Karl, Nicholls, and Gregory 2001; Holland 1995). Moreover, some climate scientists argue that ocean circulation changes associated with climate change may counter the effects of added warmth (Hileman 1995). In its most recent regional assessment, the IPCC concluded that "[t]here is no consensus regarding the conclusions of studies related to the behavior of tropical cyclones in a warmer world . . . current information is insufficient to assess current trends, and confidence in understanding and models is inadequate to make firm projections" (IPCC 2001a). However, the IPCC did conclude, with "moderate confidence," that the intensity of tropical cyclones is likely to increase by 10–20 percent in the Pacific region when atmospheric levels of carbon dioxide reach double pre-industrial levels (IPCC 2001a).

Potential Impacts of Climate Change on PIDC Freshwater Resources

Sea-Level Rise

As indicated earlier, sea level in PIDCs may rise by an average of 50 centimeters over the next century. Researchers have expressed concern that sea-level rise could result in intrusion of saltwater into the freshwater lenses of coral islands and atolls in the region (Watson, Zinyowera, and Moss 1998). Yet recent research indicates that a rise in sea level of 40–50 centimeters would have virtually no effect on groundwater supplies, or might even raise their volume, because the top of the freshwater lens would rise while its base would remain relatively unaffected (East-West Center 2001, Falkland 1999a). A small rise in sea levels might also be salutary if it raises freshwater lenses to more permeable layers on some islands (Falkland 1999a).

However, this assumes that the width of small islands is not reduced by inundation or erosion in the future. If these phenomena occur as a consequence of climate change, which as indicated earlier is anticipated, groundwater lenses would shrink beneath larger islands and "virtually disappear" under smaller islands (Roy and Connell 1991). For example, a recent study of the impacts of climate-induced sea-level rise on the Bonriki freshwater lens in Tarawa, Kiribati, concluded that the likely reduction of the width of the island by inundation could reduce the thickness of the freshwater lens by 29 percent (East-West Center 2001). The study also concluded that a concomitant rise of sea levels and reduction of rainfall could seriously reduce the volume of freshwater lenses. The study found that a 50-centimeter rise in sea-level in conjunction with a 25 percent reduction in rainfall could reduce the Bonriki freshwater lens by 65 percent, though the study emphasized that most of this impact was attributable to reductions in precipitation (Falkland 1999a).

Moreover, there is the danger that rising sea levels will result in water tables climbing close to or above the land surface, potentially resulting in full evapotranspiration of the resource (Burns 2000). This could ultimately result in groundwater becoming no longer potable on many islands, making human habitation impossible (Roy and Connell 1991). Also, the intrusion of saltwater into freshwater lenses could also result in severe reductions in several subsistence crops in PIDCs, including taro (Wilkinson and Buddemeier 1994; Watson, Zinyowera, and Moss 1998), breadfruit, coconuts (Lobban and Schefter 1997), and sugarcane (Nunn 1997) as well as lowland forests (SPREP 1999).

Precipitation Changes

As indicated earlier, climate change will likely visit changes in precipitation patterns on many PIDCs, with substantial increases in rainfall in some parts of the region and declines in others. Increased rainfall in the central and eastern tropical Pacific will likely result in a substantial increase in flooding, resulting in the intrusion of seawater through the aquifer recharge zones of islands, substantially reducing potable water supplies, and threatening crops grown in coastal regions (UNEP 1998). If increases in precipitation occur as shorter, but more intense events, recharge of freshwater lenses could drop as water runs off into the sea (SPREP 1999).

As discussed above, climate change could result in an increased incidence of ENSO events, which in the past have been associated with massive decreases in rainfall in the western portion of the Pacific. This could substantially reduce freshwater supplies in nations such as Micronesia and the Marshall Islands, where rainwater is the primary source of supply (SPREP 1999). Decreased rainfall could also diminish the volume of groundwater supplies and contribute to drought conditions in PIDCs, which in the past have been associated with devastating reductions in production of major agriculture export commodities such as sugar cane in Fiji, squash in Tonga (Tutangata 1996), and copra and giant taro in Kiribati (East-West Center 2001). Additionally, serious outbreaks of cholera in PIDCs have been associated with inadequate water supplies during ENSO events (Lobban and Schefter 1997).

Other Impacts

Freshwater resources may be further imperiled if the incidence and/or intensity of storms increase in the Pacific region as a consequence of climate change. Storms can generate waves that result in seawater inundation of groundwater resources, though they can also bring heavy rains that recharge these resources (Falkland 1999a).

Higher temperatures may result in an increase in potential evaporation (atmospheric water demand) rates in tropical regions (Watson, Zinyowera, and Moss 1998). This may accelerate the drying out of soil and vegetation, increasing water demand. Additionally, streamflow on some PIDCs could be affected by the cumulative effects of increased evaporation from watersheds (SPREP 1999).

Recommendations and Conclusions

The impacts described in this chapter are predicated on the assumption that greenhouse gas emissions will reach double pre-industrial levels or higher by the end of this century. Given the rather disheartening record of the parties to the United Nations Framework Convention on Climate Change (UNFCCC) in reducing greenhouse gas emissions, such levels are probably inevitable, and such projections could even prove conservative (Box 5.1).

Thus, it is incumbent upon PIDCs to focus on adaptation strategies to protect and conserve freshwater resources given the increasing threats to such resources that likely will occur during this century and beyond as a consequence of climate change.

Among the strategies that should be pursued by PIDCs are the following:

- A more comprehensive assessment of current water demand and projections of future demand in PIDCs needs to be conducted.
- A sustained research program to assess available fresh water resources on PIDCs is critical to monitor the performance of these resources under natural stresses, such as drought, and stresses associated with human activities such as surface water diversions and pumping. This should include the use of electronic data logging equipment, including hydrological data processing software, to ensure regular reporting and monitoring of water trends in the region. Such research will help to establish baseline information for water managers seeking to assess the additional threats

that may be posed by climate change. Additional training of local staff and island councils in this context should also be a priority.

- More emphasis needs to be placed on rehabilitation and maintenance of water catchment and distribution systems, including efforts to fortify such infrastructure from the possible impacts of climate change, such as violent storms.

- More attention needs to be focused on the impacts of development projects, including tourism infrastructure, on the vulnerability of coastal freshwater resources.

- More attention needs to be placed on controlling leaks from existing water systems, which as indicated earlier, result in the loss of as much as 70 percent of the freshwater resources in some delivery systems. Leakage control programs in nations such as the Seychelles, Malta, and the Bahamas have substantially reduced water loss (UNEP 1998).

There are a number of regional and international organizations that provide funding for research and adaptation programs in the region, including the World Meteorological

---

Box 5.1   The UN Framework Convention on Climate Change (UNFCCC)

The 186 parties to the UNFCCC, signed on May 9, 1992 (UNFCCC 1992; 31 I.L.M. 849), pledge themselves to "achieve . . . stabilization of greenhouse gas concentrations in the atmosphere at a level that would prevent dangerous anthropogenic interference with the climate system." In the face of continued increases in emissions, the parties adopted the Kyoto Protocol in 1997 (Kyoto Protocol 1997). The Protocol calls on industrialized parties to reduce their greenhouse gas emissions by an average of 5.2 percent below 1990 levels in the commitment period of 2008–12. However, despite the fact that the United States, which is responsible for 25 percent of the world's greenhouse gas emissions, has announced its opposition to the Protocol, the Protocol is a tepid commitment that can have little impact on climate trends. Parry et al. (1998) concluded that full implementation of the Protocol would reduce warming by 2050 by only one-twentieth of one degree Celsius and result in virtually no change in projected sea-level rise. To stabilize atmospheric emissions of CO_2 at double pre-industrial levels (550 parts per million) by the middle of this century would require reductions twenty times as large as those contemplated under Kyoto. It is very difficult to be sanguine that such a commitment by the global community will be forthcoming in the face of evidence that emissions continue to grow rapidly in both developed and developing nations (Arts 2000, IEA 2001). Overall, the International Energy Agency projects that CO_2 emissions will rise 60 percent between 1997 and 2020, with developing countries accounting for the majority of these emissions by the end of this period.

Organization, the United Nations Environment Programme, the South Pacific Applied Geoscience Commission, the South Pacific Regional Environment Programme (SPREP), and the Pacific Island Climate Change Assistance Program of the United Nations Development Program/SPREP. However, funding for these programs is extremely limited and in some cases is declining (Falkland 1999b). While the UNFCCC calls for funding of programs to assist those nations most vulnerable to the impacts of climate change, the parties have been slow to provide for adaptation assistance in developing nations, limiting themselves largely to funding the preparation of reports required under the agreement (Burton 2000). It can be hoped that more funding for such programs will be forthcoming now that several industrialized parties to the UNFCCC have announced their intention to provide $410 million annually by 2006 for funding climate change programs in developing nations (International Institute for Sustainable Development 2001). However, even this sum is far from adequate to address the vast array of needs of developing countries that will arise from climate change.

While Chinese mythology portrays heaven as a group of rock islands where immortals live blissfully, the existence of Pacific islanders by the end of this century is likely to be far from idyllic. The world's response to the specter of climate change's impacts on these nations will speak volumes as to our commitment to equity and the lot of some of the world's most vulnerable peoples.

REFERENCES

Anonymous. 1995. *Scientists Remain Unanimously Concerned over Climate Change, Eco-Log Wk,* Vol. 23, July 14, available at 1995 WL 2406417.

Aplin, G. 1999. *Global Environmental Crises* (2d ed.), p. 222.

Arts, B. 2000. "New arrangements in climate policy." *Change,* Vol. 47, pp. 1–2.

Asian Development Bank (ADB). 2001. "Geographical variability in water resources." Available at http://www.adb.org/Documents/Reports/Water/geographical_variability.asp.

Barron, E.J. 1995. "Climate models: How reliable are their predictions?" *Consequences,* pp. 17–18. Available at http://www.gcrio.org/CONSEQUENCES/introCON.html.

Benarde, Melvin A. 1992. *Global Warming . . . Global Warning.* John Wiley and Sons, Inc., New York, p. 45.

Bequette, J.F. 1994. "Small islands: Dreams and realities." *UNESCO Courier,* March, p. 23.

Brasseur, G. 1994. "Global warming and ozone depletion: Certainties and uncertainties." In Gary C. Bryner (ed.), *Global Warming and the Challenge of International Cooperation: An Interdisciplinary Assessment.* The David M. Kennedy Center for International Studies, Brigham Young University, Utah, pp. 29–30.

Brodie, J.E., Regina A. Prasad, and R.J. Morrison. 2001. "Pollution of small island water resources." Small Island Water Information Network. Available at http://www.siwin.org/textbase/swtb0013.pdf.

Burns, W.C.G. 2000. "The impact of climate change on Pacific island developing countries in the 21st century." In A. Gillespie and W.C.G. Burns (eds.), *Climate Change in the South Pacific: Impacts and Responses in Australia, New Zealand, and Small Island States.* Kluwer Academic, Dordrecht, pp. 233–251.

Burns, W.C.G. 2001a. "From the harpoon to the heat: Climate change and the International Whaling Commission in the 21st century." *Georgetown International Environmental Law Review,* Vol. 13, pp. 335–339.

Burns, W.C.G. 2001b. "The possible impacts of climate change on Pacific island state ecosystems." *International Journal of Global Environmental Issues,* Vol. 1, No. 1, p. 56.

Burton, I. 2000. "Adaptation to climate change: Advancing the agenda for collective global benefits." *Bridges,* Vol. 4, No. 8, pp. 9–12.

Campbell, J.R. 1996. "Contextualizing the effects of climate change in Pacific island countries." In T.W. Giambelluca and A. Henderson-Sellers (eds.), *Climate Change: Developing Southern Hemisphere Perspectives.* John Wiley and Sons, Inc., New York.

Climate Impacts Group (CIG). 2001. *CIG Research Summary.* University of Washington. Available at http://www.jisao.washington.edu/PNWimpacts/index.html. Accessed on October 12, 2001.

Commonwealth Scientific and Industrial Research Organization. 2001. "The greenhouse effect." Available at http://www.dar.csiro.au/publications/Holper_2001b.htm.

Dijon, R. 1994. "General review of water resources development in the region with emphasis on small islands." *Proceedings of the Regional Workshop on Water Resources of Small Islands,* Commonwealth Science Council, Technical Publication, Vol. 152, No. 2, pp. 25–44. Suva, Fiji.

Druyan, L.M. 1999. "A GCM investigation of global warming impacts relevant to tropical cyclone genesis." *International Journal of Climatology,* Vol. 19, pp. 607–616.

East-West Center. 2001. *Pacific Island Regional Assessment of the Consequences of Climate Variability and Change.* East-West Center, Honolulu, Hawaii, Chapter 2, p. 27.

Falkland, A. 1999a. "Impacts of climate change on water resources of Pacific islands." PACCLIM Workshop, Modelling the Effects of Climate Change and Sea Level Rise in Pacific Island Countries. Auckland, New Zealand, p. 2.

Falkland, A. 1999b. "Tropical island hydrology and water resources: Current knowledge and future needs." Second Colloquium on Hydrology and Water Management in the Humid Tropics, March 21–24, Panama.

Falkland, A., and E. Custodio. 1991. "Hydrology and water resources of small islands: A practical guide." UNESCO contribution to the International Hydrological Programme.

Fearnside, P.M. 2001. "Saving tropical forests as a global warming countermeasure: An issue that divides the environmental movement." *Ecological Economics,* Vol. 39, pp. 167–171.

Feldman, D.L. 1995. "Iterative functionalism and climate management organizations: From Intergovernmental Panel on Climate Change to Intergovernmental Negotiating Committee." In R.V. Bartlett, P.A. Kurian, and M. Malik (eds.), *International Organizations and Environmental Policy.* Greenwood Press, Westport, Connecticut, pp. 1,195–1,196.

General Assembly Resolution (G.A. Res.). 1989. G.A.Res. 43/53, U.N. GAOR, 2d Comm., 43rd Sess., Supp. No. 49, at 133, U.N. Doc. A/43/49. United Nations, New York.

Granger, S. 1996. "Geography of small tropical islands." *Coastal and Estuarine Studies,* Vol. 51, p. 166.

Haarsman, R.J. 1993. "Tropical disturbances in a GCM." *Climate Dynamics,* Vol. 8, p. 247.

Hadley Centre for Climate Prediction and Research. 1999a. *The Greenhouse Effect and Climate Change.* Hadley Centre, United Kingdom.

Hadley Centre for Climate Prediction and Research. 1999b. "Regional climate." Available at http://www.meto.gov.uk/sec5CR_div.bak/Brochure/regn_pre.html.

Hansen, J., et al. 2000. NASA Goddard Institute for Space Studies. "Global warming in the 21st century: An alternative scenario" (2000). Available at http://www.giss.nasa.gov/gpol/cites/2000.html.

Hanson, R.L. 2001. "Evapotranspiration and droughts." U.S. Geological Survey. Available at http://geochange.er.usgs.gov/sw/changes/natural/et.

Hay, J. 1999. "Small island states and the climate treaty." *Tiempo,* Vol. 33, p. 3.

Hay, J., and C. Kaluwin. 1993. *Proceedings, Second SPREP Meeting on Climate Change and Sea Level Rise in the South Pacific Region,* p. 190. South Pacific Regional Environmental Program. Apia, Western Samoa.

Hennessy, K.J. 2001. "CSIRO climate change output." Available at http://www.dar.csiro.au/pub/programs/climod/impacts/data.htm.

Herman, M.E., R.W. Buddemeier, and S.W. Wheatcraft. 1986. "A layered aquifer model of atoll island hydrology: Validation of a computer simulation." *Journal of Hydrology,* Vol. 84, pp. 303–304.

Hileman, B. 1995. "Climate observations substantiate global warming models." *Chemical and Engineering News,* November 27, p. 4.

Hoegh-Guldberg, O., H. Hoegh-Guldberg, D.K. Stout, H. Cesar, and A. Timmerman. 2000. *Pacific in Peril.* Greenpeace, Sydney, Australia, p. 36.

Holland, G.J. 1995. "The maximum intensity of tropical cyclones." *Journal of Atmospheric Science,* Vol. 54, pp. 2519–2541.

Intergovernmental Panel on Climate Change (IPCC). 1994. *Radiative Forcing of Climate Change.* Cambridge University Press, Cambridge.

Intergovernmental Panel on Climate Change. 1995. *Contribution of Working Group I to the IPCC Second Assessment Report, IPCC X1/Doc. 3.* Cambridge University Press, Cambridge.

Intergovernmental Panel on Climate Change. 1997. *An Introduction to Simple Climate Models Used in the IPCC Second Assessment Report.* Cambridge University Press, Cambridge.

Intergovernmental Panel on Climate Change. 2001a. *Working Group II to the Third Assessment Report, Climate Change 2001: Impacts, Adaptation, and Vulnerability.* Cambridge University Press, Cambridge, pp. 854–855.

Intergovernmental Panel on Climate Change. 2001b. *Climate Change 2001: The Scientific Basis, Contribution of Working Group I to the Third Assessment Report of the Intergovernmental Panel on Climate Change.* Cambridge University Press, Cambridge, p. 89.

Intergovernmental Panel on Climate Change. 2001c. *Working Group I, Third Assessment Report, Summary for Policymakers.* Cambridge University Press, Cambridge.

Intergovernmental Panel on Climate Change. 2001d. *Technical Summary of the Working Group I Report, Third Assessment Report.* Cambridge University Press, Cambridge.

Intergovernmental Panel on Climate Change. 2001e. *Working Group I, Summary for Policymakers.* Cambridge University Press, Cambridge.

International Energy Agency (IEA). 2001. *World Energy Outlook 2000.* Available at http://www.iea.org/weo/index.html.

International Institute for Sustainable Development. 2001. *Summary of the Resumed Sixth Session of the Conference of the Parties to the UN Framework Convention on Climate Change: 16–27 July 2001, Earth Negotiations Bulletin,* Vol. 12, No. 176. Available at http://www.iisd.ca/climate/cop6bis.

Jacobson, G., P.J. Hill, and F. Ghassemi. 1997. "Geology and hydrogeology of Nauru Island." Chapter 24 in *Geology and Hydrogeology of Carbonate Islands, Developments in Sedimentology,* Vol. 54, pp. 707–742.

Jardine, K. 1994. "Finger on the carbon pulse." *Ecologist,* Nov./Dec., p. 220.

Jones, R. 2000. "Climate change in the South Pacific." *Tiempo,* Vol. 35, pp. 17–18.

Kaluwin, C., and A. Smith. 1997. "Coastal vulnerability and integrated coastal zone management in the Pacific island region." *Journal of Coastal Research,* Vol. 24, p. 95.

Karl, T.R., N. Nicholls, and J. Gregory. 2001. "The coming climate." *Scientific American.* Available at http://www.sciam.com/ 0597issue/0597karl.html.

Klein, R.J.T., and R.J. Nicholls. 1999. "Assessment of coastal vulnerability to climate change." *Ambio,* Vol. 28, No. p. 182, 182.

Knutson, T.R., R.E. Tuleya, and Y. Kurihara. 1998. "Simulated increase of hurricane intensities in a CO_2-warmed climate." *Science,* Vol. 279, February 18, p. 1,018.

Kondratyev, K. Ya., and P. Cracknell. 1998. *Observing Global Climate Change.* Taylor & Francis, Philadelphia.

Kripalani, R.H., A. Kulkarni, and S.S. Sabade. 2001. "ENSO-monsoon weakening: Is global warming really the player?" *CLIVAR Exchanges,* Vol. 6, No. 3, p. 11.

Kyoto Protocol. 1997. December 10, FCCC/CP/1997/L.7/Add. 1, 37 International Legal Materials (ILM) 22.

Lauerman, J.F. 1997. "Trouble in paradise." *Environmental Health Perspectives,* Vol. 105, No. 9. Available at http://ehpnet1.niehs.nih.gov/docs/1997/105-9/focus-full.html.

Leatherman, S.P. 1997. "Sea level rise and small island states: An overview." *Journal of Coastal Research,* Vol. 24, pp. 3–4.

Lobban, C.S., and M. Schefter. 1997. *Tropical Pacific Island Environments.* Island Environments Books, Mangilao, Guam.

Meehl, G.A. 1996. "Vulnerability of freshwater resources to climate change in the tropical Pacific region." *Water, Air and Soil Pollution,* Vol. 92, pp. 203–210.

Ministerial Conference on Environment and Development in Asia and Pacific. 2001. "Review of the state of the environment of the Pacific islands." Available at http://www.unescap.org/mced2000/pacific/SoE-pacific.htm.

NASA Goddard Institute for Space Studies. 2001. "How will the frequency of hurricanes be affected by climate change?" Available at http://www.giss.nasa.gov/research/intro/druyan.02.

Nicholls, R.J., and N. Mimura. 1998. "Regional issues raised by sea-level rise and their policy implications." *Climate Research,* Vol. 11, pp. 5–6.

Nunn, P.D. 1997. "Vulnerability of South Pacific island nations to sea-level rise." *Journal of Coastal Research,* Vol. 24, pp. 133–140.

Panjabi, R.K.L. 1993. "Can international law improve the climate? An analysis of the United Nations Framework Convention on Climate Change signed at the Rio Summit in 1992." *North Carolina Journal of International Commercial Regulation,* Vol. 18, pp. 491–500.

Parry, M., N. Arnell, M. Hulme, R. Nicholls, and M. Livermore. 1998. "Buenos Aires and Kyoto targets do little to reduce climate change impacts." *Global Environmental Change,* Vol. 8, No. 4, pp. 285–289.

Parry, M., N. Arnell, T. McMichael, R. Nicholls, P. Martens, S. Kovats, M. Livermore, C. Rosenzweig, A. Iglesias, and G. Fischer. 2001. "Millions at risk: Defining critical climate change threats and targets." *Global Environmental Change,* Vol. 11, pp. 181–183.

Pearce, F. 1995. "World lays odds on global catastrophe." *New Science,* April 8, p. 4.

Pernetta, J.C. 1992. "Impacts of climate change and sea-level rise on small island states." *Global Environmental Change,* March, p. 20.

Roy, P., and J. Connell. 1991. "Climate change and the future of atoll states." *Journal of Coastal Research,* Vol. 7, No. 4, pp. 1,057–1,064.

Salinger, M.J., R.E. Basher, B.B. Fitzharris, J.E. Hay, P.D. Jones, J.P. MacVeigh, and I. Schmidely-Leleu. 1995. "Climate trends in the South-West Pacific." *International Journal of Climatology,* Vol. 15, p. 285.

Schlesinger, M.E. 1993. "Model projections of CO_2-induced equilibrium climate change." In R.A. Warrick, E.M. Barrow, and T.M.L. Wigley (eds.), *Climate Change and Sea Level Change.* Cambridge University Press, Cambridge, pp. 285–300.

Small Island Developing States Net. 2001. "The impacts of climate change on Pacific island countries." Available at http://www.sidsnet.org.

Society of Professional Well Log Analysts. 2001. "Glossary of water terms." Available at http://www.spwla.org/gloss/reference/glossary/glossz/glossz.htm.

Solman, S.A., and M.N. Nunez. 1999. "Local estimates of global climate change: A statistical downscaling approach." *International Journal of Climatology,* Vol. 19, pp. 835–836.

Solomon, S.M., and D.L. Forbes. 1999. "Coastal hazards and associated management issues on South Pacific islands." *Ocean and Coastal Management,* Vol. 42, pp. 527–529.

South Pacific Applied Geoscience Commission. 2001a. "Water woes. . . . Available at http://www.sopac.org.fj/Data/Press/Detail.html?PRID=36.

South Pacific Applied Geoscience Commission. 2001b. "Water resources assessment and sanitation." Available at http://www.sopac.org.fj/Secretariat/Units/Wru/wateract.html.

South Pacific Regional Environmental Programme (SPREP). 1999. *Analysis of the Effects of the Kyoto Protocol on Pacific Island Countries,* Part 2, *Regional Climate Scenarios and Risk Assessment Methods.* SPREP, Apia, Western Samoa.

Stevens, W.K. 1999. "1999 continues warming trend around globe." *New York Times,* Dec. 19, p. 1.

Tutangata, T.I. 1996. *Vanishing Islands, Our Planet.* Available at http://www.ourplanet.com/imgversn/103/06_van.htm.

Tutangata, T.I. 2000. "Sinking islands, vanishing worlds." *Earth Island Journal,* Summer, p. 44.

UNESCO. 2001. *Pacific Science Programme Update.* Available at http://www.unesco.org/csi/act/pacific/project2.htm.

United Kingdom Department of the Environment, Transport and the Regions. 1999. *Climate Change and Its Impacts.* United Kingdom.

United Nations Environment Programme (UNEP). 1998. *Source Book of Alternative Technologies for Freshwater Augmentation in Small Island Developing States.* UNEP, Nairobi, Kenya.

United Nations Environment Programme. 1999. *Source Book of Alternative Technologies for Freshwater Augmentation in Small Island Developing States.* UNEP, Nairobi, Kenya, p. 22.

United Nations Environment Programme. 2000. *Overview on Land-Based Pollutant Sources and Activities Affecting the Marine, Coastal, and Freshwater Environment.* Regional Seas Reports and Studies, No. 174. Available at http://www.unep.org/nova/applications/regseas/sprep/english.html.

United Nations Environment Programme, GEO-2000. 1999. *The State of the Environment-Global Issues.* Available at http://www.unep.org/geo2000/english/0034.htm.

United Nations Framework Convention on Climate Change (UNFCCC). 1992. *International Legal Materials,* Vol. 31, No. 849, May 9.

University of Capetown, Environmental and Geographical Science Department. 1999. *Climate Change—Some Basics.* Available at http://www.egs.uct.ac.za/csag/faq/climate-change/faq-doc-5.html.

U.S. Central Intelligence Agency. 2001. *The World Factbook 2000.* Available at http://www.odci.gov/cia/publications/factbook.

U.S. Naval Meteorology and Oceanography Command. 2001. *Pacific Ocean.* Available at http://oceanographer.navy.mil/pacific.html.

Warbrick, C., and D. McGoldrick. 1998. "Global warming and the Kyoto Protocol." *International and Comparative Law Quarterly,* Vol. 47, pp. 446–447.

Watson, R.T., M.C. Zinyowera, and R.H. Moss. 1998. *The Regional Impacts of Climate Change.* Cambridge University Press, Cambridge.

Whittaker, R.J. 1998. *Island Biogeography.* Oxford University Press, New York.

Wigley, T.M.L. 2001. "The Science of climate change." In E. Claussen (ed.), *Climate Change: Science, Strategies and Solutions.* Pew Center on Global Change, Washington, D.C., p. 7.

Wigley, T.M.L., and S.C.B. Raper. 2001. "Interpretation of high projections for global-mean warming." *Science,* Vol. 291, pp. 453–454.

Wilkinson, C.R. 1996. "Global change and coral reefs: Impacts on reefs, economies and human cultures." *Global Change Biology,* Vol. 2., pp. 547–549.

Wilkinson, C.R., and R.W. Buddemeier. 1994. *Global Climate Change and Coral Reefs: Implications for People and Reefs.* Report of the UNEP-IOC-ASPEI-IUCN Global Task Team on the Implications of Climate Change on Coral Reefs. United Nations Environment Programme, Nairobi, Kenya.

World Bank 2000. *Cities, Seas, and Storms.* Papua New Guinea and Pacific Island County Unit. The World Bank, Washington, D.C., Vol. 4, p. 25.

Managing Across Boundaries: The Case of the Colorado River Delta

Michael Cohen

The development of water resources has contributed to tremendous economic growth, generated relatively clean power, provided regular, dependable supplies to urban and agricultural users, and enabled settlement and agricultural development in areas where such activity would have been unimaginable without the construction of a massive hydrologic infrastructure and the institutions to drive and control it.

Yet this development has come at the cost of a profoundly altered physical landscape. Lakes and wetlands have disappeared while valleys and gorges have been inundated, rivers have been straightened and channelized, and clear streams have been polluted with all manner of contaminants, driving a host of aquatic and terrestrial species to, or even over, the edge of extinction. Perhaps the single most dramatic example of the environmental impacts of water resources development has been the desiccation of the Aral Sea, as described in Chapter 2 of this volume, but a large number of freshwater systems have suffered equally important, if less dramatic, impacts. This chapter offers an overview of the environmental impacts on and challenges to one such system, the Colorado River delta, and offers a set of recommendations to protect and preserve this resource.

The development of water resources occurs within complex legal and institutional frameworks that typically favor economic development and human uses at the expense of instream and environmental uses. In the early-twentieth-century American West—indeed, throughout the world—water left in a stream was considered "wasted"; the very language of water "development" and "reclamation" points to the perception that water was a resource to be extracted and put to use. Yet society's appreciation for the value of ecosystems is changing fundamentally (Cortner et al. 1996). In the mid-1960s, the meaning of "beneficial use" of water began to expand from a narrow economic interpretation to a broader view that reflected the growing public awareness of the value of environmental resources. Instream uses of water, for recreation, fish and wildlife habitat, and aesthetic concerns, have finally begun to receive formal protections (Hoffman-Dooley 1996; Howe 1996; Colby, McGinnis, and Rait 1991). For example, South Africa has formally recognized the right to water for natural ecosystems in their new water laws and constitution. Rigid and outdated institutions, however, often hinder these changes in values (Cortner et al. 1996).

The shift in priorities and sensibilities has translated slowly into a similar recognition of the value of transboundary and binational environmental resources. Managing transboundary water resources presents additional challenges, challenges that are instructive given the abundance of such resources across the globe. A 1978 study by the United Nations identified 214 international river basins; the most recent estimate now exceeds 260 international basins because of changes in political borders and our ability to map basins accurately (Wolf et al. 1999, Gleick 2000). One of these international river basins is the Colorado, with its headwaters in the Rocky Mountains of the United States and its remnant delta in Mexico.

The Colorado River

The delta of the Colorado River provides a valuable case study for several reasons. The Colorado River is subject to greater institutional control than perhaps any other river in the world. More than 80 major diversions move water out of the Colorado River channel and transport it to agriculture and other uses in the United States and Mexico, irrigating more than 750,000 hectares (ha) and serving some 30 million people (Pontius 1997). The delta in Mexico supports several endangered species and is itself endangered by lack of flows. Each drop of Colorado River water is used an average of 17 times. The U.S. Bureau of Reclamation (in Fradkin 1981, p. 16) reports that

> [t]he Colorado River is not only one of the most physically developed and controlled rivers in the nation, but it is also one of the most institutionally encompassed rivers in the country. There is no other river in the Western Hemisphere that has been the subject of as many disputes of such wide scope during the last half century as the Colorado River.

This control has caused tremendous changes along the river as a whole, and in particular in its delta and estuary. In many years, the Colorado River runs dry before it reaches the ocean. The combination of an extensive physical infrastructure, including more than ten major dams on the Colorado River mainstem, and a complex and detailed legal framework have irrevocably altered the physical and biological composition of the river, straightening the river channel, flattening the hydrograph, decreasing the temperature while increasing salinity and the concentration of other contaminants, and driving several endemic species to the brink of extinction.

These changes are most evident in the river's delta. Yet despite the marked reduction in inflows, the Colorado River delta is still the largest desert estuary in North America, supporting a variety of wildlife and marine species, including several threatened and endangered species listed in both the United States and Mexico. The delta is a key stopover for birds migrating along the Pacific Flyway and it supports large numbers of wintering waterfowl. The delta is also of interest because its vegetation and remaining wildlife have demonstrated a remarkable resilience and ability to regenerate in response to inadvertent releases of water from upstream dams. This suggests that modest efforts to restore the ecosystem of the delta could have substantial, positive environmental results. The following sections provide a brief review of the Colorado

River's hydrology, the institutions that developed to manage the river, the impacts that water resources development have had on the river's delta, and recommended steps to preserve the region.

Hydrology

The Colorado River basin covers 632,000 square kilometers (km^2), from snow-covered mountain ranges to arid salt flats (Pontius 1997); some 32,000 km^2 of the basin lie in Mexico (see Figure 6.1). The Colorado River basin, and particularly the lower basin, is

FIGURE 6.1 The Colorado River basin.

Source: U.S. Bureau of Reclamation

characterized by a tremendous disparity in the spatial and temporal distribution of water (Gleick 1993). The headwaters, high in the Rocky Mountains of the United States, receive most of the precipitation in the region, primarily as snow. Much of the rest of the basin is extremely arid, with less than 80 millimeters (mm) of precipitation per year. The Colorado is largely snowmelt-driven: some 70 percent of the river's annual natural flow occurs from May through July (Harding, Sangoyomi, and Payton 1995).

Marked fluctuations in the volume of flow, both between and within years, characterized the flow of the Colorado River prior to the construction of dams and diversions. The maximum reconstructed flow (i.e., the flow that would have occurred without human intervention in the form of dams and withdrawals) during the period of record occurred in 1983, with an estimated flow of 31 cubic kilometers (km^3); the minimum reconstructed flow occurred in 1934, with an estimated flow of 7.7 km^3. Researchers cite varying estimates for the river's average annual flow, ranging from an estimated long-term mean of 16.7 km^3 based on tree-ring records (Meko, Stockton, and Boggess 1995) to 18.6 km^3 for the past century of instrumental record (Owen-Joyce and Raymond 1996).

Intra-annual fluctuations on the river are also large. The flow of the river is largely dependent on the size of the snowpack in the headwaters and how quickly it melts. Peak flows typically occur in June, with the lowest flows in mid-winter (Ohmart, Anderson, and Hunter 1988). Peak season flows of 2,300 m^3/sec were common at Lee Ferry—a measurement point at the divide between the "upper" and "lower" basins (see Figure 6.1), dropping to less than 85 m^3/sec from late summer through winter (GAO 1996). For several days in 1934, no measurable discharge was recorded for the river near Yuma, Arizona (IBWC 1998).

Irrigators began diverting water from the Colorado River in the late nineteenth century, but the flood and drought cycles challenged development efforts and often washed out irrigation headgates, inundating fields and towns. Laguna Dam, the first major dam on the river, was completed in 1909, but not until the completion of Hoover Dam in 1935 was there a structure on the river capable of regulating river flows. The old adage about the Colorado River was that it was "too thick to drink, too thin to plow." One of the major effects of the river's impoundment has been the near-cessation of the river's transport of sediment, fundamentally changing the character of the river from warm, turbulent, and sediment-filled to cold, regulated, and clear.

Within a year of its construction, the Colorado had filled the reservoir behind Laguna Dam with silt. Imperial Reservoir, completed in 1938 10 kilometers upstream from Laguna Dam, had an initial storage capacity of 100 million cubic meters (mcm); it now has no useful storage, due to silting by the river (Reclamation 1996). Measurements of suspended sediment collected at Lee Ferry prior to the closure of upstream dams indicated sediment concentrations in excess of 10,000 ppm; the concentrations of samples collected since the closure of upstream dams are generally below 200 ppm (Schmidt and Graf 1990). Upstream dams, especially Hoover and Glen Canyon, have been so effective at trapping sediment that the sediment levels at Imperial Dam now average only 870 tons/day (Cox 1997), well below the Virgin River's load of 50,000 tons/day (White 1991). This loss of sediment input has caused the river's delta to reverse the normal process of accumulation, and tidal action now removes more material from the delta than the river replaces (Kowalewski et al. 2000).

Dams and diversionary structures on the Colorado now have a storage capacity exceeding four years of average river flow. This storage capacity has, with few exceptions, enabled river managers to flatten the river's pre-impoundment hydrograph, greatly reducing the flooding that previously characterized the system, particularly affecting the river delta. The pre-impoundment hydrograph has now been replaced by a predictable set of releases, timed to meet the needs of irrigators and urban areas. These releases are further modified to maximize peak power generation, so that river levels below dams rise and fall as much as 4 feet on a daily basis (Reclamation 1996).

Legal Framework: The "Law of the River"

Water in the western United States developed under institutions designed to encourage settlement and consumptive, off-stream use (NRC 1992). These institutions generated a wealth of subsidies for water, power, agriculture, and land, which led to disproportionately high water consumption and a disregard for resource conservation (Reisner and Bates 1990, Reisner 1993). Waterstone (1992) claims that western water scarcity is less the result of physical scarcity and more a result of institutional factors favoring water-intensive agriculture over more productive, less demanding uses of water. The institutional problems governing the use of water are more salient and more perplexing even than the myriad of physical, technical, and economic issues that have generated such frustration and conflict among water users in the West (Ingram et al. 1984).

A complex, dynamic legal framework has evolved over the past century to manage and control water demands and the operation of facilities on the Colorado River. Within the United States, this framework is known as the "Law of the River." The Law of the River is a multilayered, at times inconsistent, body of rules, compacts, state and federal laws, regulations, Supreme Court decisions, and treaties. The Law of the River is not explicitly defined or codified in any single form; it is a dynamic bundle of rules subject to frequent reinterpretation, revision, and expansion (Hundley 1966, 1975; Getches 1985; Bloom 1986). The Law of the River has imposed a three-tiered system for allocating Colorado River water. At the top is the United States' international obligation to Mexico to deliver 1.85 km^3 (1,850 mcm) of water annually, within a prescribed salinity range. The second tier divides the water between the upper and lower basins in the United States, and to the various states within each basin. The lowest tier is the allocation of water within each state (Wilson 1994). In Mexico, regulation of Colorado River use is centralized at the federal level (Castro 1995).

The flows of the Colorado River below Hoover Dam are controlled and regulated based on flood-control requirements, downstream diversion orders, and demands for hydroelectric power (cf. Nathanson 1980). The degree of institutional control over the Colorado River cannot be overstated: the 1983 flood has been the only instance since the construction of Hoover Dam in 1935 when discharge from the dam, and along the lower Colorado River, was not completely controlled by the Bureau of Reclamation (Holburt 1984). The concept of a major river whose flow can be turned on and off is difficult to comprehend, but it is the central fact of the lower Colorado River. Except in extremely rare instances of unusually high inflows to Lake Mead and limited storage availability (triggering U.S. Army Corps of Engineers Flood Control Release Guidelines), the flows in the lower Colorado River are released by the U.S. Bureau of Reclamation. The Bureau of Reclamation determines release rates based on a complex algorithm that integrates agricultural diversion orders, required deliveries to Mexico, storage requirements, flood control, and hydroelectric power generation contracts. The Law of the River allocates no water to the Colorado River delta: any flows that reach the delta are either the result of low-quality return flows or infrequent flood and space-building releases from upstream dams.

The flow of Colorado River water to Mexico and the delta occurs within this legal framework. The 1944 U.S.-Mexico Treaty on the Utilization of Waters of the Colorado and Tijuana Rivers and of the Río Grande ("the Treaty") commits the United States to deliver 1,850 mcm of Colorado River water to Mexico each year, of which at least 1,678 mcm is to be delivered in the mainstem above Mexico's Morelos Dam. The remainder may be delivered at the southerly land boundary near the mainstream. Article 10(b) of the Treaty allocates an additional 246.7 mcm to Mexico when the U.S. Section of the International Boundary and Water Commission determines that there exists a surplus of Colorado River water above the amount needed to supply U.S. uses. Such Treaty surpluses were declared between 1997 and 2000 but are unlikely to occur in the next 15 years due to the recent U.S. adoption of an interim set of operating guidelines (Pitt et al. 2000).

After a long dispute between Mexico and the United States about the quality of water delivered to Mexico, which had been degraded by brackish discharge from Arizona's Wellton-Mohawk Irrigation and Drainage District in the early 1960s, the two countries adopted Minute 242 of the International Boundary and Water Commission in 1973 (Wahl 1989). Minute 242 states that 1,678 mcm of annual water deliveries to Mexico at Morelos Dam would have an average salinity no more than 115 ppm (±30 ppm) greater than the salinity of the river at Imperial Dam; the remaining 173 mcm, delivered at the International Boundary with Mexico near San Luis, would have "a salinity substantially the same as that of the waters customarily delivered there" (in Hundley 1986, p. 39). Congress then passed the Colorado River Basin Salinity Control Act of 1974, authorizing measures to enable compliance with Minute 242, including the construction of the $250 million Yuma Desalting Plant and the construction of the Main Outlet Drain Extension (MODE) and its bypass extension, discharging agricultural drainage (with a salinity >2,900 ppm) into Mexico's Ciénega de Santa Clara, in the southeastern corner of the delta (Glenn et al. 1992).

The Mexican Constitution establishes the legal framework for water management in Mexico and reserves to the federal government the rights to national waters, including the Colorado River (Castro 1995). The Constitution also gives ownership of ground water to the national government (Cossio-Díaz 1995). In practice, Mexico's Comisión Nacional de Aguas (CNA) determines deliveries of Colorado River water and regulates groundwater extraction within the delta region (Mumme 1996). CNA determines cropping patterns and allocates water accordingly. In recent years, however, efforts have begun to decentralize this authority and provide for greater autonomy at the local level (Clinton et al. 2001).

A fundamental problem is that the Colorado River is over-allocated: more water is legally apportioned ("paper water") from the river than actually flows ("wet water") in most years. This unfortunate situation arose from assumptions based on a faulty analysis of the long-term flow of the river: the U.S. negotiators of the Colorado River Compact of 1922 used the information available at the time, which did not accurately reflect the average hydrology. In hindsight, we now understand that the fifteen years prior to the 1922 Compact were unusually wet: the annual flow of the river near Lee Ferry during that time was estimated at 22.3 km^3 (Morrison, Postel, and Gleick 1996); 1917 had witnessed the highest recorded flow at Lee Ferry, at 29.5 km^3 (Holburt 1984). The Colorado River was apportioned among the basin states based upon these incomplete records, allocating a total of 19.7 km³ to the states. This over-allocation has been tempered by a recent wet period and by the fact that the upper basin states have yet to fully develop their allocations, enabling the lower basin states (particularly California) to use more than their formal entitlement. As the upper basin states continue to expand their use, and in the face of average or below-average inflows to system reservoirs, pressure has increased on California to limit its use. This over-allocation greatly complicates efforts to identify and dedicate water for environmental purposes.

The Colorado River Delta

The Colorado River delta was formed by the deposition of sediment from periodic Colorado River floods (Sykes 1937). The delta reflects the impacts of upstream dams and diversions, land use changes, and management decisions. These changes have reduced the delta's extent from some 7,770 km² to a series of remnant wetlands of around 600 km² (Luecke et al. 1999). The hydrologic extent of the delta encompasses the Imperial Valley in California, the Mexicali Valley in Baja California, and the confluence of the Colorado and Gila Rivers in Arizona, extending into the lower Coachella Valley in California, the San Luis Valley in Sonora, and parts of the Laguna Salada basin in Baja California. Much of this fertile alluvial soil has been converted to irrigated agriculture, leaving a small remnant delta located along the border of the Mexican states of Baja California and Sonora (see Figure 6.3). The upper reaches of the remnant delta include a narrow riparian corridor, broadening downstream to include a series of emergent wetlands fed by distinct sources of agricultural runoff. Further downstream, tidal effects become more pronounced.

As with most river deltas, the Colorado River delta was a highly productive and diverse ecosystem prior to management and diversion of the river. The pre-1930s Colorado River delta was nearly twice the size of Rhode Island and populated by a diversity of plant and animal species, including an estimated 200 to 400 species of vascular plants. Early explorers in the delta reported jaguars, beavers, deer, and coyotes in addition to an astounding abundance of waterfowl, fish, and other marine and estuarine species (Leopold 1949, Sykes 1937, and others). Human settlements in the delta date back 15,000 years; Yuman-speaking native peoples inhabited the area some 3,000 years ago. The Cucapá people, descended from the Yumana linguistic family, have lived in the delta for a millennium and numbered about 20,000 at the arrival of the Spanish (Kelly 1950).

The transformation of the Colorado River and its delta began in the late nineteenth century, as water was diverted from the river and the river's fertile floodplain began to be cleared and irrigated. For six years in the late 1930s, as Lake Mead filled behind Hoover Dam, virtually no fresh water reached the river's delta. The dewatering of the delta was repeated again from 1963 to 1981 as Lake Powell filled behind Glen Canyon Dam in Arizona. Figure 6.2 shows the estimated amount of water that would have flowed to the Colorado River delta absent upstream dams and diversions, and the actual discharge recorded at a gauging station near the upstream limit of the delta. Those two reservoirs are now used to store and regulate the flow of the Colorado River for maximum reliable use by water-rights holders.

During the twentieth century, river flows into the delta have been reduced nearly 75 percent; in 24 of the past 40 years, less than 2 percent of the Colorado River's estimated

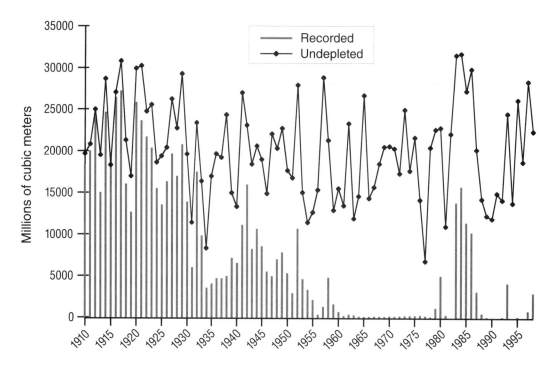

FIGURE 6.2 Measured and estimated undepleted flow through the delta region, 1920–1998.

Source: Cohen, Henges-Jeck, and Castillo-Moreno 2001

undepleted flows reached the delta (Pitt et al. 2000). The gauge at the Southerly International Boundary (SIB) (the southernmost point of the border dividing Baja California, Mexico, from Arizona in the United States) records discharge to the upstream extent of the delta. In 9 years within the most recent 30-year period of record (1969–98), annual discharge at the SIB has exceeded 1,000 mcm. Mean annual discharge at the SIB during this period measured 2,350 mcm, while median discharge was 190 mcm (Σ = 4,400 mcm). The Colorado River discharges to the delta when either or both of the following sets of conditions are satisfied: the elevation of Lake Mead on the Colorado River, or Painted Rock Reservoir on the Gila River, and projected runoff into that reservoir are both sufficiently high to trigger flood-control releases, and the timing and magnitude of such releases exceed the demands and diversion capacity of downstream diverters (Cohen, Henges-Jeck, and Castillo-Moreno 2001).

After being ignored for a century, the delta of the Colorado River is the subject of growing scientific and political interest. Much of this interest stems from the delta's resilience and recent regeneration in response to the renewed flow of water to the region, as well as the region's relative abundance of native riparian habitat and several endangered species. The recent literature on the ecology and restoration of the delta of the Colorado River emphasizes the importance of natural and anthropogenic sources of water for sustaining delta habitats (Glenn et al. 1992, 1996, 1999; Zengel et al. 1995; Morrison, Postel, and Gleick 1996; Valdés-Casillas et al. 1998; Luecke et al. 1999; Kowalewski et al. 2000; Pitt et al. 2000; see also *Journal of Arid Environments* [2001] 49:1).

Despite reports that the delta was a dead ecosystem where the Colorado River no longer reached the sea (Fradkin 1981), agricultural drainage and the occasional releases of Colorado River water from upstream reservoirs have prompted significant new growth of valuable native riparian and emergent wetland habitat, supporting the

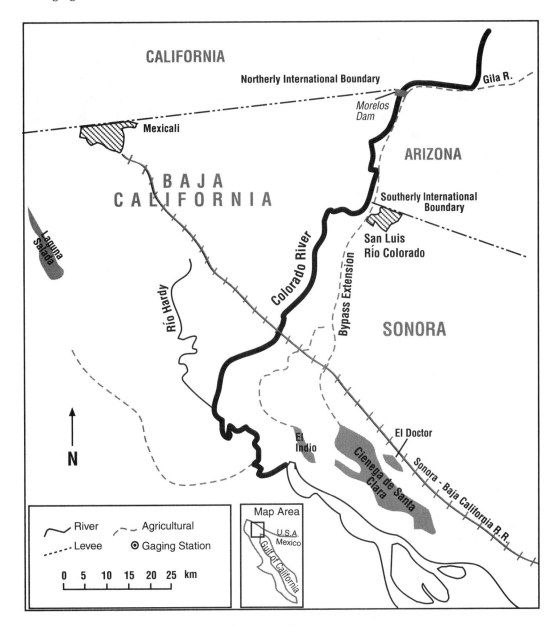

FIGURE 6.3 The Colorado River delta in Mexico.
Source: Adapted from Cohen, Henges-Jeck, and Castillo-Moreno 2001; Luecke et al. 1999

largest and most critical arid wetland in North America and sustaining avian and aquatic species of concern (Glenn et al. 1992, 1996; Luecke et al. 1999). Recent flood releases have also been strongly correlated with a rise in the shrimp catch in the upper Gulf (Galindo-Bect et al. 2000), an indication of the potential to renew the viability of an important estuary.

The remnant delta encompasses the riparian area downstream of Morelos Dam, the Río Hardy wetlands northwest of the levee on the right bank, the Ciénega de Santa Clara (4,200 ha), and the El Indio (1,900 ha) and El Doctor (750 ha) wetlands east of the levee on the left bank. The delta also includes the intertidal zone along the final 19 kilometers of the river, encompassing 440 ha (Luecke et al. 1999). Through its delta, the Colorado River runs as a low-gradient, meandering stream with no firm channel: the downstream, broader reach of the floodplain is replete with oxbows and backwaters, vestiges of former channels (Sykes 1937). The total length of the Colorado River from Morelos Dam to its mouth near Isla Montague, at the upper Gulf of California, is about 150 kilometers. The maximum elevation of the river in this stretch is approximately 32 meters above sea level, at the base of Morelos Dam. The delta lies in the Sonoran desert, characterized by low precipitation (54 mm/year) and high evaporation rates (2,046 mm/year), where maximum daily temperatures may exceed 38° Celsius more than five months each year.

The upstream extent of the delta, characterized by willow *(Salix goodingii)* thickets and cottonwood *(Populus fremontii)*/willow gallery forests (Valdés-Casillas et al. 1998), is constrained by levees that were constructed to protect the surrounding agricultural areas from flooding. The area within the levees broadens downstream near the Colorado's confluence with the Río Hardy, where wetland vegetation and non-native saltcedar *(Tamarix ramosissima)* supplant the native riparian vegetation (Luecke et al. 1999). Downstream of the confluence lies the intertidal zone, characterized by endemic saltgrass *(Distichlis palmerii)*, affected by the extreme tides (amplitude >8 meters) (Lavín, Beier, and Badan 1997) of the upper Gulf of California (Glenn et al. 1999). The delta also commonly includes three wetland areas distinct from the mainstem system: the Ciénega de Santa Clara and El Indio wetlands, characterized by dense stands of cattails *(Typha domingensis)*, common reed *(Phragmites australis)*, and bulrush *(Scirpus americanus)* (Glenn et al. 1992), and El Doctor wetlands, supporting 29 wetland plant species (Zengel et al. 1995). The Ciénega, the largest of these distinct wetlands, has a total inundated area of 12,000 ha, of which some 4,200 ha are vegetated (Luecke et al. 1999; Zamora-Arroyo et al. 2001). The Ciénega lies in a depression formed by the Cierro Prieto fault, in a former arm of the Colorado River (Glenn et al. 1999). In the 1970s, agricultural drainage from Mexico's Riito Drain and local artesian springs supported a smaller (200 ha) wetland at the site. Agricultural drainage discharging behind the levee on the left bank of the Colorado River supports El Indio wetlands (Luecke et al. 1999). Artesian springs along the eastern edge of the delta sustain El Doctor wetlands (Glenn et al. 1999).

These riparian and emergent wetland delta habitats support a variety of wildlife and aquatic species, including several threatened and endangered species listed in both the United States and Mexico. Estimates indicate that delta habitats support 68,000 resident and 49,000 nonresident summer birds. The habitat in the lower basin of the Colorado River in the United States is estimated to support fewer than half as many birds (Pitt et al. 2000). The largest known populations of two species listed as endangered in the United States—Yuma clapper rail *(Rallus longirostris yumanensis)* and desert pupfish *(Cypranodon macularius)*—are found in the delta.

The saltwater habitats of the upper Gulf of California are also affected by the lack of freshwater flows. In the 1970s, the totoaba *(Cynoscion macdonaldii)* fishery in the upper Gulf collapsed, and in the late 1980s and early 1990s the shrimp catches dropped by over 50 percent, signaling a virtual collapse in the shrimp fishery. In 1993, a year in which floods in the Gila River brought flood flows into the delta and upper Gulf, there was a resurgence of corvina *(Cynoscion xanthalus)* schools in the upper Gulf, after a 40-year virtual disappearance of this species. Several marine and estuarine species of the delta and upper Gulf are endangered, mostly notably the vaquita *(Phoceana sinus)* and the totoaba. A species of clam endemic to the delta and upper Gulf, *Mulinia colora-doensis*—once occurring at densities of about 50 clams/m^2 and now at densities of only

3 clams/m²—is endangered by the lack of freshwater input (Kowalewski et al. 2000) and is being proposed for listing under the Endangered Species Act.

Lack of freshwater flows has exacerbated water-quality problems, particularly salinity and selenium. Flows crossing the international border must meet salinity thresholds described above, but the water discharged through the MODE to the Ciénega de Santa Clara does not, and agricultural irrigation return flows in Mexico increase salinity. Selenium may be a more serious problem, since it can bioaccumulate to toxic levels for wildlife, causing high rates of embryonic mortality and deformity. Elevated levels of selenium have been found in water, sediment, and fish tissues in the Colorado River within the delta, and in the waters of the Río Hardy. These levels exceeded the EPA's criterion of 5 mg/liter for the protection of freshwater aquatic life by as much as 14 times, as well as violating Mexican limits of 8 mg/L (García-Hernández et al. 2001). Evidence shows that selenium concentrations are being magnified by the agricultural practices on the Mexicali Valley, and by evaporation in certain sites where standing water remains without being flushed (Valdés-Casillas et al. 1998).

The delta generates significant economic activity in addition to irrigated agriculture. Three communities—El Golfo de Santa Clara, San Felipe, and Puerto Peñasco—rely on fishing. Sixteen tourist camps near the Río Hardy/Río Colorado confluence are used by visitors from Mexicali and the United States for fishing, hunting, and other water-based recreation, providing employment for local residents who work as guides. Eco-tourism, especially in the region of the Ciénega de Santa Clara, is a growing economic sector for several of the local communities.

Restoration Opportunities

Interest in the ecological values of the Colorado River delta, and opportunities for preserving and enhancing these values, have grown markedly in the past decade. This is partly attributable to the diversity and abundance of habitat types and species found in the delta, and partly because the delta has demonstrated a remarkable resilience. Both the riparian corridor and the emergent wetlands areas have regenerated in response to releases of fresh and brackish water, with no outside intervention or management. Simply adding water to the system apparently is sufficient to prompt such growth. For the mainstem riparian corridor, periodic flood flows of 320 mcm (Luecke et al. 1999) with a minimum instantaneous discharge of 80 m³/sec (Zamora-Arroyo et al. 2001) appear to be sufficient to promote and sustain the recruitment of cottonwood and willow seedlings and to provide them with a competitive advantage over the invasive saltcedar. An estimated annual instream flow of 40–50 mcm annually would maintain sufficient water in the upper reaches of the channel to permit the viability of prey species for migratory passerines (Luecke et al. 1999). The size of the Ciénega de Santa Clara shows a strong correlation with the volume of brackish drainage released from the United States; inflows of roughly 120 mcm/year appear sufficient to maintain the Ciénega at its present size (Glenn et al. 1996).

Despite the small volumes of water required to preserve this remnant habitat—less than 1 percent of the river's mean annual flow—identifying and allocating such water to the delta is far from simple. Water users in the United States have strongly resisted efforts to date to dedicate Colorado River flows for delta preservation and enhancement, citing limitations of the Law of the River. This resistance reflects the tension between growing public support for environmental protection and restoration, and the water development legal framework developed almost a century ago. Ultimately, overcoming this institutional resistance may require an amendment to the 1944 Treaty with Mexico (Pitt et al. 2000). Guaranteeing existing flows to the Ciénega may be less challenging, though a legal assurance would improve the long-term prospects of this resource.

Dedicating water for the delta's habitats will require several interim steps. These include developing local capacity and linking economic development in the region to the health of the environment, developing specific policies that acknowledge the needs of water users for consistency and reliability in the system, increasing public awareness and appreciation of the delta, an expanded research effort to refine and document the delta's water needs, and greater communication among stakeholders.

The active involvement of local communities is essential to the success of any restoration effort. Several local community groups have begun to develop ecotourism operations in the lower parts of the delta and in the upper Gulf. These could be linked, better promoted, and expanded to include certified fisheries, aquaculture operations, and sport hunting. Perhaps the greatest single challenge to delta restoration is the array of real and perceived legal and institutional obstacles to the dedication of Colorado River water to environmental uses. Overcoming these obstacles will require the development and promotion of viable short- and long-term alternatives and sound solutions for delta restoration that provide reliability and predictability for users in both countries. Raising public awareness and appreciation of the region will increase pressure on elected officials and decision makers to address the ecological problems of the delta, as well as more specifically to support (and make more effective) the current binational restoration process. A coordinated and comprehensive long-term research program designed to investigate outstanding institutional, hydraulic, and ecological questions and provide long-term monitoring would provide invaluable feedback to refine management programs and maximize the utility of limited Colorado River water.

Colorado River water itself could come from a variety of sources. Perhaps the most palatable would be to identify willing sellers or leasers of water rights, in the United States and Mexico, and dedicate such waters for environmental purposes. Additional sources could include a water assessment on water reallocations, dedicating a portion of water transferred for the delta. These may be combined with conservation improvements, enabling agricultural and municipal users to improve their efficiency, and transferring the resultant saved water. Such water would require legal protections, such as an amendment to the U.S.-Mexico Treaty, to ensure that it reaches its intended destination.

Conclusions

Momentum has been building to address the needs of the delta and reverse the environmental impacts of the past century of water development. In May 2000, the U.S. Department of the Interior and Mexico's Secretaría de Medio Ambiente, Recursos Naturales y Pesca (SEMARNAP) signed the Joint Declaration to Enhance Cooperation in the Colorado River Delta. Subsequent discussions among U.S. stakeholders led the IBWC to the December 2000 adoption of Minute 306, Conceptual Framework for U.S.-Mexico Studies for Future Recommendations Concerning the Riparian and Estuarine

Ecology of the Limitrophe Section of the Colorado River and its Associated Delta. The text of Minute 306 is reproduced in this volume as a Water Brief. Shortly after the adoption of the Minute, U.S. Secretary of the Interior Babbitt addressed the annual meeting of Colorado River users, saying,

> I want to emphasize that dealing with the needs of the delta may be the single most important piece of unfinished business on the Colorado River, and I urge you, as water users on the American side of the border, to approach this issue proactively. We know there are a number of potential win/win opportunities that can and should be explored in bilateral negotiations and with the advice of stakeholders in both countries.

Such bilateral negotiations have begun, under the auspices of Minute 306. In September 2001, the governments of both nations hosted a symposium on the delta to increase understanding of the institutional, hydrologic, and ecologic issues. Water users, academics, and non-governmental organizations continue to exchange information and work toward developing common understandings of the issues involved, as a precursor toward developing sustainable solutions. Those with existing water rights seek resolution of the issue to improve their confidence in the reliability and predictability of future supplies; some also see the issue as an opportunity to expand flexibility within the Law of the River.

Efforts to address the threatened Colorado River delta demonstrate changing societal values and the growing recognition of the value of water left instream. Formal institutional arrangements and the presence of an international boundary separating the remnant delta from its river challenge these efforts, yet recent binational accords and an expanding body of research have improved the prospects for the delta. The dedication of even a small proportion of the Colorado River's mean annual flow would preserve a valuable and threatened ecosystem and would signal marked progress in reconciling water resources development with ecosystem needs. While the identification, procurement, and dedication of water for the delta remain some time off, confidence is high that actions to protect the delta and ultimately other critical ecosystems around the world may well be realized.

REFERENCES

Bloom, P.L. 1986. "Law of the River: A critique of an extraordinary legal system." In G. Weatherford and F. Brown (eds.), *New Courses for the Colorado River: Major Issues for the Next Century.* University of New Mexico Press, Albuquerque, pp. 139–154.

Castro, J.E. 1995. "Decentralization and modernization in Mexico: The management of water services." *Natural Resources Journal,* Vol. 35, pp. 461–487.

Clinton, M., M. Hathaway McKeith, J. Clark, P. Cunningham, D.H. Getches, J.L. Lopezgamez, L.O. Martínez Morales, B. Bogada, J. Palafox, and C. Valdés-Casillas. 2001. *Immediate Options for Augmenting Water Flows to the Colorado River Delta in Mexico,* May, 41 pp.

Cohen, M.J., C. Henges-Jeck, and G. Castillo-Moreno. 2001. "A preliminary water balance for the Colorado River delta, 1992–1998." *Journal of Arid Environments,* Vol. 49, pp. 35–48.

Colby, B., M.A. McGinnis, and K.A. Rait. 1991. "Mitigating environmental externalities through voluntary and involuntary water reallocation: Nevada's Truckee-Carson River basin." *Natural Resources Journal,* Vol. 31, pp. 756–783.

Cortner, H.J., M.A. Shannon, M.G. Wallace, S. Burke, and M.A. Moote. 1996. *Institutional Barriers and Incentives for Ecosystem Management: A Problem Analysis.* U.S. Forest Service General Technical Report PNW-GTR-354.

Cossio-Díaz, D.J.R. 1995. "Constitutional framework for water regulation in Mexico." *Natural Resources Journal,* Vol. 35, pp. 489–499.

Fradkin, P. 1981. *A River No More: The Colorado River and the West.* Alfred A. Knopf, New York.

Galindo-Bect, M., E. Glenn, H. Page, K. Fitzsimmons, L. Galindo-Bect, J. Hernandez-Ayon, R. Petty, J. Garcia-Hernandez, and D. Moore. 2000. "Analysis of *Penaeid* shrimp landings in the upper Gulf of California in relation to Colorado River freshwater discharge." *Fishery Bulletin,* Vol. 98, pp. 222–225.

García-Hernández, J., K.A. King, A.L. Velasco, E. Shumilin, M.A. Mora, and E.P. Glenn. 2001. "Selenium, selected inorganic elements, and organochlorine pesiticides in bottom material and biota from the Colorado River delta." *Journal of Arid Environments,* Vol. 49, pp. 65–89.

Getches, D. 1985. "Competing demands for the Colorado River." *University of Colorado Law Review,* Vol. 56, pp. 413–479.

Gleick, P.H. 1993. *Water in Crisis: A Guide to the World's Freshwater Resources.* Oxford University Press, New York.

Gleick, P.H. 2000. *The World's Water 2000–2001: The Biennial Report on Freshwater Resources.* Island Press, Washington, D.C.

Glenn, E.P., R.S. Felger, A. Burquez, and D.S. Turner. 1992. "Cienega de Santa Clara: Endangered wetland in the Colorado River delta, Sonora, Mexico." *Natural Resources Journal,* Vol. 32, pp. 817–824.

Glenn, E.P., J. Garcia, C. Congdon, and D. Luecke. 1999. "Status of wetlands supported by agricultural drainage water in the Colorado River delta, Mexico." *HortScience,* Vol. 34, pp. 16–21.

Glenn, E.P., C. Lee, R. Felger, and S. Zengel. 1996. "Effects of water management on the wetlands of the Colorado River delta, Mexico." *Conservation Biology,* Vol. 10, pp. 1,175–1,186.

Harding, B.L., T.B. Sangoyomi, and E.A. Payton. 1995. "Impacts of a severe sustained drought on Colorado River water resources." *Water Resources Bulletin,* Vol. 31, pp. 815–824.

Hoffman-Dooley, S. 1996. "Determining what is in the public welfare in water appropriations and transfers: The Intel example." *Natural Resources Journal,* Vol. 36, pp. 103–126.

Holburt, M.B. 1984. "The 1983 high flows on the Colorado River and their aftermath." *Water International,* Vol. 9, pp. 99–105.

Howe, C.W. 1996. "Water resources planning in a federation of states: Equity *versus* efficiency." *Natural Resources Journal,* Vol. 36, pp. 29–36.

Hundley, N., Jr. 1966. *Dividing the Waters: A Century of Controversy between the United States and Mexico.* University of California Press, Los Angeles.

Hundley, N., Jr. 1975. *Water and the West: The Colorado River Compact and the Politics of Water in the American West.* University of California Press, Los Angeles.

Hundley, N., Jr. 1986. "The West against itself: The Colorado River—an institutional history." In G. Weatherford and F. Brown (eds.), *New Courses for the Colorado River.* University of New Mexico Press, Albuquerque, pp. 9–49.

Ingram, H.M., D.E. Mann, G.D. Weatherford, and H.J. Cortner. 1984. "Guidelines for improved institutional analysis in water resources planning." *Water Resources Research,* Vol. 20, pp. 323–334.

International Boundary and Water Commission (IBWC). 1998. *Western Water Bulletin: Flow of the Colorado and Other Western Boundary Streams and Related Data.* Department of State, Washington, D.C.

Kowalewski, M., E.A.S. Guillermo, K.W. Flessa, and G.A. Goodfriend. 2000. "Dead delta's former productivity: Two trillion shells at the mouth of the Colorado River." *Geology,* Vol. 28, pp. 1,059–1,062.

Lavín, M.F., E. Beier, and A. Badan. 1997. "Estructura hidrográfica y circulación del Golfo de California: Escalas estacional e interanual." In M.F. Lavín (ed.), *Contribuciones a la Oceanografá Fisica en México.* Unión de Geofísica Mexicana. Monografía No. 3, pp. 41–171.

Leopold, Aldo. 1949. *A Sand County Almanac: With Other Essays on Conservation from Round River.* Oxford University Press, New York.

Luecke, D.F., J. Pitt, C. Congdon, E. Glenn, C. Valdés-Casillas, and M. Briggs. 1999. *A Delta Once More: Restoring Riparian and Wetland Habitat in the Colorado River Delta.* Environmental Defense Publications, Washington, D.C.

Meko, D., C.W. Stockton, and W.R. Boggess. 1995. "The tree-ring record of severe sustained drought." *Water Resources Bulletin,* Vol. 31, No. 5, pp. 789–801.

Morrison, J.I., S.L. Postel, and P.H. Gleick. 1996. *The Sustainable Use of Water in the Lower Colorado River Basin.* Pacific Institute Report. Pacific Institute for Studies in Development, Environment, and Security, Oakland, California.

Mumme, S.P. 1996. "Groundwater management on the Mexico–United States border." Report to the Commission on Environmental Cooperation, Montreal, Canada.

Nathanson, M.N. 1980. *Updating the Hoover Dam Documents: 1978.* U.S. Department of the Interior, Bureau of Reclamation. U.S. Government Printing Office, Denver, Colorado.

National Research Council (NRC). 1992. *Water Transfers in the West: Efficiency, Equity, and the Environment.* National Academy Press, Washington D.C.

Ohmart, R.D., B.W. Anderson, and W.C. Hunter. 1988. *The Ecology of the Lower Colorado River from Davis Dam to the Mexico–United States International Boundary: A Community Profile.* U.S. Fish and Wildlife Service Biological Report 85 (7.19).

Owen-Joyce, S.J., and L.H. Raymond. 1996. *An Accounting System for Water and Consumptive Use Along the Colorado River, Hoover Dam to Mexico.* U.S. Geological Survey Water-Supply Paper 2407. Denver, Colorado.

Pitt, J., D.F. Luecke, M.J. Cohen, E.P. Glenn, and C. Valdés-Casillas. 2000. "Two countries, one river: Managing for nature in the Colorado River delta." *Natural Resources Journal,* Vol. 40, pp. 819–864.

Pontius, D. 1997. "Colorado River basin study." Final Report to the Western Water Policy Review Advisory Commission.

Postel, S. 1993. "Water and agriculture." In P.H. Gleick (ed.), *Water in Crisis: A Guide to the World's Freshwater Resources.* Oxford University Press, New York, pp. 56–66.

Reisner, M. 1993. *Cadillac Desert: The American West and Its Disappearing Water.* Penguin Books, New York.

Reisner, M., and S. Bates. 1990. *Overtapped Oasis: Reform or Revolution for Western Water.* Island Press, Covelo, California.

Schmidt, J.C., and J.B. Graf. 1990. *Aggradation and Degradation of Alluvial Sand Deposits, 1965 to 1986, Colorado River, Grand Canyon National Park, Arizona.* U.S. Geological Survey Paper 1493. Denver, Colorado.

Sykes, G. 1937. *The Colorado Delta.* Publication No. 460. Carnegie Institution, Washington, D.C.

U.S. Bureau of Reclamation (Reclamation). 1996. *Description and Assessment of Operations, Maintenance, and Sensitive Species of the Lower Colorado River.* U.S. Bureau of Reclamation, Lower Colorado Region, Boulder City, Nevada.

Valdes-Casillas, C., O. Hinojosa-Huerta, M. Munoz-Viveros, F. Zamora-Arroyo, Y. Carrillo-Guerrero, S. Delgado-Garcia, M. Lopez-Camacho, E.P. Glenn, J. Garcia, J. Riley, D. Baumgartner, M. Briggs, C.T. Lee, E. Chavarria-Correa, C. Congdon, and D. Luecke. 1998. *Information Database and Local Outreach Program for the Restoration of the Hardy River Wetlands, Lower Colorado River Delta, Baja California and Sonora, Mexico.* Instituto Tecnologico y de Estudios Superiores de Monterrey (ITESM), Campus Guaymas, Guaymas, Sonora, Mexico.

Wahl, R.W. 1989. *Markets for Federal Water: Subsidies, Property Rights, and the Bureau of Reclamation.* Resources for the Future, Washington, D.C.

Wilson, F. 1994. "A fish out of water: A proposal for international stream flow rights in the lower Colorado River." *Colorado Journal of International Environmental Law and Policy,* Vol. 5, pp. 249–272.

Wolf, A.T., J.A. Natharius, J.J. Danielson, B.S. Ward, and J. Pender. 1999. "International river basins of the world." *International Journal of Water Resources Development,* Vol. 15, No. 4 (December).

Zamora-Arroyo, F., O. Hinojosa-Huerta, E. Glenn, and M. Briggs. 2001. "Vegetation trends in response to instream flows in the Colorado River delta, Mexico." *Journal of Arid Environments,* Vol. 49, pp. 49–64.

Zengel, S., V. Mertetsky, E. Glenn, R. Felger, and D. Ortiz. 1995. "Cienega de Santa Clara, a remnant wetland in the Rio Colorado delta (Mexico): Vegetation distribution and the effects of water flow reduction." *Ecological Engineering* 4:19–36.

CHAPTER 7

The World Commission on Dams Report: What Next?

Katherine Kao Cushing

There has been a long and growing controversy over the construction and operation of large dams. Many of the issues—their economic and environmental consequences, the impacts of dams on local communities and people, and the role that large infrastructure plays in long-term water planning and management—have been addressed in the first two volumes of *The World's Water* (Gleick 1998, 2000). In recent years, controversies over the planning, construction, and operation of dams have become more frequent and heated. Out of this conflict and the need for its resolution came the World Commission on Dams (WCD), an independent body established in 1998 and charged with

> [r]eview[ing] the development effectiveness of dams and assess[ing] alternatives for water resources and energy development, and . . . develop[ing] internationally-acceptable criteria and guidelines to advise future decision-making in the planning, design, construction, monitoring, operation, and decommissioning of dams. (WCD 1999)

The Commission was supposed to shed light on questions that perennially plagued the large dams debate, such as, What have been the projected versus actual benefits, costs, and impacts of large dams? and, What has been the distribution of project benefits and costs? These were issues for which data were scarce and unreliable. In November 2000, the World Commission on Dams concluded a groundbreaking 2-year study on the sustainability of large dams. The work of the Commission is considered by many to be the most comprehensive assessment of dams yet undertaken. This Chapter summarizes the Commission's objectives, work products, and processes, and describes initial responses to the Commission's work.

The WCD Organization

The Commission consisted of 12 Commissioners and a Secretariat based in Cape Town, South Africa. Professor Kadar Asmal, then Minister for Water Affairs and Forestry (and

TABLE 7.1   WCD Commissioners

Commissioner	Organizational Affiliation
Prof. Kadar Asmal	Minister of Education, South Africa
Lakshmi Chand Jain	Industrial Development Services
Dr. Judy Henderson	Oxfam International
Goran Lindal	ABB Ltd.
Prof. Thayer Scudder	California Institute of Technology
Joji Carino	Tebtebba Foundation
Donald Blackmore	Murray-Darling Basin Commission
Medha Patkar	Struggle to Save the Narmada River
Jose Goldemberg	University of São Paulo
Deborah Moore	Environmental Defense
Jan Veltrop	International Commission on Large Dams
Achim Steiner	IUCN

subsequently Minister of Education) for South Africa, served as Chair. The other Commissioners were selected to represent the broad spectrum of interests in the large dams debate, ranging from Medha Patkar, leader of the Struggle to Save the Narmada River (an Indian non-governmental organization, or NGO), to Jan Veltrop, honorary president of the International Commission on Large Dams (ICOLD). The full list of Commissioners is presented in Table 7.1. Dozens of consultant teams also supported the work of the Commission.

The WCD collected data and feedback from five major sources: (1) in-depth case studies of eight large dams on four continents, together with more general reviews of dam building in India and China; (2) cross-check surveys of large dams located in 52 countries; (3) thematic reviews that covered five categories of issues relevant to water and energy development;[1] (4) regional consultative meetings with stakeholders in Cairo, São Paulo, Columbo, and Hanoi; and (5) submissions from interested individuals, groups, and institutions. The outputs of these activities comprised a shared "knowledge base" upon which the Commissioners based their deliberations. In addition, the Commission held regular meetings with the WCD Forum, a 70-member multi-stakeholder group that acted as a mechanism for maintaining a dialogue between the Commission and the constituencies of Forum members (WCD 2000a).

Stakeholders challenged the WCD to produce three deliverables. First, it was to conduct a broad-brush performance review of large dams worldwide, taking into account economic, social, and environmental impacts. The Commission accomplished this task. The Commission's information-gathering activities spanned the globe and included data from over 1,000 dams. It also covered all major aspects of multipurpose water and power projects ranging from hydropower production to returns on irrigation investments to impacts on natural riparian and aquatic ecosystems. Perhaps most importantly, it gave equal attention to the social and environmental impacts of dams. The Commission's work was also interdisciplinary, applying staff and consultant expertise in fields that included economics, marine biology, sociology, history, and engineering. The WCD assessment, while imperfect and criticized by

---

1. In total, the Commission produced 17 Thematic Reviews and 130 papers that addressed five major areas: social and distributional issues, environmental issues, economic and financial issues, options assessment, and governance and institutional processes (WCD 2000a, 32).

certain groups of stakeholders, is still the most comprehensive, multidisciplinary post hoc analysis of large dams to date.

The WCD's second objective was to identify and analyze non-dam alternatives to water resources and energy development. The Commission's success in accomplishing this task was mixed. While the final report does provide descriptive information on non-dam alternatives to water and energy development (e.g., demand side management) and promotes a decision-making process that generally emphasizes the importance of considering non-dam alternatives, it fails to explicitly show how such non-dam options can be better integrated into that process. The Commission does not provide an adequate level of guidance on how, specifically, non-dam options can be better integrated into the process of assessing options. This is a key weakness of the report.

Finally, the Commission's third goal was to propose a set of criteria and guidelines to shape future decision making related to water and energy resources development. If fully implemented, these recommendations would indeed epitomize sustainable development and set an unprecedented standard for project planning, management, and evaluation. The Commission paints a well-defined vision of what is possible. While all stakeholder groups are in agreement about and supportive of the ultimate goals and principles of the Commission, how, exactly, to achieve those goals is still being hotly debated. The suggested work items put forward by the Commission are idealistic, rather than immediately realistic, and will depend heavily on the interests and ability of stakeholders to push them forward.

An additional topic of considerable confusion and debate among groups that read the Commission's report centered on the concept of "guidelines." In a number of places, the WCD's proposals for guidelines have led to confusion about the extent to which organizations should modify their existing practices to reflect the Commission's recommendations. In addition, the level of detail for many of the guidelines is not sufficient to help organizations that would be implementing the recommendations.

Findings and Recommendations

The five main "messages" of the Commission can be grouped into two categories: findings related to project performance and recommendations related to improving the planning and operation of water and energy resource development projects. The WCD's first three messages summarize project performance:

> 1. Dams have made an important and significant contribution to human development, and the benefits from them have been considerable.

> 2. In too many cases an unacceptable and often unnecessary price has been paid to secure these benefits, especially in the social and environmental terms, by people displaced, by communities downstream, by taxpayers, and by the natural environment.

> 3. Lack of equity in the distribution of benefits has called into question the value of many dams in meeting water and energy development needs (WCD 2000a).

Further description of Commission findings related to these messages is presented in Table 7.2. The WCD's last two messages (see Table 7.2) address the mismatch between the distribution of project costs and the benefits:

1. By bringing to the table all those whose rights are involved and who bear the risks associated with different options for water and energy resources development, the conditions for a positive resolution of competing interests and conflicts are created.

2. Negotiating outcomes will greatly improve the development effectiveness of water and energy projects by eliminating unfavorable projects at an early stage and by offering as a choice only those options that key stakeholders agree represent the best ones to meet the needs in question (WCD 2000a).

The foundation of these recommendations lies in creating a fundamentally different approach to assessing and managing water development activities. Traditionally, preproject water and energy development assessments have focused on weighing the projected economic benefits of a project against its economic costs. Social and envi-

TABLE 7.2   Major Findings of the WCD

Large dams display a high degree of variability in delivering predicted water and electricity services, as well as related social benefits.

[They] demonstrate a marked tendency towards schedule delays and cost overruns.

Large dams designed to deliver irrigation services have typically fallen short of physical targets, did not recover their costs, and have been less profitable in economic terms than expected. Large hydropower dams tend to perform closer to, but still below, targets for power generation. However the dams studied generally met their financial targets, with notable under- and over-achievers.

Social groups bearing the social and environmental costs and risks of large dams, especially the poor, the vulnerable, and future generations, are often not the same groups that receive water and electricity services, nor the social and economic benefits from these.

Large dams generally have a range of extensive impacts on rivers, watersheds, and aquatic ecosystems. These impacts are more negative than positive and, in many cases, have led to irreversible losses of species and ecosystems.

Efforts to date to counter the ecosystem impacts of large dams have met with limited success. This result is attributable to the lack of attention paid to anticipating and avoiding impacts, the poor quality and uncertainty of predictions, the difficulty of coping with all impacts, and only partial implementation and success of mitigation measures.

Pervasive and systematic failure to assess the range of potential negative impacts and implement adequate mitigation, resettlement, and development programs for the displaced and the failure to account for the consequences of large dams on downstream livelihoods have led to the impoverishment and suffering of millions, giving rise to growing opposition to dams by affected communities worldwide.

There are many non-dam options that can improve or expand water and energy services in all sectors, in all segments of society, and in all regions of the world. These include: DSM, supply efficiency, new supply options, recycling and recovery, and decentralization. Currently, there is considerable room for improving the adoption and performance of these alternatives.

Source: Adapted from WCD 2000a

ronmental impacts, if assessed at all, were given a much lower priority. Decision making was centralized and often lay in the hands of only a few key players, such as governments, developers, and multilateral development banks.

In contrast, the Commission proposes using concepts of "recognition of rights" and "risk assessment" as a more appropriate way to develop water and energy resources. Recognition of rights means acknowledging the various rights and entitlements that project-affected parties hold (WCD 2000a). Various rights may be relevant in the context of large dam projects. These include constitutional rights, customary rights, treaty rights, Executive Order rights, rights codified in legislation, and common law. What this would mean to the current project development process is that a much broader range of parties would need to be included in the decision-making process, most notably displaced communities, indigenous people's groups, and those representing the interests of future generations and ecosystems.

Regarding risk, the Commission distinguishes between voluntary risk *takers* (e.g., developers, governments) and involuntary risk *bearers* (e.g., indigenous people, displaced communities, future generations, and ecosystems). It argues that in the past only the risks faced by voluntary risk takers have been taken into account and that involuntary risk bearers—the ones who often bear the bulk of the project costs—have enjoyed few or no benefits. Assessing rights at risk, the Commission argues, creates a broader stakeholder group (which includes involuntary risk bearers) and will enhance everyone's ability to define the level and type of risks they wish to take and to explicitly define their boundaries and acceptability (WCD 2000a).

The Commission's fifth major message is that negotiated agreements among relevant stakeholder parties should be used to reconcile the interests of those who have a stake in the project and the risks they bear under various project options. These agreements would be legally binding. The WCD also recommends that the process used to arrive at the agreement include alternative decision-making mechanisms, such as arbitration, mediation, and judicial review, should a negotiated settlement prove infeasible. Finally, they recommend that some kind of appeal mechanism also be incorporated into the process (WCD 2000a).

Strategic Priorities, Criteria, and Guidelines

To lay a common foundation for sustainable water and energy resources development, the Commission proposes a set of seven strategic priorities, which it considers tenets of sustainable and equitable water resources development. These priorities and their explanations are listed in Table 7.3.

To provide further instruction on how to put the Commission's recommendations into action, the final report also provides a set of five "criteria" that correspond with key decision points in planning and project development in the energy and water sector. These criteria and their role in the planning and project development process are illustrated in Figure 7.1.

The Commission recommends that each of these criteria be met in a manner that satisfies relevant "strategic priorities." For example, Table 7.4 illustrates how Criteria 1—Has the need for water and energy services been validated?—might be verified.

Finally, the Commission proposes a series of 26 "guidelines" that "describe in general terms how to assess options and plan and implement dams projects" that meet

TABLE 7.3 WCD's Strategic Priorities (framed as achieved outcomes)

Priority	Key Message
Gaining Public Acceptance	Public acceptance of key decisions is essential for equitable and sustainable water and energy resources development. Acceptance emerges from recognizing rights, addressing risks, and safeguarding the entitlements of all groups of affected people, particularly indigenous and tribal peoples, women, and other vulnerable groups. Decision-making processes and mechanisms are used that enable informed participation by all groups of people, and result in the demonstrable acceptance of key decisions. Where projects affect indigenous and tribal peoples, such processes are guided by their free, prior, and informed consent.
Comprehensive Options Assessment	Alternatives to dams do often exist. To explore these alternatives, needs for water, food, and energy are assessed and objectives clearly defined. The appropriate development response is identified from a range of possible options. The selection is based on a comprehensive and participatory assessment of the full range of policy, institutional, and technical options. In the assessment process, social and environmental aspects have the same significance as economic and financial factors. The options assessment process continues through all stages of the planning, project development, and operations.
Addressing Existing Dams	Opportunities exist to optimize benefits from many existing dams, address outstanding social issues and strengthen environmental mitigation and restoration measures. Dams and the context in which they operate are not seen as static over time. Benefits and impacts may be transformed by changes in water use priorities, physical and land use changes in the river basin, technological developments, and changes in public policy expressed in environment, safety, economic and technical regulations. Management and operation practices must adapt continuously to changing circumstances over the project's life and must address outstanding social issues.
Sustaining Rivers and Livelihoods	Rivers, watersheds, and aquatic ecosystems are the biological engines of the planet. They are the basis for life and the livelihoods of local communities. Dams transform landscapes and create risks of irreversible impacts. Understanding, protecting, and restoring ecosystems at river basin level is essential to foster equitable human development and the welfare of all species. Options assessment and decision making around river development prioritizes the avoidance of impacts, followed by the minimization and mitigation of harm to the health and integrity of the river system. Avoiding impacts, good site selection, and project design is a priority. Releasing tailor-made environmental flows can help maintain downstream ecosystems and the communities that depend on them.
Recognizing Entitlements and Sharing Benefits	Joint negotiations with adversely affected people result in mutually agreed and legally enforceable mitigation and development provisions. These provisions recognize entitlements that improve livelihoods and quality of life, and affected peoples are beneficiaries of the project.

continues

TABLE 7.3 *Continued*

Priority	Key Message
Recognizing Entitlements and Sharing Benefits (*continued*)	Successful mitigation, resettlement, and development are fundamental commitments and responsibilities of the State and the developer. They bear the onus to satisfy all affected people that moving from their current context and resources will improve their livelihoods. Accountability of responsible parties to agreed mitigation, resettlement, and development provisions is ensured through legal means, such as contracts, and through accessible legal recourse at national and international levels.
Ensuring Compliance	Ensuring public trust and confidence requires that governments, developers, regulators, and operators meet all commitments made for the planning, implementation, and operation of dams. Compliance with applicable regulations, criteria, and guidelines, and project-specific negotiated agreements is secured at all critical states in project planning and implementation. A set of mutually reinforcing incentives and mechanisms is required for social, environmental, and technical measures. These should involved an appropriate mix of regulatory and non-regulatory measures, incorporating incentives and sanctions to ensure effectiveness where flexibility is needed to accommodate changing circumstances.
Sharing Rivers for Peace, Development, and Security	Storage and diversion of water on transboundary rivers has been a source of considerable tension between countries and within countries. As specific interventions for diverting water, dams require constructive cooperation. Consequently, the use and management of resources increasingly becomes the subject of agreement between States to promote mutual self interest for regional cooperation and peaceful collaboration. This leads to a shift in focus from the narrow approach of allocating a finite resource to the sharing of rivers and their associated benefits in which States are innovative in defining the scope of issues for discussion. External financing agencies support the principles of good faith negotiations between riparian States.

Source: WCD 2000: 215-251.

the Commission's Strategic Priorities. They can be thought of as sectoral "best practices." Table 7.5 lists the topics for which the Commission produced guidelines.

Reaction to the WCD Report

Response to the Commission's final report has varied widely, from fully supportive and enthusiastic to disparaging.[2] It should be noted that due to the report's complexity and wide scope, as well as the fact that the report was released initially only in English, not all the voices who want to weigh in on the report have done so. As such, it is still too

---

2. The bulk of the data presented below is based on an assessment of the WCD conducted by the World Resources Institute (2001) and is used with permission of the authors. Our thanks to the authors, particularly Mairi Dupar, for allowing use of their material.

soon to tell what the long-term impacts of the WCD will be. Nevertheless, some countries, organizations, and coalitions have made their positions on the WCD's work known, and these reactions provide at least some insight into how the Commission's recommendations will play out in the future. Box 7.1 summarizes some of the reaction and activities surrounding the final WCD Report as of mid-2001.

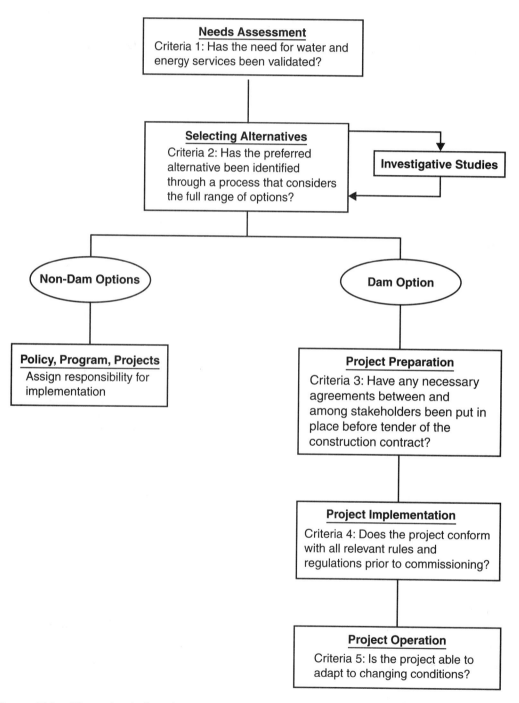

FIGURE 7.1   Five criteria for planning and project development.
Source: Adapted from WCD 2000

TABLE 7.4 Criteria Checklist for Stage 1: Needs Assessment—Validating the Needs for Water and Energy Services

Strategic Priority	Checklist
Gaining Public Acceptance	Develop a consultation plan using stakeholder analysis to identify the groups that should be involved. The plan should define mechanisms for verifying needs at the local, regional, and national level.
	Use a process of public consultation and make sure the results are disseminated to all relevant stakeholders.
	Make sure project goals reflect a basin-wide understanding of social, economic, and environmental values, requirements, functions, and impacts, and that these goals take into account synergies and potential areas of conflict.
	Use an appropriate process to address disparities between needs expressed through public consultation and the stated development objectives.
Comprehensive Options Assessment	Review legal, policy, and institutional frameworks.
	Assess and address any bias against resource conservation, efficiency, decentralized options, and any provisions that hinder an open and participatory assessment of needs and options.
Addressing Existing Dams	Evaluate and incorporate outstanding social and environmental impacts from past projects.
Sustaining Rivers and Livelihoods	Assess ecosystem baseline studies and maintenance needs at a strategic level.

Source: Adapted from WCD 2000a: 265.

Non-governmental Organizations

Overall, those organizations and their representatives opposed to new dam projects or active in attempting to incorporate social and environmental factors in decisions about dam development have reacted positively to the report. They wanted governments and lending institutions to take immediate and far-reaching action. A statement issued by the International Rivers Network and 135 other NGOs called for immediate implementation of all WCD recommendations generally (especially regarding free, prior, and informed consent of indigenous peoples), the establishment of multi-stakeholder reviews of planned and existing projects, and reparations for the cost bearers of existing and past projects. Perhaps most alarming to lending institutions, these organizations also called for a moratorium on the funding, planning, or construction of new dams until compliance with the aforementioned actions was achieved (WRI 2001).

Similarly, World Wildlife Fund International called for governments and the private sector to adopt the WCD's recommendations and for the OECD governments to publicly pledge not to construct any more large dams for at least the next 20 years (WRI 2001).

NGOs working on issues relevant to dam-affected people generally embraced the results of the report, which they felt brought to light the many injustices they had suffered under the current system as well as a strong rationale for them to take a greater part in reaping project benefits. And while some groups would like to have seen stronger language from the Commission concerning the rights of dam-affected people, they indicated that the Commission's report is "something they can work with" (WRI 2001).

TABLE 7.5   WCD's "Good Practice" Guideline Topics

Stakeholder analysis

Negotiated decision-making process

Free, prior, and informed consent

Strategic impact assessment for environmental, social, health, and cultural heritage issues

Project-level impact assessment for environmental, social, health, and cultural heritage issues

Multi-criteria analysis

Life cycle assessment

Greenhouse gas emissions

Distributional analysis of projects

Valuation of social and environmental impacts

Improving economic risk assessment

Ensuring operating rules reflect social and environmental concerns

Improving reservoir operations

Baseline ecosystems surveys

Environmental flow assessments

Maintaining productive fisheries

Baseline social conditions

Impoverishment risk analysis

Implementation of the mitigation, resettlement, and development action plan

Project benefit-sharing mechanisms

Compliance plans

Independent review panels for social and environmental matters

Performance bonds

Trust funds

Integrity pacts

Procedures for shared rivers

Source: Adapted from WCD 2000a: 278.

Funding Institutions

Arguably the most influential and widely anticipated reactions to the Commission's Report came from the World Bank and other international funding agencies.[3] As one Commissioner put it, "A lot of the 'success' of the Commission will depend on how the World Bank reacts to it. A lot of organizations look to them for guidance, so if they don't like it or adopt it, that won't bode well for success. The WCD won't be considered effective if it doesn't influence Bank behavior" (Moore 2001). While embracing the general principles behind the Commission's work, the extent to which future Bank practices will incorporate the Commission's recommendations into planning and operating activities remains unclear.

---

3. The following quotation from WRI's (2001) assessment of the WCD points out two important reasons why the World Bank's response was considered crucial. "By the time of the WCD's report release, the Bank was involved in only one percent of large dam building internationally, but its agenda setting power in the development discourse and its continuing leverage with client governments gave it considerable opportunity to influence the `international acceptability' of the WCD guidelines" (WRI 2001, 123).

Box 7.1   List of Follow-up Initiatives known to WCD Secretariat, August 2001

National Responses

France

The Ministry of Environment is conducting an internal review of WCD proposals and relevance to current dam development policy. The French Export Credit Agency (COFACE) is reviewing its own export finance guidelines in light of the WCD Report.

Germany/GTZ/KfW/BMZ Ministry of Development Cooperation

Multi-stakeholder review of WCD proposals and relevance to current policy initiated by Minister for Development Cooperation held on January 16–17, 2001, in Berlin attended by 150 representatives. The Minister and Deputy Minister chaired the meeting. NGO/Industry Working Group established to advise Minister on review of German Aid Guidelines and policy dialogue with EU/OECD/World Bank.

India

Between May and August 2001, various institutions within India hosted more than seven multi-stakeholder meetings in Mumbai, Delhi, Hyderabad, Shillong, Khedi Madhya Pradesh, Pune, Maharashtra, and Ranchi Jharkhand. The Delhi meeting was organized by Lokayan as part of the Independent Review Group and was attended by former Commissioners Medha Patkar, Jan Veltrop, and Laxmi Jain, and Secretariats Larry Haas and S. Parasuman.

Japan

Japan conducted a review of WCD Report as part of OECD environmental guidelines reform.

Lesotho

The Lesotho Transformation Resource Centre organized a two-day multi-stakeholder workshop of NGOs, government, hydrologists, the Lesotho Highlands Development Authority, and affected peoples to work toward

continues

Box 7.1 *continued*

implementation of the WCD Report in Maseru on May 14–15 2001. A follow-up meeting was held in August 2001.

Nepal

National meetings were arranged to review report and present at a regional meeting on Integrated Water Resources Management, organized by Nepal Water Partnership in Kathmandu in November 2001.

Norway

WCD Briefing meeting held on November 30, 2000. The Ministry of Foreign Affairs, including both domestic policy and overseas development officers, coordinated formal consideration of the Report and its recommendations among relevant government agencies. That report, released in June, is available on the WCD web site.

Pakistan

Presentation of WCD Report in Islamabad on November 18, 2000. Ministry of Environment is considering follow-up meetings at provincial and federal level.

Panama

Multi-stakeholder meeting organized by Fundación Natura and the Panama National Committee of IUCN on May 23, 2001. A wide range of stakeholders gathered for a presentation of the WCD Report and initial discussions on how to begin working with the Report. The CICH (an inter-institutional commission created to coordinate the management of the water resources of the Canal watershed) joined Fundación Natura and the Panama National Committee of IUCN to assist in gathering a representative group to steer the development of a work plan and specific follow-up initiatives to further disseminate and begin the "Panamanización" of the WCD Report and consider appropriate implementation mechanisms.

Poland

On July 6, 2001, in Warsaw, a partnership of the World Wide Fund for Nature (WWF), the Global Water Partnership (GWP), academics, and civil engineers, in cooperation with the people, businesses, and government of

continues

Poland, released the Overview Report of the World Commission on Dams (WCD) in their own language. According to officials, the Polish version of the Report makes the findings and recommendations of the WCD available and useful to a wider constituency of more than 2,000 individuals planning to use the Report.

Philippines

A one-day workshop on the recommendations of the World Commission of Dams was hosted by the Asian Development Bank and held in Manila on August 30, 2001.

South Africa

Briefing meeting held with Minister of Water Affairs and Forestry in January 2001. The Minister and representatives of the Department of Water Affairs and Forestry made presentations at the WCD Forum in February 2001. DWAF conducted pilot exercise on proposed Skuifraam dam project with WCD guidelines. The delegates to the July 23–24 Symposium adopted a resolution that "declares itself to be broadly supportive of the strategic priorities outlined in the WCD Report, but believes that the guidelines need to be contextualised in the South African situation." The full resolution can be found at www.dams.org.

GLOBE Southern Africa Parliamentary Services hosted a WCD presentation in Cape Town on March 20, 2001. The audience was 40 members of South Africa's water and environmental affairs committees, who sought more feedback from its Ministers on how the Report could be used.

Spain

On July 13, WCD launched the Spanish language WCD Report in Madrid with the Centro de Estudios Hidrograficos, Madrid, and the International Water Resources Association (Spanish chapter). WCD provided hard copies of the Final Report to the 55 participants and hundreds of Spanish CD-ROMs. The Director General of CEDEX, Snr. Manuel L. Martin Anton, welcomed the WCD Report as a constructive contribution to the debate over water resources.

Sri Lanka

Presentation of WCD Report was made in Colombo on November 21, 2000.

continues

Box 7.1 *continued*

Sweden/SIDA

The Swedish International Development Agency's (SIDA) view is that the conclusions drawn in the WCD Report are largely in line with their existing regulations and guidelines and indicated that the WCD's report will constitute important reference material prior to deliberations and collective assessments on support to dam projects in the future. The WCD Report will also form part of the basis for the forthcoming review of SIDA's guidelines on environmental impact assessments. SIDA intends to give financial support to disseminate the Report among their cooperation partners and to continue to support the development of laws, regulations, institutions, and so on in partner countries so that dam projects can be discussed in a democratically, socioeconomically, and environmentally acceptable and transparent fashion. (For their full response, see www.dams.org.)

Switzerland

Swiss development agencies hosted a multi-stakeholder briefing and WCD Report on January 17, 2001.

UK/Department for International Development (DFID)

DFID cosponsored a British Dams Society meeting held in London on February 1, 2001, at which the Parliamentary Under-Secretary of State, Chris Mullins, gave an address.

The UK government has established a cross-departmental group to review the WCD Report and develop a coordinated and consistent UK approach. This will include consideration of the implications for the Export Credit Guarantees Department and dams in the UK. DFID have indicated that it will consider providing support to governments in developing countries wanting to implement elements of the WCD Report.

Multi-lateral Development Banks
African Development Bank

A letter from AfDB sent to WCD in January 2001 indicated active support for implementation of report. AfDB plans to incorporate the criteria and guidelines during the development of Bank's technical guidelines to support their recently completed policy on Integrated Water Resources Management.

continues

AfDB will soon embark upon revision of its Environmental Policy, which will give a stronger emphasis on social issues typically relevant for large infrastructure projects. The Policy will also incorporate the recent international trends addressing issues related to Resettlement and Involuntary Displacements such as those typically associated with large dam projects.

The overall conclusion of a Bank seminar held on July 17, 2001, was that the AfDB would not place a moratarium on large dams that are consistent with Bank's IWRM policy. Africa needs dams, but they need to be developed in economically viable and environmentally and socially sound manner. It was explained that at minimum, the Bank's new policy on Environment (under revision) and new policy on Involuntary Resettlement will provide appropriate opportunities to bring in the WCD's risk and rights approach (see the text) and the Strategic Principles. The Guidelines will be used as a reference and "best practice" report on a case-by-case basis.

Asian Development Bank

A regional workshop was held in Manila February 19–20, 2001, with participants from 14 governments and NGO representatives. ADB will reexamine its own procedures, including its environment and social development policies, and determine the extent to which the Report's recommendations may necessitate changes in these procedures. ADB is planning to hold a series of national-level consultations.

ADB also hosted a Philippines in-country workshop on the recommendations of the World Commission of Dams in Manila on August 30, 2001. A report is available on the ADB website.

US-EXIM Bank

Revision of the EXIM Bank's environmental guidelines is expected to incorporate recommendations from the OECD Working Group. The draft revision includes reference to the WCD recommendations.

Inter-American Development Bank

WCD presented a report on its findings to the IADB on May 14, 2001. They established an internal Task Force to review recommendations and implication for Bank policies and procedures.

World Bank

The WCD Report was presented to the Board of the World Bank on January 14, 2001. The WB sent out a consultation mission in January 2001 to seven

continues

Box 7.1 *continued*

key member countries to get feedback on the Report. This mission was controversial. Member countries offered feedback to the World Bank Committee on Development Effectiveness on February 15, 2001. The WB prepared a response on March 31, which outlines general approach to working with the Report. Additional activities are being prepared.

International Development Organizations
World Conservation Union (IUCN)

IUCN is working on the design of projects to support implementation of WCD recommendations on the ground. A Task Force was established in response to a request from IUCN General Assembly to monitor WCD Report implementation.

IUCN position statement on the Report has been released and can be read online at www.dams.org.

OECD Environmental Development Committee

WCD presented the WCD Report in Paris on March 9, 2001. Many bilateral agencies indicated review in process.

Incorporated a review of WCD recommendations as part of the OECD program to harmonize environmental guidelines. References were made to WCD guidelines in the communiqué of the G8 in Trieste.

OECD Export Credit Agency Working Party

The WCD Report was presented at the OECD workshop on December 14, 2000.

UNEP

UNEP Financial services initiative held on November 17, 2000, in Bonn; briefing by WCD at meeting.

Briefing by WCD on April 5–6, 2001, in Nairobi.

UNEP hosted a two-day meeting in Manila to explore how the finance sector in Asia-Pacific can promote sustainable development.

UNEP will host a two-year follow-up in a new Dams and Development Unit. They will also be responsible for further dissemination of the Report.

continues

World Water Council (WWC)

The WWC supports the core values and strategic priorities proposed in the Report and wishes to contribute positively to follow-up dialogue. The WWC Task Force will continue operation to facilitate the Council's contribution to this process.

Regional Groups

European Parliament

The Henrich Böll Foundation hosted a meeting for European Parliament in Brussels on December 19, 2000, to introduce and discuss the WCD Report.

Mekong River Commission (MRC)

IUCN and MRC Secretariat are considering follow-up program for Thailand/Vietnam/Laos and Cambodia. MRC is organizing translation of Part II sections of the Report into four languages. A regional workshop was held in Phnom Penh on June 28–29, 2001.

Greater Mekong Sub-region Power Interconnection and Trade Group

WCD presentation was made at the GMSR Group workshop in Vientiane December 5–6, 2000, organized by the Asian Development Bank.

Mesoamerica

IUCN organized a regional meeting for Central American countries in Costa Rica, March 2001, to discuss reaction and follow-up.

SADC

SADC considered the Report at the 10th Water Resource Technical Committee Sector Meeting on May 8, 2001, Swaziland. The meeting brought together about 30 high-level government water resources experts from various countries of Southern Africa.

continues

Box 7.1 *continued*

International Conventions
Convention on Biological Diversity

Report results were presented to the sixth meeting of the Subsidiary Body on Scientific, Technical, and Technology Advice at The Hague on April 19, 2001.

WCD briefed a meeting of SBSTTA in Montreal, Canada, on March 12–16, 2001. A meeting resolution requires that the Conference of the Parties take note of the recommendations contained in the Report of the WCD in regard to the implementation of the program of work on biological diversity of inland water ecosystems.

Ramsar Convention

Briefing note is being prepared for the scientific and technical committee and recommendations for Conference of the Parties (COP) 8, 2002. The Ramsar Secretariat to follow up.

Private Sector
Balfour Beatty

Balfour Beatty has welcomed the Report produced by the WCD and declared its support to establish a framework for promoting appropriate practice in the development and execution of hydroelectric schemes. Balfour Beatty has also committed itself to taking the WCD principles, criteria, and guidelines into account in determining whether and how it should be involved in any future hydroelectric projects. A shareholder meeting on May 2, 2001, rejected a motion from Friends of the Earth that would require the Company to adopt and be bound by the 26 guidelines set out in the WCD Report. See http://www.dams.org/media.

The Hydro Equipment Association

The Hydro Equipment Association (HEA), consisting of Voith Siemens, Alstom, and VAHydro, has been formed to represent the hydro equipment industry in the follow-up dialogue and actions with other stakeholders resulting from the WCD Report. It will assist in the review and development of standards and guidelines relating to the hydropower industry. The HEA plans to contact the principal WCD stakeholders to inform them of the Association's objectives, discuss the status of follow-up activities to the

continues

WCD Report, and explain the role of the Association as a partner in energy strategy development.

Skanska

Skanska intends to apply the WCD guidelines for major hydropower projects. Although most of Skanska's companies already work with a certified environmental management system in accordance with ISO 14001, by January 2002 all operations will be certified. See http://www.dams.org/report/reaction_skanska.htm.

Sustainable Asset Management (SAM) Sustainability Group/Dow Jones

The SAM Sustainability Group in Zurich has a joint venture with Dow Jones Indexes to undertake research on top companies for the Dow Jones Sustainability Index. They have formulated a questionnaire for companies involved in the dam-building industry based on WCD recommendations.

Source: Modified from http://www.dams.org/report/followups.htm.

At first vaguely positive, the Bank's response has since become modest and cautious. According to the WRI's assessment of events that transpired after the final report was released, "prospects for *significant* change within the Bank based on the Commission's recommendations seem fairly remote" (WRI 2001: 124). Excerpts from the Bank's response to the Commission's report are presented in Table 7.6.

In response to the request of international NGOs that they immediately and comprehensively adopt the Commission's recommendations, the Bank responded by saying, "Not all WCD guidelines apply in all cases and should not be taken literally, but rather as guiding principles. The decision on whether to adopt the [Commission's] guidelines lies with the respective developing countries." The World Bank will not "comprehensively adopt the 26 WCD guidelines, but will use them as a reference point when considering investments in dams." This response is, in fact, consistent with the Commission's interpretation of the intent of the guidelines and the extent to which organizations should adopt their recommendations.[4] The Bank rejected the request that they institute a moratorium on dam projects until certain conditions are met,

---

4. The intent of the Commission's recommendations is more fully elaborated in the following excerpts from its final report: "[T]he Commission proposes a set of *guidelines* . . . to describe how its policy principles can be realized. These guidelines add to existing decision-support instruments and should be incorporated by governments, professional organizations, financing agencies, civil society, and others as they continue to improve their own relevant guidelines and policies over time" (WCD 2000, p. 260); "The 26 guidelines add to the wider range of technical, financial, economic, social and environmental guidelines. They are advisory tools to support decision-making and need to be considered within the framework of existing international guidance and current good practice" (WCD 2000, 278); and "Nobody can, of course, simply pick up the report and implement it in full. It is not a blueprint" (WCD 2000, p. 311).

TABLE 7.6 The World Bank's Response to the WCD's Final Report

The World Bank believes that the World Commission on Dams Report has made a great contribution in framing many of the major issues in this contentious and broad-ranging debate. It fully shares the Commission's goals of equity, efficiency, sustainability, participation, and accountability. It believes that the Report is a valuable guide, and it is putting in place an action plan, and is working with its partners in implementing the Report's good recommendations.

The Bank welcomes the WCD Report and will use it as a valuable reference to inform its decision-making process when considering projects that involve dams.

The Bank will continue to support dams that are economically well justified and environmentally and socially sound.

In support of our borrowers, and at their request, the Bank will:

a) Support strategic planning processes by borrowers for decisions concerning water and energy to enhance the evaluation of options and alternatives, and will support borrowers in financing the priority investments emerging from such processes.

b) Work with borrowers on new projects to map these against the Strategic Priorities articulated in the WCD Report and assess their applicability in the specific setting.

c) Review how the principles of the twenty-six WCD criteria may be put into individual use in the context of specific projects.

The Bank will continue to strengthen implementation of its safeguard policies, use of consultation and disclosure of information for all projects including those with dams.

The Bank will undertake measures in the form of a Preliminary Action Plan to strengthen its work in the water and energy sectors and to improve the evaluation, implementation and operation of dams when they are the appropriate development option.

The Bank will continue to disseminate and discuss the WCD Report with its borrowers, recognizing its broad interest for development issues.

The Preliminary Action Plan comprises:

• working with borrowers to move "upstream"
• effectively implement the World Bank's existing "safeguard policies"
• continue to support borrowers in improving the performance of existing dams
• continue to stress institutional reform for more efficient use of water and energy
• develop a more proactive and development-oriented approach to international waters
• continue to support innovative approaches for dealing with complex dam-related management and technical issues

Source: Adapted from World Bank (2001).

stating that dams must remain an important development option and that a blanket moratorium would be just another means of paralyzing the dams debate. Finally, with regard to reparations on past projects, the Bank's interpretation of the WCD's report is that this area is a state responsibility, and they would only consider dealing with legacy issues if and when asked to do so by a borrower (World Bank 2001).

The Bank did identify areas in which its existing policies differed from those of the WCD's recommendations, particularly in the areas of resettlement and indigenous peoples. In response, it stated that it is currently "[r]eviewing how and where WCD recommendations can be reflected in our current work," and that it would "continue to strengthen implementation of its safeguard policies, use of consultation, and disclosure of information for all projects, including those with dams" (World Bank 2001).

Other lending institutions echoed the Bank's view, saying that they had already incorporated many of the Commission's recommendations into existing policies and that duplication did not make sense. In general, however, the banks felt that many of the messages of the report were valid (Eden 2001). The Asian Development Bank and the African Development Bank have been more receptive to the Commission's findings. For example, in a letter to Commission Chair Kadar Asmal, the President of ADB stated that

> ADB will reexamine its own procedures, including our environment and social development policies, and determine the extent to which the report's recommendations may necessitate changes in these procedures. We will also encourage our member countries to do the same. (Chino 2000)

ADB also held a meeting in February 2001 with government representatives from numerous countries it works with, as well as development consultants, NGOs, and regional institutions, to discuss the report and implementation issues.

International Industry and Trade Associations

Like all other interest groups, those associated with the dam construction and hydropower industries publicly applauded the effort and intent of the Commission. Industry associations also agreed with the Commission's larger principles and affirmed their intent to work within an "improved framework" that will lead to better projects, more water-derived energy, more dependable water supplies, and a livable environment for all people (Eden 2001). However, some members of this stakeholder group were "not pleased" with some of the significant findings and recommendations made in the Commission's report.

From the industry's perspective the Commission failed to adequately consider the benefits created by dams, focusing instead on the negative social and environmental impacts. Some industry representatives expressed concerns that the Commission's 26 guidelines would be seen as setting up an expectation that the criteria *must* be adopted worldwide, thus creating new regulations at an international level unsanctioned by an international governing body. Another key industry concern is that WCD-suggested procedures (such as comprehensive stakeholder analysis) could bog down a project in pre-approval red tape. This could stop projects, not on merit, but simply due to the weight of policy and procedure (Eden 2001).

Industrial representatives also disagreed with some of the Commission's recommendations. For example, some felt that the WCD's call for "comprehensive, integrated, cumulative, and adaptive periodic [project] evaluations at intervals of five to 10 years" was an inappropriate use of resources. They contend that these kinds of reviews are already conducted as part of relicensing processes, albeit at 35- to 50-year time intervals, or as part of ongoing environmental monitoring programs. In addition, they argue that the non-dams options assessment emphasized by the Commission is already incorporated into feasibility assessments of potential projects. Finally, they voiced concern over the negotiated decision-making process proposed by the Commission, saying that it would impose a structure giving one constituency effective veto power over a project (Eden 2001).

At the beginning of the process, industry had sought clearer rules for their involvement in dam-related projects. Many hoped, like their counterparts in financial institutions, that the WCD would provide a way to streamline the development of proposals by defining straightforward criteria for either rejecting or accepting projects with minimal conflict. They expressed disappointment that the WCD had not provided the simple solutions they were looking for and perceived that the WCD report introduced further uncertainty in timing and outcome to the water resources development process (WRI 2001). Forum members from industry were also unhappy with the final stages of the WCD process and complained that the lack of an interim final report and consultation prevented conflicts over points of emphasis in the report and the word *guidelines* from being addressed ahead of time. They felt that these kinds of conflicts could have been ironed out and made more palatable had the Commission better consulting with Forum members prior to release of the final report (WRI 2001).

Governments

A positive reaction from governments is critical to the success of the WCD's efforts. The WCD's final report emphasizes that governments have a pivotal role and bear a major responsibility for ensuring that water and energy resources development in their countries are sustainable and equitable. Reactions from governments are difficult to categorize as a whole because governments have many different departments and ministries and their opinions can, and often do, conflict. By early 2002, relatively few national governments had issued statements on the WCD report.

The governments of Germany and the U.K. had a positive response to the report. For example, the British Dams Society hosted a multi-stakeholder meeting to discuss the report, which was attended by over 100 representatives from government departments, the private sector, NGOs, and the academic community (WCD 2001a). The British government has already established a cross-department review to coordinate a new approach to water and energy development, and announced that its foreign aid agency will offer support to governments in developing countries wanting to implement the Commission's report. Similarly, Germany's Minister for Development Cooperation convened a "dialogue forum" on the WCD process. One hundred and fifty people representing all sides of the dams debate attended the Forum. At the meeting, the Minister pledged to undertake a review of German guidelines and to promote consideration of the Commission's recommendations to the World Bank and other financing institutions (WCD 2001a).

An analysis of available responses from Southern governments conducted by WRI (2001), based mainly on responses from water ministries, characterizes initial reaction as "defensive" and states that, "this defensiveness stemmed partly from a misreading of the WCD's recommendations for governments" and that "The considerable length and sometimes lack of clarity in the final report may have also added to the confusion."

The governments of India and China, the two largest dam-building nations in the world, have either chosen not to participate in the WCD process and/or rejected the validity of the Commission and its findings. China, which initially had a Chinese Commissioner participating in the process, eventually withdrew her and its support of the WCD's investigative efforts. Some analysts suggest that China's withdrawal was due to concerns that the Commission might be "anti-dam" and that its findings might jeopardize the monumental Three Gorges Dam project, an endeavor of critical importance to the Chinese government (WRI 2001). The Indian government's participation in the WCD process was soured by the appointment of two Indian Commissioners who were viewed as "relatively anti-dam" by the Indian government (WRI 2001). They criticized the Commission for failing to acknowledge the benefits that dams bring to large sections of society "for whom the dam and water supply flowing from the dam are nothing short of a lifeline" and stated that the "WCD's obsessive concern for preserving the rights of affected local people makes it distrust the entire public set up, even the legal framework of the country to which these people belong" (WRI 2001).

While praising the core values of the report, other developing country governments, "in a misinterpretation of the WCD's stated intent, argued that it was inappropriate for them to import the guidelines wholesale and that they had already incorporated many, if not all, procedures put forward by the Commission in their current policies" (WRI 2001). This is not what the WCD intended. Rather, the Commission called on governments to begin national dialogues and establish review panels to see where and how the Commission's guidelines could be integrated into existing systems (WRI 2001).

These reactions illustrate the need to better clarify the WCD's recommendations and their implications for government action. In addition, they suggest that governments need to be more involved in any kind of intergovernmental policy that might stem from the WCD's report. Box 7.1 summarizes most of the follow-up activities that have been initiated by national governments and non-governmental organizations as of mid-2001. For regular updates on these activities, see www.dams.org.

The End of the Beginning

In February 2001, WCD Forum members met for a final time in Cape Town, South Africa, to discuss next steps. Once the Commission had officially disbanded, it was time for Forum members and their constituencies to carry on the initiative.

At the meeting, Forum members decided to create a Dams and Development Unit (DDU), to be hosted by the United Nations Environmental Program (UNEP). The DDU's purpose is threefold: to ensure widespread dissemination and understanding of the report, its findings, and its recommendations; to promote implementation of the Commission's proposed guidelines (keeping in mind the need for testing, refinement, and adaptation of the guidelines to practical contexts); and to promote dialogue, information exchange, and networking in working with the WCD report (WCD 2001b). The DDU has a mandate to operate for two years and is governed by a multi-stakeholder steering committee, which includes representatives from industry, the World Bank, UNEP, donor countries and organizations, environmental NGOs, and water development agencies.

The accomplishments of the WCD have been substantial. They have pushed the debate forward by conducting the largest and most comprehensive study of large dams to date and by proposing an equally broad-ranging set of guidelines and recommendations to guide future decision making on water and energy development. The Commission's work shows how concepts of sustainable development can be put into action within the context of dam planning, construction, and operation and paints a bold vision of what appropriate development can achieve. The fact that its final report is blessed by the consensus of twelve Commissioners who represent many, if not all, sides of the large dams debate lends significant weight to its validity.

The road ahead will not be easy, however, and many challenges exist to making the Commission's vision a reality. The WCD's findings and conclusions do not clearly come down on one side of the dams debate or the other. Rather, their report reflects the complexity of the projects they analyzed—projects that often affect many groups and ecosystems, serve multiple purposes, require interdisciplinary skills to manage, and last for several generations.

As the following statement from Nelson Mandela on the Commission's work makes clear, it will always be a struggle to balance interests in the large dams debate. Yet the solution to dam-related problems lies not in determining whether dams can or cannot meet all interests but in determining how people with differing interests can work together to meet common goals, goals more clearly elucidated by the WCD's work:

> Political freedom alone is still not enough if you lack clean water. Freedom alone is not enough without light to read at night, without time or access to water to irrigate your farm, without the ability to catch fish to feed your family. For this reason, the struggle for sustainable development nearly equals the struggle for political freedom. . . . It is one thing to find fault with an existing system. It is another thing altogether, a more difficult task, to replace it with an approach that is better. . . . Rather than single out dams for excessive blame or credit, we must learn the answer: it is all of us! (Mandela 2001 as quoted in WCD 2000b)

REFERENCES

Chino, T. 2000. President, Asian Development Bank. Letter to Prof. Kadar Asmal, Chair, World Commission on Dams. December 22.

Dubash, N.K., M. Dupar, S. Kothari, and T. Lissu. 2001. *A Watershed in Global Governance? An Independent Assessment of the World Commission on Dams. Draft Report.* July 5. World Resources Institute, Washington, D.C.

Eden, L. 2001. "Responding to the report of the World Commission on Dams." *Hydro Review Worldwide,* Vol. 9, No. 2, pp. 28–31.

Mandela, N. 2000. Excerpts from speech made at the launching of the WCD's final report, November 16, London, England.

Moore, D. 2001. "The World Commission on Dams: An experiment in global policy making." World Commission on Dams Commissioner. Statement made at Berkeley Energy and Resources Group Spring 2001 Colloquium, March 21, Berkeley, California.

World Bank. 2001. "The World Bank & the World Commission on Dams report Q&A." Available at http://www.worldbank.org. March. Accessed June 19, 2001.

World Commission on Dams (WCD). 2000a. *Dams and Development: A New Framework for Decision-Making.* Earthscan Publications, London.

World Commission on Dams. 2000b. "Dateline dispatches: Around the world in ten days." *DAMS: Official Newsletter of the World Commission on Dams,* No. 8 (December).

World Commission on Dams. 2001a. "Gone . . . But Not Forgotten." *DAMS: Official Newsletter of the World Commission on Dams,* No. 9 (March).

World Commission on Dams. 2001b. "Changing of the guard: New director appointed for transition period." WCD Press Release, April 2. Available at http://www.dams.org/press/default.php?article=368. Accessed July 24, 2001.

World Resources Institute (WRI). 2001. "A watershed in global governance? An independent assessment of the World Commission on Dams." Available at http://www.wri.org/governance/wcdassessment.html.

Water Briefs

The Texts of the Ministerial Declarations from The Hague (March 2000) and Bonn (December 2001)

In the last two years, a number of major international water conferences have been held to discuss the science, economics, and politics of solving water problems. Two of these meetings included parallel ministerial-level meetings where water and environment ministers met to discuss actions that might be taken and commitments that might be made at the official governmental level.

The first of these Ministerial Conferences took place on March 21 and 22, 2000, in The Hague as part of the Second World Water Forum. H.M. Queen Beatrix of the Netherlands officially opened the Ministerial Conference. Prior to the meeting, senior officials were asked to prepare a rough draft of the final declaration, outlining key water challenges. Ministers then participated in round-table discussions in parallel sessions. The Ministerial Conference was an initiative of the Dutch government and aimed to mobilize political support for countering global water problems. Delegations from 130 countries participated, including 114 ministers. The product was a formal Ministerial Declaration. Conference materials for The Hague meeting can be found at http://www.worldwaterforum.net/index2.html.

In December 2001, the German government hosted a follow-up meeting as preparation for the tenth anniversary of the United Nations Conference on Environment and Development held in Rio de Janeiro in 1992. This International Conference on Freshwater also had a ministerial meeting, with ministers from 46 countries participating. A formal Ministerial Declaration was also prepared from this meeting. Information, documents, and descriptions of the Bonn Conference can be found at http://www.water-2001.de/.

These Declarations should be read very carefully. They are cautious documents, reflecting the noncontroversial elements of international water meetings. They also, however, indicate the growing importance of water at the international level and the willingness of senior government officials to join the broader international community in taking steps to solve water problems.

The full texts of both Declarations are included here.

Declaration of The Hague

Ministerial Declaration of The Hague on Water Security in the 21st Century Agreed to on Wednesday 22 March, 2000, in The Hague, The Netherlands

1. Water is vital for the life and health of people and ecosystems and a basic requirement for the development of countries, but around the world women, men and children lack access to adequate and safe water to meet their most basic needs. Water resources, and the related ecosystems that provide and sustain them, are under threat from pollution, unsustainable use, land-use changes, climate change and many other forces. The link between these threats and poverty is clear, for it is the poor who are hit first and hardest. This leads to one simple conclusion: business as usual is not an option. There is, of course, a huge diversity of needs and situations around the globe, but together we have one common goal: **to provide water security in the 21st Century.**[1] This means ensuring that freshwater, coastal and related ecosystems are protected and improved; that sustainable development and political stability are promoted; that every person has access to enough safe water at an affordable cost to lead a healthy and productive life and that the vulnerable are protected from the risks of water-related hazards.

2. These threats are not new. Nor are attempts to address them. Discussions and actions started in Mar del Plata in 1977, continued through Dublin and were consolidated into Chapter 18 of Agenda 21 in Rio in 1992. They were reaffirmed in Paris 1998, CSD-6 and in the Second World Water Forum and Ministerial Conference. The process will continue in the meeting in Bonn in 2002 ("Dublin+10"), through the 10-year review of implementation of Agenda 21, and beyond. These and other international meetings have produced a number of agreements and principles that are the basis upon which this and future statements should be built. The goal of providing water security in the 21st Century is reflected in the unprecedented process of broad participation and discussion by experts, stakeholders and government officials in many regions of the world. This process has profited from the important contributions of the World Water Council, who launched the World Water Vision process at the First World Water Forum in Marrakech, from the formation of the World Commission on Water in the 21st Century and from the development of the Framework for Action by the Global Water Partnership.

The Main Challenges

3. To achieve water security, we face the following main challenges:

Meeting basic needs: to recognise that access to safe and sufficient water and sanitation are basic human needs and are essential to health and well-being, and to empower people, especially women, through a participatory process of water management.

---

1. All bold text here is bold in the original statement.

Securing the food supply: to enhance food security, particularly of the poor and vulnerable, through the more efficient mobilisation and use, and the more equitable allocation of water for food production.

Protecting ecosystems: to ensure the integrity of ecosystems through sustainable water resources management.

Sharing water resources: to promote peaceful co-operation and develop synergies between different uses of water at all levels, whenever possible, within and, in the case of boundary and trans-boundary water resources, between states concerned, through sustainable river basin management or other appropriate approaches.

Managing risks: to provide security from floods, droughts, pollution and other water-related hazards.

Valuing water: to manage water in a way that reflects its economic, social, environmental and cultural values for all its uses, and to move towards pricing water services to reflect the cost of their provision. This approach should take account of the need for equity and the basic needs of the poor and the vulnerable.

Governing water wisely: to ensure good governance, so that the involvement of the public and the interests of all stakeholders are included in the management of water resources.

Meeting the Challenges

4. We, the Ministers and Heads of Delegation, recognise that our gathering and this Declaration are part of a wider process, and are linked to a wide range of initiatives at all levels. We acknowledge the pivotal role that governments play in realising actions to meet the challenges. We recognise the need for institutional, technological and financial innovations in order to move beyond "business as usual" and we resolve to rise to meet these challenges.

5. The actions advocated here are based on **integrated water resources management,** that includes the planning and management of water resources, both conventional and non-conventional, and land. This takes account of social, economic and environmental factors and integrates surface water, groundwater and the ecosystems through which they flow. It recognises the importance of water quality issues. In this, special attention should be paid to the poor, to the role, skills and needs of women and to vulnerable areas such as small island states, landlocked countries and desertified areas.

6. Integrated water resources management depends on collaboration and partnerships at all levels, from individual citizens to international organisations, based on a political commitment to, and wider societal awareness of, the need for water security and the sustainable management of water resources. To achieve integrated water resources management, there is a need for coherent national and, where appropriate, regional and international policies to overcome fragmentation, and for transparent and accountable institutions at all levels.

7. We will further advance the process of collaboration in order to turn agreed principles into action, based on partnerships and synergies among the government, citizens and other stakeholders. To this end:

A. We will establish targets and strategies, as appropriate, to meet the challenges of achieving water security. As part of this effort, we support the development of indicators of progress at the national and sub-national level. In carrying this forward, we will take account of the valuable work done for the Second World Water Forum.

B. We will continue to support the UN system to re-assess periodically the state of freshwater resources and related ecosystems, to assist countries, where appropriate, to develop systems to measure progress towards the realisation of targets and to report in the biennial World Water Development Report as part of the overall monitoring of Agenda 21.

C. We will work together with other stakeholders to develop a stronger water culture through greater awareness and commitment. We will identify best practices, based on enhanced research and knowledge generation capacities, knowledge dissemination through education and other channels and knowledge sharing between individuals, institutions and societies at all appropriate levels. This will include co-ordination at regional and other levels, as appropriate, to promote arrangements for coping with water-related disasters and for sharing experiences in water sector reform. It will also include international co-operation in technology transfers to, and capacity building in, developing countries.

D. We will work together with stakeholders to increase the effectiveness of pollution control strategies based on polluter pays principles and to consider appropriate rules and procedures in the fields of liability and compensation for damage resulting from activities dangerous to water resources.

E. Against the background of the preparatory work for and discussions in The Hague, we will work within multilateral institutions, particularly the UN system, International Financial Institutions and bodies established by Inter-Governmental Treaties, to strengthen water-related policies and programmes that enhance water security, and to assist countries, as appropriate, to address the major challenges identified in this Declaration.

F. We call upon the Secretary General of the United Nations to further strengthen the co-ordination and coherence of activities on water issues within the UN system. We will adopt consistent positions in the respective governing bodies to enhance coherence in these activities.

G. We call upon the Council of the Global Environmental Facility (GEF) to expand activities that are within the mandate of the GEF in relation to freshwater resources by catalysing investments in national water management issues that have a beneficial impact on international waters.

H. We welcome the contribution of the World Water Council in relation to the Vision and of the Global Water Partnership with respect to the development of the Framework for Action. We welcome follow-up actions by all relevant actors in an open, participatory and transparent manner that draws upon all major groups in society.

I. We note the statements (attached to this declaration) made by the representatives of the major groups and welcome them as a clear reflection of their readiness to work with us towards a secure water future for all.

8. Recognising that the actions referred to in paragraph 7, including progress on targets and strategies, are important and ambitious, we will review our progress periodically at appropriate fora, including the meeting in Bonn in 2002 and the 10-year review of the implementation of Agenda 21.

9. The Ministerial Conference acknowledges with appreciation that a range of issues were discussed during the Second World Water Forum, and that the Chair of the Forum presented these issues to the Ministerial Conference. The importance of these issues is unquestionable; we will raise them for further consideration in relevant fora in the future and will consider their implications for our individual national situations.

10. The challenges are formidable, but so are the opportunities. There are many experiences around the world that can be built on. What is needed is for us all to work together, to develop collaboration and partnerships, to build a secure and sustainable water future. We will, individually and acting together, strive to achieve this and stimulate and facilitate the contributions of society as a whole. To this end, we note with appreciation that pledges were made at The Hague (attached to our declaration). This Declaration reflects the determination of our governments and represents a critical step in the process of providing water security for all.

11. We, the Ministers and Heads of Delegation, thank the government and people of The Netherlands for their vision and for their hospitality in hosting this conference and forum.

The full text of the Declaration can be found at www.worldwaterforum.net/Ministerial/declaration.html.

Ministerial Declaration

Adopted by the Ministers Meeting in the Ministerial Session of the International Conference on Freshwater, Bonn, Germany Agreed to on 4 December 2001

We, ministers with responsibilities for water affairs, environment and development from 46 countries throughout the world, have assembled in Bonn to assess progress in implementing Agenda 21 and to discuss actions required to increase water security and to achieve sustainable management of water resources.

We consider that the World Summit for Sustainable Development, scheduled for August 2002 in Johannesburg, needs to demonstrate renewed commitment to sustainable development and political will to action.

We consider the equitable and sustainable use and the protection of the world's freshwater resources a key challenge facing governments on the road to a safer, more peaceful, equitable and prosperous world. Combating poverty is the main challenge for achieving equitable and sustainable development, and water plays a vital role in relation to human health, livelihood, economic growth as well as sustaining ecosystems. The outcome of the World Summit on Sustainable Development must include decisive action on water issues.

We express our deep concern that at the beginning of the 21st century 1.2 billion people live a life in poverty without access to safe drinking water, and that almost 2.5 billion have no access to proper sanitation. Safe and sufficient water and sanitation are basic human needs. The worldwide struggle to alleviate poverty must bring safe and decent living conditions to those who are deprived of these basic requirements.

We confirm our resolve to reach the International Development Targets agreed by the UN Millennium Summit, in particular the target to halve, until the year 2015, the proportion of people living in extreme poverty and to halve the proportion of people who suffer from hunger and are unable to reach or to afford safe drinking water. We also confirm our resolve to stop the unsustainable exploitation of water resources by developing water management strategies at regional, national and local levels.

Water is needed in all aspects of life. For sustainable development, it is necessary to take into account water's social, environmental and economic dimensions and all of its varied uses. Water management therefore requires an integrated approach.

We emphasise that ten years after the UN Conference on Environment and Development and the Dublin Conference, and several years after the global water conferences in Paris and The Hague, there is still a need for greater commitment to implement commonly agreed principles on water resource management. Pressures on the world's scarce freshwater resources and aquatic systems have increased. Water pollution and unsustainable patterns of water consumption are among the causes. Water use efficiency needs to improve.

We agree that governments, the international community, the private sector, the nongovernmental organisations and all other stakeholders need to base their actions on the following:

(Governance)

The primary responsibility for ensuring the sustainable and equitable management of water resources rests with the governments.

Each country should have in place applicable arrangements for the governance of water affairs at all levels and, where appropriate, accelerate water sector reforms.

We urge the private sector to join with government and civil society to contribute to bringing water and sanitation services to the unserved and to strengthen investment and management capabilities. Privately managed service delivery should not imply private ownership of water resources. Service providers should be subject to effective regulation and monitoring.

We encourage riparian states to cooperate on matters related to international water-courses.

(Funding Gap)

There is an enormous gap in funding investments in water infrastructure, maintenance, training and capacity building, research, and data generation.

It is urgent to close this gap using existing resources more efficiently and with additional financial resources from all sources: public investment budgets, capital markets, and community based finance, user and polluter charges; as well as increased international development financing from public and private sources particularly for developing countries to reflect the acute needs in the water sector.

The lack of financial resources for water infrastructure investment, operations and maintenance is particularly hurting the poor in Least Developed Countries and in other countries with people living in extreme poverty.

Critical actions for closing the financial gap are poverty alleviation and the improvement of opportunities for trade and income generation for developing countries.

Resources also need to be made available to assist developing countries to mitigate the effects of natural disasters and to assist in adapting to the impacts of climate change.

Water development programmes, to be successful, should be based on a good understanding of the negative impact desertification causes to people living in affected areas.

(Role of the International Community)

We call on the international community to strengthen its commitment and its efforts to enable developing countries to manage water sustainably and to ensure an equitable sharing of benefits from water resources.

We call upon the Secretary General of the United Nations to strengthen the co-ordination and coherence of activities within the UN system on water issues in an inclusive manner.

We recall the agreed UN target for official development assistance of 0.7% of GDP. Developed countries which have not yet reached the target should exert their best efforts to do so.

(Capacity Building and Technology Transfer)

We recognise that capacity building and innovative technologies including the improvement of indigenous technologies are needed to efficiently utilise water, control pollution and develop new and alternative water sources in water stressed countries.

We will support capacity building programmes and information exchange to ensure the effective use of human, financial, and technical resources for water management.

We will facilitate technology transfer initiatives to enable technologically less developed countries to acquire capacity to manage water with the best available knowledge and equipment.

We need improved and coherent assessments of state and trends in the world water situation.

(Gender)

Water resources management should be based on a participatory approach. Both men and women should be involved and have an equal voice in managing the sustainable use of water resources and sharing of benefits. The role of women in water related areas needs to be strengthened and their participation broadened.

(Next Steps)

We urge the World Summit on Sustainable Development to take account of the outcome of this International Conference on Freshwater.

We expect that the International Year of Freshwater in 2003, and the 3rd World Water Forum in Japan will be a good opportunity to further discuss on the roles and actions for all players in international society on the issues of sustainable development of freshwater.

We thank the government of Germany for its hospitality and its determination to promote dynamic action on water issues.

The full text of the Bonn Declaration is available at www.water-2001.de/outcome/Ministerial_declaration.asp. Versions are available in English, Spanish, French, Arabic, Chinese, and Russian.

The Southeastern Anatolia Project (GAP) and Archaeology

Amar S. Mann

Situated on the broad plains of upper Mesopotamia, the southeastern Anatolia region was the cradle of civilization in ancient times and the northernmost extension of the "Fertile Crescent." This region, located in what is now southern Turkey, played a pivotal role in the history of the Near and Middle East, as some of the world's earliest agricultural settlements, followed by great civilizations, arose in the valleys between the Euphrates and Tigris Rivers. Major trade routes developed along these rivers, such as the Silk Road, and splendid cities were erected, their palaces richly adorned with the work of sculptors, potters, and metalsmiths. Presently, as part of its ongoing development project for southeastern Anatolia (the Southeastern Anatolia Project, or GAP from the Turkish), the Turkish government has constructed multiple dams in the river valleys of the Euphrates and Tigris, which has led to the inundation of hundreds of archaeological sites, many dating from prehistoric times. Since archaeological surveys of dam sites were often rushed or inadequately funded, the full extent of these losses will never be known. In the coming decades, at least 10 more dams are scheduled for construction in southeastern Turkey and many more archaeological remnants will be destroyed. In addition, lost in the focus on the dams is the potential destruction of sites on the plains now being irrigated or to be irrigated in the future.

Overview of GAP

In one of the largest public works undertakings in the world, and the biggest development project ever undertaken by Turkey, GAP is a multi-sectoral and integrated development plan encompassing a wide array of physical, social, and economic infrastructure, that is conceptually analogous to the American Tennessee Valley Authority or the Mekong Valley Program. Following the economic and political collapse of the Ottoman Empire in 1922, the newly created Turkish state had to make many difficult financial choices in rebuilding the country, and the southeastern Anatolia region received little attention. Because of limited funds and a greater state of industrialization already in place in the western portion of the country, Mustafa Kemal Ataturk, the founder of the Turkish Republic, invested heavily in the continued development of western Turkey during the initial five-year plans. While the western portion of the country developed industrially, the eastern portion remained stagnant with little growth. This condition was made worse by a chronically high rate of unemployment, steady out-migration of workers from east and southeast Turkey, and high rates of illiteracy (Nestor 1995).

Following several decades of neglect, the Turkish government finally turned its attention to developing southeastern Anatolia. For nearly two millennia the region had served as a remote backwater, even though it contained better-than-average land, abundant water from two world-class river systems, and rich mineral deposits (Kolars and Mitchell 1991). Early studies carried out by Turkey's State Hydraulic Works Administration (DSI) in the 1960s examined the irrigation and energy potential for the lower

Euphrates basin and Tigris basin, and in 1977, the government decided to unite planned activities in both basins under the title of Southeastern Anatolia Project. By 1990 the scope of GAP grew into a multi-sectoral, socioeconomic project, and the first version of the Southeast Anatolia Project (GAP) Master Plan was completed outlining development objectives, strategies, and a schedule for implementation of the massive program. The total cost is expected to exceed $US 30 billion (Government of Turkey 2001). Located within the Turkish portions of the Euphrates and Tigris river basins, GAP is advertised by the Turkish government to bring civilization and greatness back to upper Mesopotamia.

With expected completion in approximately 2010, the master plan for GAP envisages the construction of 22 dams and 19 hydroelectric power plants on the Euphrates and Tigris Rivers and their tributaries, of which 12 dams and 6 hydroelectric power plants are completed. At full development, a system of concrete-lined canals, tunnels, and pumps will carry water to 1.7 million hectares of land for irrigated cultivation—a dramatic increase considering that Turkey's total irrigable land area is roughly 8.5 million hectares. The GAP's network of dams and canals will regulate the flow on the Euphrates and Tigris Rivers, which together average 50 billion cubic meters of water annually, and harness 28 percent of Turkey's total water potential.

The hydroelectric portion of the project is expected to generate around 27 billion kilowatt hours (kWhr) of electricity annually, with an installed capacity of about 7.5 gigawatts (Government of Turkey 2001). The annual electricity generation from GAP accounts for 22 percent of the country's economically viable hydropower potential (118 billion kWhr), however, 75 percent of GAP's installed hydroelectric power production capacity is already installed and producing power. The Ataturk and Karakaya Dams are fully operational and alone produce 60 percent of GAP's hydroelectric power. The construction of new hydroelectric plants, such as the one at Birecik on the lower Euphrates, has not kept pace with growing demands for electric power. Turkey already imports 3 billion kWhr annually from Bulgaria, and from 1991 to 1997 nationwide electricity consumption increased by about 8 percent annually (from 60 to 102 billion kWhr) (Foreign Economic Relations Board 1998). Rapid growth in electricity consumption (9–10 percent annually) is projected to continue over the next 15 years (EIA 2001).

The GAP includes not only dams for hydroelectricity and irrigation but also infrastructure supporting agriculture, mining, telecommunications, tourism, and other economic and social quality-of-life improvements, such as transportation and improved education and health services. Given that GAP is intended to create broad economic and social changes once energy and irrigation schemes come on line, the Turkish government views the project as a comprehensive "integrated regional development project." Some of the major objectives of the GAP project are summarized below (Government of Turkey 2001):

- To transform southeastern Anatolia into an export center for agriculture with sustained economic growth, through efficient utilization of the region's land and water resources;

- To raise the income level in the GAP region by improving the economic structure in order to narrow the income disparity between the region and other regions;

- To increase productivity and employment opportunities in the area, minimizing migration of people out of rural areas and decreasing unemployment;

- To create an infrastructure in order to increase the effectiveness of the cultural institutions through integrating subcultures in the region, that is, the Kurds, into the national culture; and
- To take measures to increase literacy and the level of education, particularly in favor of women and female children.

Archaeology in the GAP Region

Geographically, southeastern Anatolia is a barren plateau at the foot of the rugged Taurus Mountains and is drained by the Tigris and Euphrates Rivers. Like the present-day population, most archaeological heritage is concentrated in the protected interior enclaves of rich valleys and plains in the region. Yet these are the very areas that are flooded by the dam reservoirs or altered by irrigation systems in the GAP. A more immediate threat to the archaeological record is that the very process of constructing each of the dams—the new roads, the extraction of soil, gravel, and rock for construction—imposes a heavy toll on the material remaining in the vicinity of the dams. The reservoirs created by the dams will flood (and are already flooding) not just the valley of the Euphrates and Tigris upstream for many miles but also numerous tributaries of all sizes. In each instance, the areas flooded include some of the soils that have been regarded as the most attractive for farming for thousands of years (Kennedy 1998).

Despite long being recognized as a home to some of the first human settlements, intensive archaeological excavations in the GAP region have only begun in recent decades. As early as 1882, *Encyclopedia Brittanica* noted that perhaps no region of the East deserved a more careful scrutiny and promised a richer harvest to the antiquarian explorer than this region (9th edition). However, the Turkish Antiquity Service, which controls all archaeological activity in Turkey, has historically focused most of its resources on other areas and is criticized for being more interested in maintaining bureaucratic control than preventing the destruction of archaeological sites, which has discouraged teams and universities from joining dam salvage projects (WCD 2000b). As a result, the Antiquity Service has not coped well with the need for archaeological work on hundreds, possibly thousands, of sites that may be flooded or disturbed by the GAP. For its part, the Turkish government has never allowed a reservoir of a completed dam to be left unfilled for the sake of any archaeological work, beginning with construction of the Keban Dam in 1970 (Izady 1996).

Developments on the Euphrates River

Although not technically considered part of GAP, the Keban Dam's presence upstream on the Euphrates is an integral element in its management (Kolars 1991). In 1967, while the dam was in the process of being built, researchers from Ankara University and the University of Michigan discovered rich remains of the ancient Hurrian Kingdom from about 2300 B.C., a culture and kingdom previously known only through occasional mention in early Mesopotamian tablets (Izady 1996). Beneath the Hurrian layers were found remains from the Neolithic and colorful pottery from the Chalcolithic (2500–3500 B.C.) epochs, while above it were the remains of subsequent cultures encompassing the Roman, Byzantine, and Ottoman Empires (Whallon 1979). The Ankara team's suggestion that a small, inexpensive diversionary earthen dam be erected to protect some of the richest sites was ignored. Another site in the dam's flood zone, Taskun Kale, was found to contain remains of mud-brick and stone churches and buildings, as well as a medieval stone fortress that dated back to about A.D. 1300, along with smaller finds dating back to 3000 B.C. (McNicoll 1983). However, the rescue dig at this site was limited to 15 weeks over three seasons before rising waters from Keban halted work. Aside from these archaeological sites, many other standing monuments, such as ancient churches and mosques, were also flooded by Keban Reservoir, which now covers roughly 670 square kilometers.

The hasty and rudimentary excavation of sites doomed by floodwaters hinders the ability of archaeologists to study sites in detail, limiting their understanding of discoveries and often making interpretation mere guesswork. Furthermore, local museums, such as the one in Elazig near the Keban Dam, are not equipped for the sudden collection, storage, or protection of newly uncovered finds from these rescue operations.

The experience at Keban was repeated with the construction downstream on the Euphrates of Karakaya and Ataturk Dams. The Karakaya Dam, about 100 miles downstream from Keban, was completed in 1986, and its reservoir backs up nearly all the way to the foot of Keban, in effect drowning many historic valleys and their archaeological heritage continuously for over 160 kilometers along the Euphrates and many of its tributaries. Since no archaeological inventory of these areas was ever made, the extent of the losses is unknown.

Of all the dams in GAP, however, the losses due to the construction of the Ataturk stand out for their enormity. Constructed about 180 kilometers downstream of Karakaya, the Ataturk Dam is the world's ninth largest rock-filled dam and the largest in Turkey, with a reservoir covering over 1,300 square kilometers. Following broad but superficial surveys in the late 1970s, archaeologists from around the world were allowed into the area in 1983, the same year construction began. By the late 1980s a lake was forming a few kilometers upstream of the dam, inundating several world-class archaeological sites and standing monuments.

The modern village of Samsat, halfway between the provincial capitals of Adiyaman and Sanliurfa, was among the first to be evacuated. The valley at Samsat is especially broad and fertile, and in antiquity the fortified city guarded an important crossing point of the Euphrates on the east-west trade route. As such it enjoyed considerable commercial and strategic importance. Before the waters of Ataturk submerged it in 1989, an imposing 50-meter-high mound at Samsat, or Samosata as it was called, was uncovered. The elongated mound represented an accretion of cultures that had rebuilt the site, as diverse as Hittites (c. 2000 B.C.), Assyrians (700 B.C.), and Persians (500 B.C. to A.D. 100), Roman legionaries (A.D. 72), the Abbasids of Aleppo (A.D. 640), the Byzantines of Constantinople (c. A.D. 900), and, finally, the Seljuk Turks (c. A.D. 900) (Ward 1990). Archaeologists found 29 levels of remains, included in which were well-preserved Roman and late Hittite walls. The deepest level found, at the base of the mound on the plain below, was from the prehistoric Uruk period, around 3500 B.C. High on one side of the mound stood the walls of a medieval Seljuk villa hovering over remains of a Roman palace. The remains of a Roman aqueduct, a broken line of arches and channels, was also unearthed. Potsherds, blue and violet and emerald green, littered the ground, some a thousand years old, some even older (Ward 1990).

In the first century A.D., Samosata was an important Roman base posed at the border of the mighty Parthian empire, with a legion of 5,000 soldiers placed there and a population of 50,000 (Ward 1990). The Commagenian king Antiochus I, allied to the Roman Empire, and his successors built palaces and monuments in Samosota, adorned by the most skilled artisans of the day. In the excavations, unique remains of the palace of the Commagenian kings were uncovered, which display the influence of imperial Rome.

Although a number of Turkish and foreign teams did carry out surveys and excavations, for so large an area they were modest (Kennedy 2000). None of the foreign institutes undertook any major project at Samosata itself, which was left to the valiant but limited resources of Turkish archaeologists. Samosata was revived from the earth for all of 11 years before its ultimate destruction by water. "Under normal conditions," stated Nimet Ozguc, an archaeologist from University of Ankara who directed the excavation at Samsat, "it would take at least 100 years to understand this place. But we had to make a fast program" (Izady 1996, Ward 1990).

Uncounted other archaeological sites were lost in the construction of the Ataturk Dam. At Nevali Cori, a rich 9,000-year-old temple, which contained numerous small limestone sculptures and clay figures as well as life-size limestone figures, provided for the first time an idea of how people in the area worshiped thousands of years before the birth of Christ. The larger work is animistic, some of it featuring humans and animals in carvings resembling totem poles. The masterpiece of the site is a sculpture of a female head grasped in the talons of a bird. Another male head is shaved, with a snake positioned at the back like a braid (Time Europe 1999). Astonishingly, the temple at Nevali Cori proved to be the world's oldest methodically built structure with a preconceived floor plan (Izady 1996). The world treasure and landmark monument now lies beneath 120 meters of water, trapped behind the Ataturk Dam.

Hundreds of other sites, many containing Greco-Roman remains, were recorded in surveys of the region in the Ataturk dam zone just prior to inundation (Kennedy 1998). In Adiyaman province alone, hundreds of sites were surveyed before the advance of floodwaters. The most exhaustive work in the province was done at Tille Hoyuk, where a team from the British Institute of Archaeology in Ankara stripped the entire top of a mound and, removing Iron Age levels, discovered and salvaged an Assyrian courtyard from the eighth and ninth centuries with alternating squares of black and white pebbles (Ward 1990). Other studies in Adiyaman revealed evidence of human occupation during Neolithic times (about 8,000 years ago) (Blaylock et al. 1990). Aside from Tille, however, most of the sites were given cursory attention and are now submerged by the Ataturk.

The GAP project involves more than the construction of dams. Other facilities include the construction of the Sanli-Urfa tunnels and concrete-lined irrigation canals, which will deliver water from the Ataturk. All of these may disturb or destroy archaeological remains in the area. Lands being brought into production around the Harran and Sanliurfa plains will be difficult to survey once they are irrigated. Once an extension of the Fertile Crescent, Harran is now a tiny village of a few dozen beehive-domed mud-brick houses clustered around the ruins of a medieval castle. It had been a magnificent city, a trade emporium where routes all over the ancient Near East converged, and a center for the worship of Babylonian moon god Sin. Believed to be the same ancient city mentioned in the Old Testament, Harran was a place Abraham spent several years of his life (Ward 1990). Furthermore, some historians have argued that Urfa, northeast of Harran and located in the GAP region, was Abraham's actual birthplace. Regardless, at the present time no archaeological survey has been made of the plain around Harran, which will be irrigated soon through the GAP (Kennedy 2001).

Downriver from Ataturk, the Birecik Dam threatens the magnificent ruins of the twin classical cities of Apamea-Seleucia, known to Greco-Roman authors simply as Zeugma, or "bridge." This name referred to the only masonry bridge that spanned the mighty Euphrates River in its entire course. Constructed in the fourth century B.C. during the reign of Seleucus I, the founder of the Seleucid Empire, Zeugma was the location of the first and only permanent bridge over the Euphrates between the Taurus Mountains and Babylonia, several hundred kilometers away, for several centuries (Izady 1996). In the first century B.C., like Samsat, it too passed under Roman rule and received a legion in garrison, one of only eight in all of Rome's Asian provinces between the Black Sea and the Red (Zeugma and Samsat, though just 70 kilometers apart, had two of them).

Zeugma, in modern times overlaid by the small Kurdish villages of Belkis and Tilamusa, flourished as a fortress city, trade center, garrison, nodal point of several key routes, and meeting point of East and West trade activities (Kennedy 2000). During Roman times, the twin towns of Zeugma were at least as large as Samosata, twice the size of Roman London, and three and a half times the size of Pompeii (Kennedy et al. 1995). The city sank into obscurity after it was sacked by Persian Sassanians around A.D. 252 and then leveled by an earthquake in the late third century A.D.

Western scholars have known the existence of ruins from the Classical period around the village of Belkis for over two centuries, but identification of the site as Zeugma was only established in the 1970s (Meyerson 2000). The tombs of Zeugma's ancient necropolis have always been visible from Belkis, and Roman coins, seals, and ancient debris are often found in the surrounding countryside (Kennedy et al. 1995, Meyerson 2000). Despite the identification of Zeugma some twenty years earlier, it was not until 1992, when construction on the Birecik Dam had already begun, that archaeologists from local Gaziantep Museum, the University of Western Australia, and the University of Glasgow began rescue excavations. Following a tunnel originally dug by looters, they excavated a tomb and two ancient villas with ornate floor mosaics, wall paintings, and small bronze figurines. The floor itself of one of the houses was a spectacular mosaic in almost perfect condition, approximately 8 meters by 4 meters, depicting the marriage of Dionysus and Ariadne (Kennedy et al. 1995). Due to a lack of security at the site, this mosaic was subsequently stolen by looters (Meyerson 2000). Zeugma is renowned for the mosaics that have been found there over the years. In the nineteenth century looters carried off superb mosaics and other works of fine art, some of which have turned up in private collections and museums in Berlin, St. Petersburg, and New York (Kennedy et al. 1995, Kennedy 2000). The most famous, a floor depicting the sea god Poseidon around whom were medallions showing personifications of Roman provinces, was discovered a century ago.

Difficulties in obtaining funding from Turkish or foreign sources for a major dig meant that by the end of 1999, Zeugma still had not been thoroughly investigated, with completion of the dam expected in less than a year. A small team of archaeologists from the Gaziantep Museum and the University of Nantes, in France, embarked on a last-ditch effort to excavate the lowest levels of the city. In mid-April of 2000, they dazzled the archaeological world with their discovery of two more palatial Roman villas, both of them remarkably intact and crammed full of artifacts, frescoes, and mosaics (Meyerson 2000). The last-minute excavations at Zeugma also turned up a 2-meter-tall bronze statue of Mars, about 3,700 Roman coins, and an archive of more than 60,000 seals. These excavations also increased the inventory of remarkable floor mosaics recovered from the site to 16 (Kennedy 2001, Meyerson 2000).

By May 2000 construction of the Birecik Dam was complete and water began rising around the remains of the ancient city. Amid press reports and public criticism (Kinzer

2000), the Turkish government ordered a ten-day suspension in the operation of the dam in early June but cited heavy financial losses as reason why postponement could not continue. With the aid of a $5 million grant from the Packard Humanities Institute in California, Turkish and foreign archaeologists continued to work on excavating Zeugma through October 2000, concentrating on those sections of the site in immediate danger of inundation. At last count, the number of unearthed mosaics from Zeugma has increased dramatically to 65 and excavation of the threatened sections is complete. The Gaziantep Museum has not been able to handle all of the newly uncovered artifacts, and some mosaics rest in the museum's garden, where they await restoration and conservation processes. Since much of Zeugma was built into a hill, a large section of the ancient city, containing official buildings and temples, will actually remain above water, and plans are underway to convert the section into an open-air museum.

Developments Along the Tigris River

Development of GAP projects on the Tigris River is at a far less advanced stage than those on the Euphrates, as four smaller projects (Kralkizi, Dicle, Devegecidi, Batman) have been completed to date, each of much lower magnitude than Ataturk, Karakaya, or Birecik. Similar to projects on the Euphrates, however, archaeological excavations in areas along the Tigris to be flooded by the GAP have usually been conducted only after construction of a dam begins. Little scientific digging was done for sites already flooded around the Tigris and its tributaries; however, some historically significant discoveries were made in the short time archaeologists worked at Hallan Cemi, a site threatened by the Batman Dam. In 1994, an international team led by Michael Rosenberg of University of Delaware, Richard Redding of University of Michigan, and Mark Nesbitt of University College, London, discovered the world's oldest evidence for domestication of pigs, dating back more than 10,000 years (Izady 1996).

The salvage project at the site revealed a hitherto unknown complex society replete with a rich material culture, including decorated stone bowls, stone sculptures, including those of pigs and goats, and notched stone tallies. Freshwater clamshells indicate that the mountain site was occupied year round and evidence of long distance trade in obsidian, copper, and Mediterranean shellfish exists, but the team did not find the expected domesticated goats and sheep and cereals. Instead the researchers found gathered nuts and seeds and the butchered remains of domesticated pigs (University of Delaware 1994). This discovery runs contrary to all early agricultural models, which predict the intensive use of cereals as a crucial first step in the transition from foraging to farming. Sedentism in the society found at Hallan Cemi appears to have developed without this primary condition. With the site at Hallan Cemi and neighboring mounds destined to be flooded in 1995, there was little time (or money) to do a systematic excavation of the 80-square-mile area to be flooded. Most of the work that was done consisted of surface collection rather than digging, as there were essentially last minute "search and seize" operations (Izady 1996).

The next scheduled dam for construction on the Tigris is Ilisu, by far the largest dam on the Tigris and the third largest in the GAP, dwarfed only by its counterparts on the Euphrates of Karakaya and the meta-dam Ataturk. The Ilisu Dam reservoir, with a planned completion date of 2007, would flood over 300 square kilometers of the upper Tigris Valley basin, including the historic town of Hasankeyf, the only town in Anatolia that has survived since the Middle Ages without destruction. The ruins and rock cave houses visible at Hasankeyf today tell of a history going back at least 2,000 years, although excavations may prove the original settlement dated back to the late Assyrians in the seventh century B.C., and some suggest there is evidence of habitation stretching back 11,000 years (BBC 2000, Young 2000). After years of informal survey by Turkish archaeologists starting in 1986, all work was stopped at Hasankeyf in 1991 after two archaeologists were killed and fighting between the Turkish Army and members of the Kurdish Workers Party (PKK) escalated. Excavation work at Hasankeyf did not resume until 1999, and Professor Olus Arik, the head of the excavation, has estimated that the site will require 50 years of work at a minimum (Swiss Federal Institute of Technology 2000).

Hasankeyf was a staging post on the Silk Road, and historical documents indicate that among others the Romans, who built a fortress to protect the frontier town from the Persians, Byzantines, Seljuk Turks, and Mongols, all occupied it. During the twelfth century Hasankeyf enjoyed its golden age as the capital of a Seljuk vassal state, and a massive but richly decorated stone bridge across the Tigris was built that made extensive traffic between Turkey and northern Mesopotamia possible. The ancient bridge was a marvel of the medieval world, and its remains are still visible across the Tigris in Hasankeyf today (Ward 1990, Young 2000). Atop the steep limestone cliff adjacent to the river are ruins of the citadel from which communication between Anatolia, high Mesopotamia, and northern Syria could be controlled. The remnants of palaces, castles, and a mosque built by the Artuqids and Ayyubids in the twelfth to fourteenth centuries, as well as other buildings built within the walls of the fortress, remain on their perch overlooking the town. Below these ruins, and carved into the cliff and surrounding hills, are cave dwellings that have been inhabited for at least 2,000 years and provide shelter to this day for residents of Hasankeyf.

Other threatened historic monuments in the lower town that have borne the test of time are several grand mosques of medieval age with tall, richly ornamented minarets and domes, such as Sulayman (thirteenth century) and Camii-Rizk (A.D. 1409), carved stone and colorfully tiled mausoleums, such as Zeynel Bey's tomb, and a Roman-style hippodrome (for chariot or horse racing), which are remnants of a prestigious past. Also at Hasankeyf is the tomb of Imam Abdullah, grandson of Cafer-I Tayyar, Mohammed's uncle, which draws 30,000 pilgrims each year.

All of the lower sites and most of the citadel will be inundated, and the malleable rock of the cliff, with countless cave dwellings representing thousands of years of human history, will quickly erode and topple those parts remaining above water once the Ilisu reservoir is filled. Although some of the citadel will initially remain above water, these parts will inevitably crumble away swiftly as the stone below absorbs water (British House of Commons 2001). Few of the ruins can be moved, and even their excavation, preservation, and transfer would take eight years according to Professor Arik (Swiss Federal Institute of Technology 2000). Additionally, buried underneath the threatened town and its standing monuments, are several levels that have yet to be excavated. The few short seasons of digging at Hasankeyf have turned up ancient hamams (Turkish baths), water systems, mosques, monasteries, churches, and public and private buildings at every level. An ancient city center and three districts with several levels of early Christian, early Islam, Seljuk, and Ottoman were discovered several years ago, with at least six large buildings from the Seljuk period in the lower city alone. Excavations of this and other sites, such as the palatial remains from the

Artuqids in the southern part of the town, have just begun to scratch the surface of what may be found (Young 2000). It is estimated that only 15 percent of all relics can be saved by rescue evacuation (Bosshard 1998), and prevention of the dam is the only realistic archaeological strategy for preserving this part of Turkish national heritage (British House of Commons 2001, Kitchen and Ronayne 2001).

Although losses at Hasankeyf have been the center of attention, the Ilisu Dam reservoir's flooding of some 520 square kilometers of the upper Tigris Valley basin will selectively destroy precisely those areas that have seen the most concentrated settlement in the region since at least the Paleolithic period, more than 10,000 years ago (Kitchen and Ronayne 2001). Only one-fifth of the area currently at risk of inundation has seen any form of archaeological survey work at all, and even this work has been beset with significant methodological and logistical difficulties (Algaze et al. 1991, Kitchen and Ronayne 2001, Tuna 1999). It is already known that hundreds, probably thousands of archaeological sites will be destroyed, including sites with several mounds that are comparable to Catal Hoyuk, the world's oldest discovered city in western Turkey. Some of the sites date from at least the pre-pottery Neolithic (about 8000 B.C.), when humans were shifting from hunting and gathering to agriculture, and may extend through into the post-medieval period (one such example is a 40-meter-high mound). Also at risk are large fortified sites dating to the 'Ubaid, Assyrian, Roman, and Byzantine periods, in one example enclosing an area of up to 30 hectares, and in certain cases preserving cultural deposits several meters deep. Additionally there are an unquantifiable number of smaller settlements and structures dating from every period of human history (Kitchen and Ronayne 2001).

Some of the principal sites at risk and under investigation are:

- Gre Dimse, where in the first season of digging in 1999, what may be the first complete early Iron Age (c. 1000 B.C.) painted vessel was discovered (British House of Commons 2001);
- Ziyaret Tepe, which has been described as a "national treasure," where settlements from the Late Bronze (c. 1500 B.C.) and Iron Age were recently discovered, as well as city walls of possibly one of the three historic, but hitherto undiscovered, Assyrian empire (c. 700 B.C.) border towns along the Tigris River (British House of Commons, 2001, Center for Research and Assessment 2001); and
- Salat Tepe, a mound spanning the time period from 5000 B.C. to the medieval period, with pottery remains suggesting the site was an important center controlling the Tigris Valley (British House of Commons 2001, Center for Research and Assessment 2001).

Excavation work at these sites is generally in the early stages, and funding and staffing constraints make thorough excavation difficult. In addition, hundreds of smaller sites have little or no prospect of exploration before inundation (Center for Research and Assessment 2001).

Although archaeologists are rightfully attempting to rescue as much of the cultural heritage threatened by Ilisu dam as possible, the proposed Cizre Dam project, just 50 kilometers downstream, has received much less attention. Planned for construction shortly after construction of Ilisu Dam begins, the Cizre Dam reservoir combined with the Ilisu Dam reservoir will flood a sizable portion of the Tigris basin, extending a linear distance of about 140 kilometers from Cizre almost up to the town of Bismil.

Associated irrigation projects from the Cizre Project will irrigate over 120,000 hectares on the Cizre-Silopi plain (Republic of Turkey 1996). These also threaten to damage or otherwise disturb countless other archaeological sites that will not be submerged (Algaze et al. 1991).

Preliminary survey of the area affected by the Cizre dam revealed many mounds and smaller occupations within them, as well as remnants of towns, villages, and smaller remains encompassing 14 periods of human history from the late Neolithic (c. 6000 B.C.), through the Assyrian periods, the Hellenistic/Parthian periods, Roman, and Byzantine periods, and eventually to the medieval Seljuk/Artuqid and Ottoman periods (c. eleventh to fifteenth centuries) (Algaze 1989, Algaze et al. 1991). Painted ceramics with bold careless designs and dozens of small settlements of Halaf age (c. 5000 B.C.) and 'Ubaid age (c. 4500 B.C.) were discovered in the Cizre plain. Evidence of migrations and periods of significant human settlement, as well as periods with conspicuously few traces of human settlement, were also detected in the area (Algaze et al. 1991), which should attract interest from not only archaeologists but also anthropologists and paleo-climatologists. However, with most time and financial resources devoted to major sites in imminent danger, such as Zeugma, Hasankeyf, and Carchemish, the area affected by Cizre Dam remains unexplored, except for a few hasty surveys performed in the late 1980s and early 1990s.

Some critics have characterized archaeological studies in the GAP region as non-scientific and superficial, and still others have gone so far as to call these studies "organized looting parties" (Izady 1996). Indeed, out of necessity, excavations have been quick and dirty, focused on the recovery of works of art and other remains of the wealthy. However, archaeology is not the retrieval of objects, but rather the retrieval of information. What is found, in terms of the architecture, the plan, and the structural sequence of a site, are as important as the objects that come out of it. The end result of thorough exploration of a site is a good understanding of the socioeconomic history, as well as the cultural and material history, along with the archaeology (Ward 1990). This understanding is being lost in areas affected by the GAP, as only a few years of study have been given to major sites that would ordinarily take decades to fully explore, and little or no attention has been paid to countless other sites.

Responses and Conclusions

Turkey's GAP project has raised concerns by many groups on several different grounds. Aside from archaeologists and art historians worried about the flooding of ancient cultural sites, local and foreign environmentalists, human rights activists, and Turkey's downstream neighbors Syria and Iraq have all expressed opposition to the project. The major archaeological losses from the Ataturk, Karakaya, and Batman Dams received scant coverage in the American and European media. Only when spectacular Greco-Roman pieces faced imminent destruction, at sites such as Zeugma and Samsat, did the Western media briefly turn its spotlight to the issue. Other, perhaps, more important historic sites that reveal information about human civilization's roots and man's early experiments with religion, architecture, and agriculture disappeared with much less protest.

Concerns both nationally and internationally about the feasibility, practicability, and efficiency of GAP have created financing problems. The World Bank has refused to support any GAP projects since 1984 without a formal water-sharing agreement between Turkey, Syria, and Iraq in place (Kolars 1991). Despite the World Bank's non-support of GAP projects, export credit agencies (ECAs), which typically lack strict formal policies on environmental and social issues, have been willing to finance the initial phases of GAP and to guarantee loans from private banks to Turkey. ECAs do not necessarily adhere to internationally accepted standards and guidelines; hence Turkey's rejection of the UN Convention on the Law of the Non-Navigational Uses of International Watercourses (Turkey was one of only three no votes; the other two were China and Burundi) has not affected ECA support (WCD 2000a).

Currently, the most controversial GAP project is Ilisu Dam, and the construction consortium charged with building it is seeking export credits and investment insurance guarantees from the ECAs of Austria, Germany, Italy, Japan, Portugal, Sweden, Switzerland, the United Kingdom, and the United States. Growing public pressure from the public, NGOs, and parliamentarians regarding Ilisu finally led ECAs to attach four conditions that the Turkish government must meet before export credits are issued for the project. The conditions, announced in December 1999, reflect concerns over the archaeological losses, resettlement program, degraded water quality, and effects on downstream users by the Ilisu Dam. However, the conditions set by the ECAs, such as a requirement for a detailed plan to preserve as much of the archaeological heritage of Hasankeyf as possible, do not stipulate standards against which compliance with their conditions will be measured.

Since the ECAs set out their conditions, the World Commission on Dams (WCD), an international body sponsored by the World Bank, established new guidelines for the dam industry (see Chapter 7 in this book). The WCD standards are widely regarded as setting a new benchmark for best practice in dam projects and hence for international standards. Although the WCD standards are not legally binding, export credit agencies in the United Kingdom and United States have implied that Ilisu will have to meet some or all of the WCD standards in order to gain export credit support. An independent and "definitive assessment" of the project commissioned by Britain's Department of Trade and Industry reached the conclusion that the Ilisu project in its current state fails miserably to meet international standards, and for the time being Britain has abandoned its support for the dam (Ahmed 2001). Other analyses of the Ilisu Dam project and its compliance with both the ECA conditions and WCD standards by a coalition of NGOs, reported that Ilisu violates each condition set by the ECAs, and each of the WCD guidelines representing international best practice (KHRP 2001, WEED 2000). The WCD has also reported that only about 25 of the 298 dam projects in Turkey have been surveyed at all for cultural heritage, and of these, only 5 have had organized, systematic rescue work conducted (WCD 2000b).

GAP represents one of the last huge water-development projects to be designed in the twentieth century, and perhaps one of the last ever to be built on such a scale. It also represents the old way of doing business: we now know that projects must be designed and built to far different standards than in the past. For the GAP project, this especially applies to efforts to protect and preserve the vast archaeological record and treasures that will be affected by the many different pieces of the project. Even with the minimal archaeological work done so far, we know that countless priceless and unique treasures have already been lost and many more are threatened. Turkey has the opportunity to do the remaining pieces of the project properly, preserve part of its own history, and show the world community how to do it right.

REFERENCES

Ahmed, K. 2001. "UK drops Turkish dam plan." *Observer,* July 1. Available at http://www.observer.co.uk/international/story/0,6903,515248,00.html.

Algaze, G. 1989. "A new frontier: First results of the Tigris-Euphrates Archaeological Reconnaissance Project, 1988." *Journal of Near Eastern Studies,* Vol. 48, No. 4, pp. 241–81.

Algaze, G., R. Breuninger, G. Lightfoot, and M. Rosenburg. 1991. "The Tigris-Euphrates Archaeological Reconnaissance Project: A preliminary report of the 1989–1990 seasons." *Anatolica,* Vol. 17, pp. 175–240.

Blaylock, S.R., D.H. French, G.D. Summers. 1990. "The Adiyaman Survey: An interim report." Anatolian Studies, Vol. 40, pp. 81–135.

Bosshard, P. 1998. "Ilisu—A test case of international policy coherence." Available at http://www.rivernet.org/turquie/ilisu.htm#Archaeological.

British Broadcasting Company (BBC) News. 2000. "Turkish dam controversy." January 22. Available at http://news.bbc.co.uk/hi/english/world/europe/newsid_614000/614235.stm.

British House of Commons Select Committee on Trade and Industry. 2001. "Twelfth report: Ilisu Dam." Available at http://www.parliament.the-stationeryoffice.co.uk/pa/cm200001/cmselect/cmtrdind/512/51204.htm.

Center for Research and Assessment of the Historic Environment (TACDAM). 2001. "Salvage project of the Ilisu and Carchemish Dam reservoirs: Activities in 1999–2000." Available at http://www.metu.edu.tr/home/wwwmuze/Tacdam2000/html_e/menu.htm.

Energy Information Administration (EIA). 2001. "Turkey: Country analysis brief." Available at http://www.eia.doe.gov/emeu/cabs/turkey.html.

Foreign Economic Relations Board (DEIK). 1998. "Energy production and consumption in Turkey." Available at http://www.deik.org.tr/turkiye/1998/sec1202.html.

Government of Turkey. 2001. "Overall information on GAP." http://www.gap.gov.tr/English/Frames/fr1.html.

Izady, M.R. 1996. *On the Drowning of the Kurdish Historical and Artistic Heritage Behind Dams.* Kurdish Worldwide Resources, New York.

Kennedy, D., R. Ergec, and P. Freeman. 1995. "Mining the mosaics of Roman Zeugma." *Archaeology,* Vol. 48, No. 2, pp. 54–55.

Kennedy, D. 1998. "The twin towns of Zeugma on the Euphrates: Rescue work and historical studies." *Journal of Roman Archaeology,* No. 27, pp. 19–29. Portsmouth, Rhode Island.

Kennedy, D. 2000. "Double tragedy on the Euphrates." Available at http://www.arts.uwa.edu.au/Classics/archaeology/Z2.html.

Kennedy, D. 2001. Personal communication with author. July 18.

Kinzer, S. 2000. "Dam in Turkey may soon flood a 2nd Pompeii." *New York Times,* May 7, pp. 1, 10.

Kitchen, W., and M. Ronayne. 2001. "The Ilisu Dam in southeast Turkey: Archaeology at risk." *Antiquity,* Vol. 75, pp. 37–38.

Kolars, J.F., and W.A. Mitchell. 1991. *The Euphrates River and Southeast Anatolia Development Project.* Southern Illinois University Press, Carbondale.

Kurdish Human Rights Project (KHRP). 2000. "The Ilisu Dam: The World Commission on Dams and export credit reform." Available at http://www.khrp.org/publish/p2000/IlisuReport2000.htm.

McNicoll, A. 1983. *Taskun Kale: Keban Rescue Excavations Eastern Anatolia.* British Institute of Archaeology at Ankara, Oxford, England.

Meyerson, J.A. 2000. "After the flood!" *Archaeology Odyssey,* November/December, pp. 18–23.

Nestor, C.E. 1995. "Dimensions of Turkey's Kurdish question and the potential impact of the Southeast Anatolian Project (GAP)." *International Journal of Kurdish Studies,* Vol. 8, No. 1, pp. 35–78. (Two parts.)

Republic of Turkey Prime Ministry Southeastern Anatolia Project (GAP). 1996. *An Innovative Approach to Integrated Sustainable Regional Management.* Southeastern Anatolia Project Regional Development Administration, Ankara, Turkey.

Swiss Federal Institute of Technology Zurich (ETH Zurich). 2000. "Sustainable management of international rivers—case study: Southeastern Anatolia Project in Turkey—GAP." Available at http://www.eawag.ch/research_e/apec/seminars_e/GAP.pdf.

Time Europe. 1999. "Archaeology: Ever nearer the past." Available at http://www.time.com/time/europe/magazine/1999/126/archaeology2.html.

Tuna, N. 1999. *Salvage Project of the Archaeological Heritage of the Ilisu and Carchemish Dam Reservoirs: Activities in 1998.* N. Tuna and J. Ozturk (ed.). METU Center for Research and Assessment of the Historic Environment, Ankara, Turkey.

University of Delaware. 1994. "10,000 year old pig probably earliest domesticated animal." Available at http://www.udel.edu/PR/Messenger/94/4/1.html.

Ward, D.R. 1990. "In Anatolia, a massive dam project drowns traces of an ancient past." *Smithsonian,* Vol. 21, No. 8, pp. 28–41.

WEED. 2000. "Ilisu Dam and export credit reform." Available at http://www.weedbonn.org/hermes/ilisu_report_e.rtf.

Whallon, R. 1979. "An archaeological survey of the Keban Reservoir area of east-central Turkey." *Memoirs of the Museum of Anthropology, University of Michigan,* No. 11, pp. 1–307.

World Commission on Dams (WCD). 2000a. "Dams and development—A new framework for decision making. Available at http://www.dams.org.

World Commission on Dams (WCD). 2000b. "Cultural heritage and dam projects in Turkey: An overview." Available at http://www.dams.org/docs/html/contrib/soc21281.htm.

Young, P. 2000. "Hasankeyf: A city in peril." *History Today,* Vol. 50, No. 11, pp. 3–4.

Environment and Security Water Conflict Chronology Version 2002

Peter H. Gleick

Beginning with the first volume of *The World's Water,* we have regularly published a Chronology of water-related conflicts and violence. This feature has become a highly valued tool for people interested in water disputes, and it is one of the most heavily visited sections of our web site at www.worldwater.org. The electronic version will be updated monthly.

As we have consistently noted, water resources have rarely been the sole source of violent conflict or war. But this fact has led some international security "experts" to ignore or belittle the complex and real relationships between water and security. It is easy to draw a narrow definition of "security" in a way that excludes water (or other resources) from the debate over international security, or to require that security threats be narrow, single-issue factors. Such an approach both misunderstands the connections between water and security and misleads policy makers and the public seeking ways of reducing tensions and violence. In fact, there is a long and highly informative history of conflicts and tensions over water resources, the use of water systems as weapons during war, and the targeting of water systems during conflicts caused by other factors.

We are continuing our efforts, begun in the late 1980s, to understand the connections between water resources, water systems, and international security and conflict. In 2001, the Institute completed a report on the links between environment and terrorism, and incorporated the issue of terrorism into the Chronology.

The interest in this Chronology has been strong. New information is regularly sent to me by historians, water experts, and readers to update, correct, and expand the listings. Recent world events in Afghanistan, Pakistan, the Middle East, and other regions have, unfortunately, also added several new entries. As a result, an updated Chronology is presented here, with new entries and a range of corrections and modifications to the older ones. In addition, we continue to make changes in how several of these entries are categorized. The heading "Basis of Conflict" now includes terrorism and a more clear set of categories than the previous listing. The current categories or types of conflict now include:

> **Control of water resources** (state and nonstate actors): where water supply or access to water is at the root of tensions.

> **Military tool** (state actors): where water resources, or water systems themselves, are used by a nation or state as a weapon during a military action.

> **Political tool** (state and nonstate actors): where water resources, or water systems themselves, are used by a nation, state, or nonstate actor for a political goal.

Terrorism (nonstate actors): where water resources, or water systems, are either targets or tools of violence or coercion by nonstate actors. This is a new category.

Military target (state actors): where water resource systems are targets of military actions by nations or states.

Development disputes (state and nonstate actors): where water resources or water systems are a major source of contention and dispute in the context of economic and social development.

It will be clear to even the casual reader that these definitions are imprecise and that single events can fall into more than one category, depending on perception and definitions. For example, intentional military attacks on water-supply systems can fall into both the **Targets** and **Tools** categories, depending on one's point of view. Disputes over control of water resources may reflect either political power disputes or disagreements over approaches to economic development, or both. I believe this is inevitable and even desirable—international security is not a clean, precise field of study and analysis. It is evolving as international and regional politics evolves and as new factors become increasingly, or decreasingly, important in the affairs of humanity. In all this, however, one factor remains constant: the importance of water to life means that providing for water needs and demands will never be free of politics. As social and political systems change and evolve, this Chronology and the kinds of entries and categories will change and evolve. I continue to look forward to contributions and comments from readers. Please email contributions, with full citations and supporting information, to pgleick@pipeline.com.

WATER CONFLICT CHRONOLOGY

Peter H. Gleick

Pacific Institute for Studies in Development, Environment, and Security

Date	Parties Involved	Basis of Conflict	Violent Conflict or in the Context of Violence?	Description	Source
1503	Florence and Pisa warring states	Military tool	Yes	Leonardo da Vinci and Machiavelli plan to divert Arno River away from Pisa during conflict between Pisa and Florence.	Honan 1996
1642	China, Ming Dynasty	Military tool	Yes	The dikes of the Huang He (Yellow) River have been breached for military purposes. In 1642, "toward the end of the Ming dynasty (1368–1644), General Gao Mingheng used the tactic near Kaifeng in an attempt to suppress a peasant uprising."	Hillel 1991
1672	French; Dutch	Military tool	Yes	Louis XIV starts the third of the Dutch Wars in 1672, in which the French overran the Netherlands. In defense, the Dutch opened their dikes and flooded the country, creating a watery barrier that was virtually impenetrable.	Columbia 2000
1863	U.S. Civil War	Military tool	Yes	General U.S. Grant, during the Civil War campaign against Vicksburg, cut levees in the battle against the Confederates.	Grant 1885, Barry 1997
1898	Egypt; France; Britain	Military and political tool; control of water resources	Military maneuvers	Military conflict nearly ensues between Britain and France in 1898, when a French expedition attempted to gain control of the headwaters of the White Nile. While the parties ultimately negotiate a settlement of the dispute, the incident has been characterized as having "dramatized Egypt's vulnerable dependence on the Nile and fixed the attitude of Egyptian policy-makers ever since."	Moorhead 1960
1907–1913	Owens Valley, Los Angeles, California	Political tool; control of water resources; terrorism; and development dispute	Yes	The Los Angeles Valley aqueduct/pipeline suffers repeated bombings in an effort to prevent diversions of water from the Owens Valley to Los Angeles.	Reisner 1993
1915	German Southwest Africa	Military tool	Yes	Union of South African troops capture Windhoek, capital of German Southwest Africa (May). Retreating German troops poison wells, "a violation of the Hague convention."	Daniel 1995

Date	Parties	Category	Military maneuvers	Description	Source
1935	California; Arizona development dispute	Political tool		Arizona calls out the National Guard and militia units to the border with California to protest the construction of Parker Dam and diversions from the Colorado River; dispute ultimately is settled in court.	Reisner 1993
1938	China; Japan	Military tool; military target	Yes	Chiang Kai-shek orders the destruction of flood-control dikes of the Huayuankou section of the Huang He River to flood areas threatened by the Japanese Army. West of Kaifeng dikes are destroyed with dynamite, spilling water across the flat plain. The flood destroyed part of the invading army and its heavy equipment was mired in thick mud, though Wuhan, the head-quarters of the Nationalist government, was taken in October. The waters flooded an area variously estimated as between 3,000 and 50,000 square kilometers, and killed Chinese estimated in numbers between "tens of thousands" and "one million."	Hillel 1991; Yang Lang 1994
1940–1945	Multiple parties	Military target	Yes	Hydroelectric dams routinely bombed as strategic targets during World War II.	Gleick 1993
1943	Britain; Germany	Military target	Yes	British Royal Air Force bombed dams on the Mohne, Sorpe, and Eder Rivers, Germany (May 16, 17). Mohne Dam breech killed 1,200 and destroyed all downstream dams for 50 km.	Kirschner 1949
1944	Germany; Italy; Britain; United States	Military tool	Yes	German forces used waters from the Isoletta Dam (Liri River) in January and February to successfully destroy British assault forces crossing the Garigliano River (downstream of Liri River). The German Army then dammed the Rapido River, flooding a valley occupied by the U.S. Army.	Corps of Engineers 1953
1944	Germany; Italy; Britain; United States	Military tool	Yes	German Army flooded the Pontine Marches by destroying drainage pumps to contain the Anzio beachhead established by the Allied landings in 1944. Over 40 square miles of land were flooded; a 30-mile stretch of landing beaches was rendered unusable for amphibious support forces.	Corps of Engineers 1953
1944	Germany; Allied forces	Military tool	Yes	Germans flooded the Ay River, France (July), creating a lake two meters deep and several kilometers wide, slowing an advance on Saint Lo, a German communications center in Normandy.	Corps of Engineers 1953
1944	Germany; Allied forces	Military tool	Yes	Germans flooded the Ill River Valley during the Battle of the Bulge (winter 1944–45), creating a lake 16 kilometers long, 3–6 kilometers wide, and 1–2 meters deep, greatly delaying the U.S. Army's advance toward the Rhine.	Corps of Engineers 1953

Date	Parties Involved	Basis of Conflict	Violent Conflict or in the Context of Violence?	Description	Source
1947 onward	Bangladesh; India	Development disputes; control of water resources	No	Partition divides the Ganges River between Bangladesh and India; construction of the Farakka barrage by India, beginning in 1962, increases tension; short-term agreements settle dispute in 1977–82, 1982–84, and 1985–88, and thirty-year treaty is signed in 1996.	Butts 1997, Samson and Charrier 1997
1947–1960s	India; Pakistan	Development disputes; control of water resources; and political tool	No	Partition leaves Indus basin divided between India and Pakistan; disputes over irrigation water ensue, during which India stems flow of water into irrigation canals in Pakistan; Indus Waters Agreement reached in 1960 after 12 years of World Bank–led negotiations.	Bingham, Wolf, and Wohlegenant 1994; Wolf 1997
1948	Middle East	Military tool	Yes	Arab forces cut of West Jerusalem's water supply in first Arab-Israeli War.	Wolf 1995, 1997
1950s	Korea; United States; others	Military target	Yes	Centralized dams on the Yalu River serving North Korea and China are attacked during Korean War.	Gleick 1993
1951	Korea; United Nations	Military tool and military target	Yes	North Korea released flood waves from the Hwachon Dam, damaging floating bridges operated by UN troops in the Pukhan Valley. U.S. Navy planes were then sent to destroy spillway crest gates.	Corps of Engineers 1953
1951	Israel; Jordan; Syria	Political tool; military tool; development disputes	Yes	Jordan makes public its plans to irrigate the Jordan Valley by tapping the Yarmouk River; Israel responds by commencing drainage of the Huleh swamps located in the demilitarized zone between Israel and Syria; border skirmishes ensue between Israel and Syria.	Wolf 1997, Samson and Charrier 1997
1953	Israel; Jordan; Syria	Development dispute; military target; political tool	Yes	Israel begins construction of its National Water Carrier to transfer water from the north of the Sea of Galilee out of the Jordan basin to the Negev Desert for irrigation. Syrian military actions along the border and international disapproval lead Israel to move its intake to the Sea of Galilee.	Samson and Charrier 1997

Date	Parties	Basis of conflict	Violent conflict or in the context of violence	Description	Sources
1958	Egypt; Sudan	Military tool; political tool; control of water resources	Yes	Egypt sends an unsuccessful military expedition into disputed territory amid pending negotiations over the Nile waters, Sudanese general elections, and an Egyptian vote on Sudan-Egypt unification; Nile Water Treaty signed when pro-Egyptian government elected in Sudan.	Wolf 1997
1960s	North Vietnam; United States	Military target	Yes	Irrigation water supply systems in North Vietnam are bombed during Vietnam War. 661 sections of dikes damaged or destroyed.	BRP Foundation 1971, Gleick 1993, Zemmali 1995
1962–1967	Brazil; Paraguay	Military tool; political tool; control of water resources	Military maneuvers	Negotiations between Brazil and Paraguay over the development of the Paraná River are interrupted by a unilateral show of military force by Brazil in 1962, which invades the area and claims control over the Guaira Falls site. Military forces were withdrawn in 1967 following an agreement for a joint commission to examine development in the region.	Murphy and Sabadell 1986
1963–1964	Ethiopia; Somalia	Development dispute; military tool; political tool	Yes	Creation of boundaries in 1948 leaves Somali nomads under Ethiopian rule; border skirmishes occur over disputed territory in Ogaden desert where critical water and oil resources are located; cease-fire is negotiated only after several hundred are killed.	Wolf 1997
1964	Cuba; United States	Military weapon	No	On February 6, 1964, the Cuban government ordered the water supply to the U.S. naval base at Guantánamo Bay cut off.	Guantánamo Bay Gazette 1964.
1965	Zambia; Rhodesia; Great Britain	Military target	No	President Kenneth Kaunda calls on British government to send troops to Kariba Dam to protect it from possible saboteurs from Rhodesian government.	Chenje 2001
1965–1966	Israel; Syria	Military tool; political tool; control of water resources; development dispute	Yes	Fire is exchanged over "all-Arab" plan to divert the Jordan River headwaters and presumably preempt Israeli National Water Carrier; Syria halts construction of its diversion in July 1966.	Wolf 1995, 1997
1966–1972	Vietnam; United States	Military tool	Yes	U.S. tries cloud seeding in Indochina to stop flow of materiel along Ho Chi Minh trail.	Plant 1995

WATER CONFLICT CHRONOLOGY *Continued*

Date	Parties Involved	Basis of Conflict	Violent Conflict or in the Context of Violence?	Description	Source
1967	Israel; Syria	Military target and tool	Yes	Israel destroys the Arab diversion works on the Jordan River head-waters. During Arab-Israeli War, Israel occupies Golan Heights, with Banias tributary to the Jordan; Israel occupies West Bank.	Gleick 1993; Wolf 1995, 1997; Wallenstein and Swain 1997
1969	Israel; Jordan	Military target and tool	Yes	Israel, suspicious that Jordan is overdiverting the Yarmouk, leads two raids to destroy the newly built East Ghor Canal; secret negotiations, mediated by the United States, lead to an agreement in 1970.	Samson and Charrier 1997
1970s	Argentina; Brazil; Paraguay	Control of water resources; development dispute	No	Brazil and Paraguay announce plans to construct a dam at Itaipu on the Paraná River, causing Argentina concern about down-stream environmental repercussions and the efficacy of their own planned dam project downstream. Argentina demands to be consulted during the planning of Itaipu, but Brazil refuses. An agreement is reached in 1979 that provides for the construction of both Brazil and Paraguay's dam at Itaipu and Argentina's Yacyreta Dam.	Wallenstein and Swain 1997
1972	North Vietnam	Military target	Yes	United States bombs dikes in the Red River delta, rivers, and canals during massive bombing campaign.	Columbia Electronic Encyclopedia 2000
1974	Iraq; Syria	Military target; military tool; political tool; development dispute	Military maneuvers	Iraq threatens to bomb the al-Thawra Dam in Syria and massed troops along the border, alleging that the dam had reduced the flow of Euphrates River water to Iraq.	Gleick 1994
1975	Iraq; Syria	Development dispute; military tool; political tool	Military maneuvers	As upstream dams are filled during a low-flow year on the Euphrates, Iraqis claim that flow reaching its territory is "intolerable" and asks the Arab League to intervene. Syrians claim they are receiving less than half the river's normal flow and pull	Gleick 1993, 1994; Wolf 1997

				out of an Arab League technical committee formed to mediate the conflict. In May Syria closes its airspace to Iraqi flights and both Syrian and Iraq reportedly transfer troops to their mutual border. Saudi Arabia successfully mediates the conflict.	
1975	Angola; South Africa	Military control of water resources	Yes	South African troops move into Angola to occupy and defend the Ruacana hydropower complex, including the Gové Dam on the Kunene River. Goal is to take possession of and defend the water resources of southwestern Africa and Namibia.	Meissner 2000
1978–onward	Egypt; Ethiopia	Development dispute; political tool	No	Longstanding tensions over the Nile, especially the Blue Nile, originating in Ethiopia. Ethiopia's proposed construction of dams on the headwaters of the Blue Nile leads Egypt to repeatedly declare the vital importance of water. "The only matter that could take Egypt to war again is water"(Anwar Sadat, 1979). "The next war in our region will be over the waters of the Nile, not politics" (Boutrous Ghali, 1988).	Gleick 1991, 1994
1978–1984	Sudan	Development dispute; military target	Yes	Demonstrations in Juba, Sudan, in 1978 opposing the construction of the Jonglei Canal led to the deaths of two students. Construction of the Jonglei Canal in the Sudan was forcibly suspended in 1984 following a series of attacks on the construction site.	Suliman n.d., Keluel-Jang 1997
1980s	Mozambique; Rhodesia/ Zimbabwe; South Africa	Military target/ terrorism	Yes	Regular destruction of power lines from Cahora Bassa Dam during fight for independence in the region. Dam targeted by RENAMO.	Chenje 2001
1981	Iran; Iraq	Military target and tool	Yes	Iran claims to have bombed a hydroelectric facility in Kurdistan, thereby blacking out large portions of Iraq, during the Iran-Iraq War.	Gleick 1993
1980–1988	Iran; Iraq	Military tool	Yes	Iran diverts water to flood Iraqi defense positions.	Plant 1995
1986	Lesotho, South Africa	Development goal; access to resources	Yes	Bloodless coup by Lesotho's defense forces, with support from South Africa, leads to immediate agreement with South Africa for water from the Highlands of Lesotho, after 30 years of unsuccessful negotiations. There is disagreement over the degree to which water was a motivating factor for either party.	Mohamed 2001
1988	Angola; South Africa; Cuba	Military goal; military target	Yes	Cuban and Angolan forces launch an attack on Calueque Dam via land and then air. Considerable damage inflicted on dam wall; power supply to dam cut. Water pipeline to Owamboland cut and destroyed.	Meissner 2000

WATER CONFLICT CHRONOLOGY *Continued*

Date	Parties Involved	Basis of Conflict	Violent Conflict or in the Context of Violence?	Description	Source
1982	Israel; Lebanon; Syria	Military tool	Yes	Israel cuts off the water supply of Beirut during siege.	Wolf 1997
1982	Guatemala	Development dispute	Yes	177 civilians killed in Rio Negro over opposition to Chixoy hydroelectric dam.	Levy 2000
1986	North Korea; South Korea	Military tool	No	North Korea's announcement of its plans to build the Kumgansan hydroelectric dam on a tributary of the Han River upstream of Seoul raises concerns in South Korea that the dam could be used as a tool for ecological destruction or war.	Gleick 1993
1986	Lesotho, South Africa	Military goal; control of water resources	Yes	South Africa supports coup in Lesotho over support for ANC and anti-apartheid, and water. New government in Lesotho then quickly signs Lesotho Highlands water agreement.	American University 2000b
1990	South Africa	Development dispute; control of water resources	No	Pro-apartheid council cuts off water to the Wesselton township of 50,000 blacks following protests over miserable sanitation and living conditions.	Gleick 1993
1990	Iraq; Syria; Turkey	Development dispute; military tool; political tool	No	The flow of the Euphrates is interrupted for a month as Turkey finishes construction of the Ataturk Dam, part of the Grand Anatolia Project. Syria and Iraq protest that Turkey now has a weapon of war. In mid-1990 Turkish president Turgut Ozal threatens to restrict water flow to Syria to force it to withdraw support for Kurdish rebels operating in southern Turkey.	Gleick 1993, 1995
1991–present	Karnataka; Tamil Nadu (India)	Development dispute; control of water resources	Yes	Violence erupts when Karnataka rejects an Interim Order handed down by the Cauvery Waters Tribunal, empaneled by the Indian Supreme Court. The tribunal was established in 1990 to settle two decades of dispute between Karnataka and Tamil Nadu over irrigation rights to the Cauvery River.	Gleick 1993, Butts 1997, American University 2000a
1991	Iraq; Kuwait; United States	Military target	Yes	During the Gulf War, Iraq destroys much of Kuwait's desalination capacity during retreat.	Gleick 1993
1991	Iraq; Turkey; United Nations	Military tool	Yes	Discussions are held at the United Nations about using the Ataturk Dam in Turkey to cut off flows of the Euphrates to Iraq.	Gleick 1993

Date	Parties	Basis of conflict	Violent conflict or in the context of violence	Description	Sources
1991	Iraq; Kuwait; United States	Military target	Yes	Baghdad's modern water supply and sanitation system are intentionally targeted by allied coalition.	Gleick 1993
1992	Czechoslovakia; Hungary	Political tool; development dispute	Military maneuvers	Hungary abrogates a 1977 treaty with Czechoslovakia concerning construction of the Gabcikovo/Nagymaros project based on environmental concerns. Slovakia continues construction unilaterally, completes the dam, and diverts the Danube into a canal inside the Slovakian republic. Massive public protest and movement of military to the border ensue; issue taken to the International Court of Justice.	Gleick 1993
1992	Bosnia; Bosnian Serbs	Military tool	Yes	The Serbian siege of Sarajevo, Bosnia, and Herzegovina includes a cutoff of all electrical power and the water feeding the city from the surrounding mountains. The lack of power cuts the two main pumping stations inside the city despite pledges from Serbian nationalist leaders to United Nations officials that they would not use their control of Sarajevo's utilities as a weapon. Bosnian Serbs take control of water valves regulating flow from wells that provide more than 80 percent of water to Sarajevo; reduced water flow to city is used to "smoke out" Bosnians.	Burns 1992, Husarska 1995
1993–present	Iraq	Military tool	No	To quell opposition to his government, Saddam Hussein reportedly poisons and drains the water supplies of southern Shi'ite Muslims, the Ma'dan. The European Parliament and UN Human Rights Commission deplore use of water as weapon in region.	Gleick 1993, American University 2000c
1993	Yugoslavia	Military target and tool	Yes	Peruca Dam intentionally destroyed during war.	Gleick 1993
1995	Ecuador; Peru	Military and political tool	Yes	Armed skirmishes arise in part because of disagreement over the control of the headwaters of Cenepa River. Wolf argues that this is primarily a border dispute simply coinciding with location of a water resource.	Samson and Charrier 1997, Wolf 1997
1997	Singapore, Malaysia	Political tool	No	Malaysia supplies about half of Singapore's water and in 1997 threatened to cut off that supply in retribution for criticisms by Singapore of policy in Malaysia.	Zachary 1997
1998	Tajikistan	Terrorism; political tool	Potential	On November 6, a guerrilla commander threatens to blow up a dam on the Kairakkhum channel if political demands are not met. Col. Makhmud Khudoberdyev made the threat, reported by the ITAR-Tass News Agency.	WRR 1998

WATER CONFLICT CHRONOLOGY *Continued*

Date	Parties Involved	Basis of Conflict	Violent Conflict or in the Context of Violence?	Description	Source
1998	Angola	Military and political tool	Yes	In September 1998, fierce fighting between UNITA and Angolan government forces broke out at Gové Dam on the Kunene River for control of the installation.	Meissner 2001
1998	Democratic Republic of Congo	Military target/ terrorism	Yes	Attacks on Inga Dam during efforts to topple President Kabila. Disruption of electricity supplies from Inga Dam and water supplies to Kinshasa.	Chenje 2001, Human Rights Watch 1998
1999	Lusaka, Zambia	Terrorism; political tool	Yes	Bomb blast destroyed the main water pipeline, cutting off water for the city of Lusaka, population 3 million.	FTGWR 1999
1999	Yugoslavia	Military target	Yes	Belgrade reported that NATO planes had targeted a hydro-electric plant during the Kosovo campaign.	Reuters 1999a
1999	Bangladesh	Development dispute; political tool	Yes	50 hurt during strikes called to protest power and water shortages. Protest led by former prime minister Begum Khaleda Zia over deterioration of public services and in law and order.	Ahmed 1999
1999	Yugoslavia	Military target	Yes	NATO targets utilities and shuts down water supplies in Belgrade. NATO bombs bridges on Danube, disrupting navigation.	Reuters 1999b
1999	Yugoslavia	Political tool	Yes	Yugoslavia refuses to clear war debris on Danube (downed bridges) unless financial aid for reconstruction is provided; European countries on Danube fear flooding due to winter ice dams will result. Diplomats decry environmental blackmail.	Simons 1999
1999	Kosovo	Political tool	Yes	Serbian engineers shut down water system in Pristina prior to occupation by NATO.	Reuters 1999c
1999	Angola	Terrorism/ political tool	Yes	100 bodies were found in four drinking water wells in central Angola.	*International Herald Tribune* 1999
1999	Puerto Rico, United States	Political tool	No	Protesters blocked water intake to Roosevelt Roads Navy Base in opposition to U.S. military presence and U.S. Navy's use of the Blanco River, following chronic water shortages in neighboring towns.	*New York Times* 1999
1999	East Timor	Military tool; political tool; terrorism	Yes	Militia opposing East Timor independence kill pro-independence supporters and throw bodies in water well.	BBC 1999

Date	Parties Involved	Basis of Conflict	Violent Conflict or in the Context of Violence	Description	Sources
1999	Kosovo	Terrorism/political tool	Yes	Contamination of water supplies/wells by Serbs disposing of bodies of Kosovar Albanians in local wells.	CNN 1999
1999–2000	Namibia; Botswana; Zambia	Military goal: control of water resources	No	Sedudu/Kasikili Island, in the Zambezi/Chobe River. Dispute over border and access to water. Presented to the International Court of Justice.	ICJ 1999
2000	Ethiopia	Development dispute	Yes	One man stabbed to death during fight over clean water during famine in Ethiopia.	Sandrasagra 2000
2000	Kyrgyzstan; Kazakhstan; Uzbekistan	Political tool	No	Kyrgyzstan cuts off water to Kazakhstan until coal is delivered; Uzbekistan cuts off water to Kazakhstan for nonpayment of debt.	Pannier 2000
2000	Gujarat, India	Development dispute	Yes	Water riots reported in some areas of Gujarat to protest against authority's failure to arrange adequate supply of tanker water. Police are reported to have shot into a crowd at Falla village near Jamnagar, resulting in the death of 3 and injuries to 20 following protests against the diversion of water from the Kankavati dam to Jamnagar town.	FTGWR 2000
2000	China	Development dispute	Yes	Civil unrest erupted over use and allocation of water from Baiyangdian Lake, the largest natural lake in northern China. Several people died in riots by villagers in July 2000 in Shandong after officials cut off water supplies. In August 2000, six died when officials in the southern province of Guangdong blew up a water channel to prevent a neighboring county from diverting water.	Pottinger 2000
2001	Israel; Palestinians	Terrorism/control of water resources	Yes	Palestinians destroy water supply pipelines to West Bank settlement of Yitzhar and to Kibbutz Kisufim. Agbat Jabar refugee camp near Jericho disconnected from its water supply after Palestinians looted and damaged local water pumps. Palestinians accuse Israel of destroying a water cistern, blocking water tanker deliveries, and attacking materials for a wastewater treatment project.	Israel Line 2001a, 2001b; ENS 2001a
2001	Pakistan	Development dispute	Yes	Civil unrest over severe water shortages caused by the long-term drought. Protests began in March and April and continued into summer. Riots, 4 bombs in Karachi (June 13), 1 death, 12 injuries, 30 arrests. Ethnic conflicts as some groups "accuse the government of favoring the populous Punjab province [over Sindh province] in water distribution."	Nadeem 2001, Soloman 2001

Water Conflict Chronology *Continued*

Date	Parties Involved	Basis of Conflict	Violent Conflict or in the Context of Violence?	Description	Source
2001	Macedonia	Terrorism/ control of water resources	Yes	Water flow to Kumanovo (population 100,000) cut off for 12 days in conflict between ethnic Albanians and Macedonian forces. Valves of Glaznja and Lipkovo Lakes damaged.	AFP 2001, Macedonia Information Agency 2001
2001	Philippines	Terrorism/ political tool	No	Philippine authorities shut off water to six remote southern villages yesterday after residents complained of a foul smell from their taps, raising fears Muslim guerrillas had contaminated the supplies. Abu Sayyaf guerrillas, accused of links with Saudi-born militant Osama bin Laden, had threatened to poison the water supply in the mainly Christian town of Isabela on Basilan Island if the military did not stop an offensive against them.	*World Environment News* 2001

Sources:

AFP. 2001. "Macedonian troops fight for water supply as president moots amnesty." June 8. Available at http://www.balkanpeace.org/hed/archive/june01/hed3454.shtml.

Ahmed, A. 1999. "Fifty hurt in Bangladesh strike violence." Reuters News Service, Dhaka, April 18. Available at http://biz.yahoo.com/rf/990418/3.html.

American University (Inventory of Conflict and the Environment; ICE). 2000a. Cauvery river dispute. Available at http://www.american.edu/projects/mandala/TED/ice/CAUVERY.htm.

American University (Inventory of Conflict and the Environment; ICE). 2000b. Lesotho "water coup." Available at http://www.american.edu/projects/mandala/TED/ice/LESWATER.htm.

American University (Inventory of Conflict and the Environment; ICE). 2000c. Marsh Arabs and Iraq. Available at http://www.american.edu/projects/mandala/TED/ice/MARSH.htm.

Barry, J.M. 1997. *Rising Tide: The Great Mississippi Flood of 1927 and How It Changed America.* Simon and Schuster, New York, p. 67.

Bingham, G., A. Wolf, and T. Wohlegenant. 1994. "Resolving water disputes: Conflict and cooperation in the United States, the Near East, and Asia." U.S. Agency for International Development (USAID). Bureau for Asia and the Near East. Washington, D.C.

BBC. 1999. "World: Asia-Pacific Timor atrocities unearthed." September 22. Available at http://news.bbc.co.uk/hi/english/world/asia-pacific/newsid_455000/455030.stm.

BRP Foundation. 1971. Bertrand Russell Peace Foundation. "Some facts on bombing of dikes." Available at http://www.homeusers.prestel.co.uk/littleton/v1117kan.htm.

Burns, J.F. 1992. "Tactics of the Sarajevo siege: Cut off the power and water." *New York Times,* September 25, p. A1.

Butts, K., ed. 1997. *Environmental Change and Regional Security.* Asia-Pacific Center for Security Studies, Center for Strategic Leadership, U.S. Army War College, Carlisle, Pennsylvania.

Cable News Network (CNN). 1999. "U.S.: Serbs destroying bodies of Kosovo victims." May 5. Available at www.cnn.com/WORLD/europe/9905/05/kosovo.bodies.

Chenje, M. 2001. "Hydro-politics and the quest of the Zambezi River Basin Organization." In M. Nakayama (ed.), *International Waters in Southern Africa.* United Nations University, Tokyo, Japan.

Columbia Electronic Encyclopedia. 2000. "Vietnam: History." Available at http://www.infoplease.com/ce6/world/A0861793.html.

Columbia Encyclopedia. 2000. "Netherlands." 6th ed. *Columbia Encyclopedia* available at http://www.bartleby.com/65/ne/Nethrlds.html

Corps of Engineers. 1953. *Applications of Hydrology in Military Planning and Operations and Subject Classification Index for Military Hydrology Data.* Military Hydrology R&D Branch, Engineering Division, Corps of Engineers, Department of the Army, Washington, D.C.

Daniel, C. (ed.). 1995. *Chronicle of the 20th Century.* Dorling Kindersley Publishing, New York.

Drower, M.S. 1954, "Water-supply, irrigation, and agriculture." In C. Singer, E.J. Holmyard, and A.R. Hall (eds.) *A History of Technology*. Oxford University Press, New York.

Environment News Service (ENS). 2001. "Environment a weapon in the Israeli-Palestinian conflict." February 5. Available at http://www.ens-news.com/ens/feb2001/2001L-02-05-01.html.

Financial Times Global Water Report. 1999. "Zambia: Water Cutoff." *FTGWR* Issue 68, p. 15 (March 19, 1999).

Financial Times Global Water Report. 2000. "Drought in India comes as no surprise." *FTGWR* Issue 94, p. 14 (April 28, 2000).

Gleick, P.H. 1991. "Environment and security: The clear connections." *Bulletin of the Atomic Scientists*, April, pp. 17–21.

Gleick, P.H. 1993. "Water and conflict: Fresh water resources and international security." *International Security*, Vol. 18, No. 1, pp. 79–112.

Gleick, P.H. 1994. "Water, war, and peace in the Middle East." *Environment*, Vol. 36, No. 3, p. 6ff. Heldref Publishers, Washington, D.C.

Gleick, P.H. 1995. "Water and conflict: Critical issues." Presented to the 45th Pugwash Conference on Science and World Affairs, Hiroshima, Japan, 23–29 July.

Gleick, P.H. 1998. "Water and conflict." In *The World's Water 1998–1999*. Island Press, Washington, D.C.

Grant, U.S. 1885. *Personal Memoirs of U.S. Grant*. C.L. Webster, New York.

Green Cross International. *The Conflict Prevention Atlas*. Available at http://dns.gci.ch/water/atlas.

Guantanamo Bay Gazette. 1964. *The History of Guantanamo Bay: An Online Edition*. Available at http://www.gtmo.net/gazz/hisidx.htm. Chapter 21, "The 1964 water crisis." Available at http://www.gtmo.net/gazz/HISCHP21.HTM.

Hillel, D. 1991. "Lash of the dragon." *Natural History*, August, pp. 28–37.

Honan, W.H. 1996. "Scholar sees Leonardo's influence on Machiavelli." *New York Times*, December 8, p. 18.

Human Rights Watch. 1998. "Human Rights Watch condemns civilian killings by Congo rebels." Available at http://www.hrw.org/press98/aug/congo827.htm

Husarska, A. 1995. "Running dry in Sarajevo: Water fight." *New Republic*, July 17 and 24.

International Court of Justice. 1999. International Court of Justice Press Communiqué 99/53, Kasikili Island/Sedudu Island (Botswana/Namibia). The Hague, Holland, 13 December, p. 2 (www.icj-cij.org).

International Herald Tribune. 1999. "100 bodies found in well." *International Herald Tribune*, August 14–15, p. 4.

Israel Line. 2001a. "Palestinians loot water pumping center, cutting off supply to refugee camp." Israel Line (http://www.israel.org/mfa/go.asp?MFAH0izu0), January 5, http://www.mfa.gov.il/mfa/go.asp?MFAH0iy50.

Israel Line. 2001a. "Palestinians vandalize Yitzhar water pipe." Israel Line, January 9, http://www.mfa.gov.il/mfa/go.asp?MFAH0izu0.

Keluel-Jang, S.A. 1997. "Alier and the Jonglei Canal." *Southern Sudan Bulletin*, Vol. 2, No. 3. From www.sufo.demon.co.uk/poli007.htm.

Kirschner, O. 1949. *Destruction and Protection of Dams and Levees*. Military Hydrology, Research and Development Branch, U.S. Corps of Engineers, Department of the Army, Washington District. From Schweizerische Bauzeitung, March 14, 1949, translated by H.E. Schwarz, Washington, D.C.

Levy, K. 2000. "Guatemalan dam massacre survivors seek reparations from financiers." *World Rivers Review*, December 2000, pp. 12–13. International Rivers Network, Berkeley, California.

Macedonia Information Agency. 2001. "Humanitarian catastrophe averted in Kumanovo and Lipkovo." Republic of Macedonia Agency of Information Archive, June 18. Available at http://www.sinf.gov.mk/PressRoomEN/2001/05/n0618.htm.

Meissner, R. 2000. "Hydropolitical hotspots in Southern Africa: Will there be a water war? The case of the Kunene river." In H. Solomon and A. Turton (eds.), *Water Wars: Enduring Myth or Impending Reality?* Africa Dialogue Monograph Series No. 2. Accord, Creda Communications, KwaZulu-Natal, South Africa, pp. 103–31.

Meissner, R. 2001. "Interaction and existing constraints in international river basins: The case of the Kunene River Basin." In M. Nakayama (ed.), *International Waters in Southern Africa*. United Nations University, Tokyo, Japan.

Mohamed, A.E. 2001. "Joint development and cooperation in international water resources: The case of the Limpopo and Orange River Basins in Southern Africa." In M. Nakayama (ed.), *International Waters in Southern Africa*. United Nations University, Tokyo, Japan.

Moorehead, A. 1960. *The White Nile*. Penguin Books, England.

Murphy, I.L., and J. E. Sabadell. 1986. "International river basins: A policy model for conflict resolution." *Resources Policy*, Vol. 12, No. 1, pp. 133–44. Butterworth, United Kingdom.

Nadeem, A. 2001. "Bombs in Karachi kill one." *Associated Press*, June 13. Available at http://dailynews.yahoo.com/h/ap/20010613/wl/pakistan_strike_3.html.

New York Times. 1999. "Puerto Ricans protest Navy's use of water." *New York Times*, October 31, p. 30.

Pannier, B. 2000. "Central Asia: Water becomes a political issue." Radio Free Europe. Available at www.rferl.org/nca/features/2000/08/F.RU.000803122739.html.

Plant, G. 1995. "Water as a weapon in war." *Water and War, Symposium on Water in Armed Conflicts*, Montreux, November 21–23, 1994, Geneva, ICRC.

Pottinger, M. 2000. "Major Chinese lake disappearing in water crisis." Reuters Science News, http://dailynews.yahoo.com/h/nm/20001220/sc/environment_china_dc_1.html.

Reisner, M. 1993. *Cadillac Desert: The American West and its Disappearing Water*. Penguin Books, New York.

Reuters. 1999a. "Serbs say NATO hit refugee convoys." April 14. Available at http://dailynews.yahoo.com/headlines/ts/story.html?s=v/nm/19990414/ts/yugoslavia_192.html.

Reuters. 1999b. "NATO keeps up strikes but Belgrade quiet." June 5. Available at http://dailynews.yahoo.com/headlines/wl/story.html?s=v/nm/19990605/wl/yugoslavia_strikes_129.html.

Reuters. 1999c. "NATO builds evidence of Kosovo atrocities." June 17. Available at http://dailynews.yahoo.com/headlines/ts/story.html?s=v/nm/19990617/ts/yugoslavia_leadall_171.html.

Samson, P., and B. Charrier. 1997. "International freshwater conflict: Issues and prevention strategies." Green Cross International. Available at http://www.dns.gci.ch/water/gcwater/study.html.

Sandrasagra, M.J. 2000. "Development Ethiopia: Relief agencies warn of major food crisis." Inter Press Service, April 11.

Simons, M. 1999. "Serbs refuse to clear bomb-littered river." *New York Times*, October 24.

Soloman, A. 2001. "Policeman dies as blasts rock strike-hit Karachi." Reuters, June 13, http://dailynews.yahoo.com/h/nm/20010613/ts/pakistan_strike_dc_1.html.

Suliman, M. n.d. "Resource access: A major cause of armed conflict in the Sudan: The case of the Nuba Mountains." Institute for African Alternatives, London. From http://srdis.ciesin.org/cases/Sudan-Paper.html.

Wallenstein, P., and A. Swain. 1997. "International freshwater resources-Conflict or cooperation?" In *Comprehensive Assessment of the Freshwater Resources of the World*, Stockholm Environment Institute, Stockholm, Sweden.

Yang Lang. 1994. "High Dam: The sword of Damocles." In Dai Qing (ed.), *Yangtze! Yangtze!* Probe International, Earthscan Publications, London, pp. 229–40.

Wolf, A.T. 1995. *Hydropolitics along the Jordan River: Scarce Water and its Impact on the Arab-Israeli Conflict*. United Nations University Press, Tokyo, Japan.

Wolf, A. T. 1997. "'Water wars' and water reality: Conflict and cooperation along international waterways." NATO Advanced Research Workshop on Environmental Change, Adaptation, and Human Security. Budapest, Hungary, 9–12 October.

World Environment News. 2001. "Philippine rebels suspected of water 'poisoning.'" Available at http://www.planetark.org/avantgo/dailynewsstory.cfm?newsid=12807.

World Rivers Review (WRR). 1998. *Dangerous Dams: Tajikistan*, Vol. 13, No. 6, p. 13.

Zachary, G.P. 1997. "Water pressure: Nations scramble to defuse fights over supplies." *Wall Street Journal*, December 4, p. A17.

Zemmali, H. 1995. "International humanitarian law and protection of water." *Water and War, Symposium on Water in Armed Conflicts*, Montreux November 21–23, 1994, Geneva, ICRC.

Note: Conflicts may stem from the drive to possess or control another nation's water resources, thus making water systems and resources a *political or military goal*. Inequitable distribution and use of water resources, sometimes arising from a water development, may lead to *development disputes*, heighten the importance of water as a strategic goal, or may lead to a degradation of another's source of water. Conflicts may also arise when water systems are used as instruments of war, either as *targets* or *tools*. These distinctions are described in detail in Gleick (1993, 1998). In 2001, the Institute began including incidents involving water and *terrorism*.

Water and Space

Elizabeth L. Chalecki

> *I am convinced that He does not play dice.*
> ALBERT EINSTEIN

> *God not only plays dice, he also sometimes throws the dice where they cannot be seen.*
> STEPHEN HAWKING

In a book about water on Earth, it is perhaps unusual to describe the search for water in outer space. When we think of water, we think of the blood of all life on Earth. Space, on the other hand, is a vast, unimaginably empty vacuum, dotted with stars and inherently inimical to life on Earth. Why would we look for water in outer space?

What would it mean to find extraterrestrial water? In learning about water in space, can we perhaps learn something about our own water? Water itself is a fascinating thing, not just because of its importance to life here, but because of where it can take us. Water in space is a necessary precursor to humans in space. And if we find water in space, we begin to wonder what else we might find out there, including life.

Surprisingly, we have found water, a simple unmistakable molecule, to be widely distributed in space. Earth's moon, the driest of bodies, may have frozen water at the poles. Mars, with its red sands and carbon-dioxide atmosphere, has water frozen in ice caps and may have subterranean ice. Several moons of Jupiter, circling their forbidding gas giant, may have large quantities of ice. Europa, in particular, looks promising as a source of liquid water, kept from freezing by the heat of its volcanic core. We have even found traces of water in far stellar clouds and galaxies.

The Origin of the Earth's Water

Scientists have asked for years how water came to this world. Many Earth scientists argue that water vapor from volcanic eruptions collected on the surface. Many space scientists have argued that water, and the complex carbon-based molecules necessary to produce life, arrived here in a mix of comets and meteors.

Cosmic Snowballs

The first volume of *The World's Water* discussed Dr. Louis Frank's novel theory that cosmic snowballs augmented terrestrial water supply (Gleick 1998). First hypothesized in 1986, Frank proposed that small comets, the size of a house or even smaller and made of pure water, strike the earth's atmosphere 20 times per minute. They do not show up as typical meteors with a fiery tail because they are made of uncompacted snow rather than rock. As a result, they are not cohesive enough to strike the earth's surface; they vaporize thousands of kilometers above the atmosphere (Kerr 1988).

Frank faced ridicule from his colleagues for this novel, and apparently counterintuitive, idea, until eleven years later, when some observations from NASA's POLAR satellite's Visible Imaging System appeared to support his hypothesis. The satellite's sensors showed what could be water from comets breaking up at high altitude, between 960 and 24,000 kilometers above the planetary surface. The suspected trail of one such cosmic snowball vaporizing over the Atlantic Ocean and Western Europe at an altitude of 8,000 to 24,000 kilometers is seen below (Figure 1). As it descends, wind disperses it; the water vapor condenses in the atmosphere and falls to Earth as precipitation (Broad 1997a). More recently, Frank and his coauthors published a new paper reporting the results of a ground-based telescope survey, which they suggest also provides supporting visual evidence for such cosmic snowballs (Frank and Sigwarth 2001). This celestial precipitation is thought to add approximately 2 to 3 centimeters of water over the entire surface of the planet every 20,000 years (Kluger 1997). If such a process has continued throughout Earth's history, as Frank hypothesizes, the water accumulation would be enough to fill all the earth's oceans (Frank et al. 1986). This hypothesis is still highly controversial, and the ultimate origin of Earth's water is unresolved.

Water-Bearing Meteorites

Snowball comets are not the only way water has arrived on Earth from space. Space scientists have uncovered the existence of liquid water older than solar system contained in two meteorites that fell to Earth in 1998, one in Monahans, Texas, and one in Zag, Morocco (Zolensky et al. 1999). Thanks to an alert group of children playing nearby, the Monahans meteorite made it to NASA's Johnson Space Center a record 46 hours after it landed. This allowed scientists there to examine the meteorite before the water it contained evaporated. They saw what appeared to be bubbles of liquid in blue-purple crystals inside the meteorite (Figures 2 and 3). Further testing indicated the presence of saltwater inside crystals that were 4.6 billion years old (Cowen 1999).

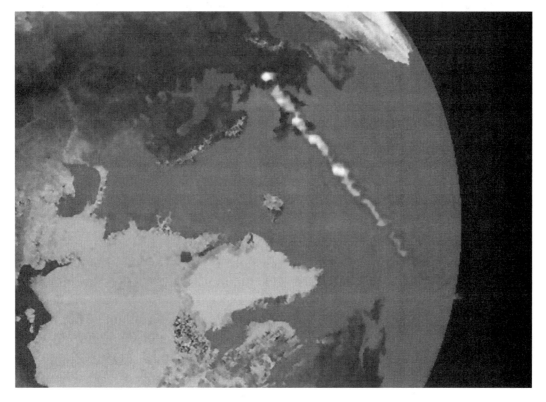

FIGURE WB.1   A water comet vaporizing in high atmosphere.
Source: L. Frank, U. Iowa

The Zag meteorite, which fell in a remote part of Morocco that same year, also contains water droplets inside blue-purple salt crystals. As in the Monahans meteorite, the salt crystals are tinged blue-purple by the cosmic radiation to which it was exposed

FIGURE WB.2   Monahans halite crystals.
Source: NASA JPL

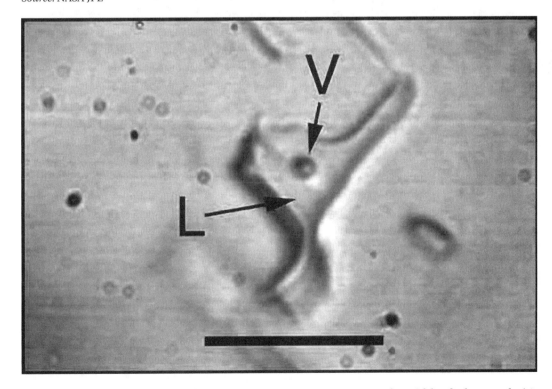

FIGURE WB.3   V = Water bubble 15 microns across (one-quarter the width of a human hair).
Source: NASA JSC

before falling to Earth. Preliminary dating of the Zag meteorite by scientists from the University of Manchester showed that the water had formed within 100 million years after the formation of the solar system, around 4.5 billion years ago. A subsequent analysis of the proportion of xenon and iodine isotopes contained in the rock showed that it and the water it contains were formed within 2 million years of the birth of the solar system (Whitby et al. 2000), making it even older than previously thought.

The fact that meteorites can contain primordial water is exciting because it has far-reaching implications for theories on the origin of life. Water is one of the essential compounds of life, and its existence outside Earth would suggest that there are other places where conditions could be favorable for the development of life.

The Moon

The search for extraterrestrial water focused early on the Moon. Efforts to find water on the Moon underlie a significant opportunity for human exploration of space. If the Moon or other celestial body has water, it could provide critical resources for human use. NASA has undertaken a series of missions to search for water on the Moon. In 1994, radar measurements from *Clementine*, a Defense Department satellite launched

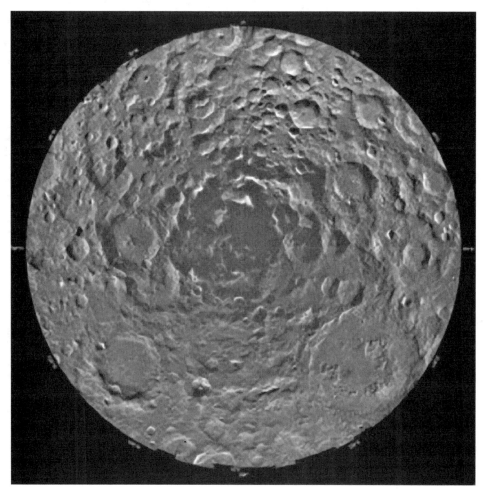

FIGURE WB.4   Lunar South Pole.
Source: NASA

to test space and missile technology, suggested the presence of permanently frozen ice at the shaded areas of the lunar poles (Figure 4), possibly deposited when water-rich comets and meteorites struck the surface (Wilford 1998).

The possibility of water on the Moon intrigued scientists, and on January 6, 1998, NASA launched the *Lunar Prospector*, designed to map the Moon's surface composition and its gravity and magnetic fields (Binder et al. 1998). Its neutron spectrometer (NS) (see Sidebar 1) scanned for hydrogen and oxygen at the Moon's north and south poles, and from its findings, NASA inferred the existence of water. Two months later, they announced the existence of anywhere from 11 to 330 million tons of frozen water mixed in with the lunar soil (Hall 1998).

In an unusual attempt to confirm this finding, NASA crashed the 160-kilogram *Lunar Prospector* on July 31, 1998, into a crater at the Moon's south pole in an effort to kick up ice-bearing rock and dust. Earth-bound observers of the crash used sensitive spectrometers to search for the ultraviolet emissions of hydroxyl (OH) molecules, the expected signature of water (NASA 1999). The crash of the *Lunar Prospector* did not yield any observable water signature, although scientists and astronomers pointed out that the chances of positive detection were judged to be less than 10 percent. To date, NASA still considers open the question of whether frozen water exists in usable quantities on the Moon.

In situ water could yield oxygen for a breathable atmosphere and hydrogen for rocket fuel, making water not only biologically necessary but also economically desirable. Currently, spacecraft must carry with them all the fuel needed for the

---

Box WB.1   Neutron Spectrometer

An NS searches for water ice by detecting the element hydrogen. Since water molecules contain atoms of hydrogen and oxygen, hydrogen is a good marker for water. If lunar water is present in usable quantities, the amount of hydrogen locked up in water molecules will dwarf the amount that may be present for any other reason. Even if only 0.5 percent of the surface material at a given location is water, Lunar Prospector is capable of finding it.

Because it flies 100 kilometers above the Moon's surface, the NS does not detect hydrogen directly. Instead, it looked for what scientists call "cool" neutrons—neutrons that have bounced off a hydrogen atom somewhere on the lunar surface. The only effective way to slow down a speeding neutron is to have it collide with something approximately its own size, such as a hydrogen atom. If the Moon's crust contains a lot of hydrogen at a certain location, any neutron that bounces around in the crust before heading out to space will cool off rapidly. When Lunar Prospector flew over such a crater, the NS could detect a surge in the number of cool ("thermal") neutrons, and a dropoff in the number of warm ("epithermal") neutrons. Soil containing ice, for instance, should yield more cool neutrons and fewer warm neutrons than soil devoid of ice. Because the spacecraft carrying the NS passed over the lunar poles every orbit, the NS produced the most accurate data precisely where the water is thought to be.

outbound and return trips. Such off-world sources of fuel would make space travel much more productive by allowing spacecraft to expand the payload weight. In addition, launching rockets from the Moon rather than Earth would extend the range of future missions. The Moon's gravity is a fraction of Earth's, and the escape velocity—the speed that a spaceship has to reach to escape its gravitational pull—is a mere 2.38 kilometers (1.4 miles) per second, compared to Earth's 11.2 km (about 7 miles) per second.

Mars

NASA's search for extraterrestrial water has not been limited to just the Moon. Mars may once have been a very wet place. Recent astronomical research from a number of institutions has uncovered what appears to be evidence of various forms of water on Mars, in the past and perhaps presently. However, Mars, like the Moon, is cold and dry. Since the average surface temperature on Mars is -81 degrees Fahrenheit and the atmospheric pressure is only one-sixth that of Earth, the existence of liquid water becomes more than problematic. Even if some internal heat source warmed up the planet enough to permit any surface ice to melt, the atmosphere is so thin that the ice would change directly to water vapor (NASA Astrobiology Institute 2001).

Since the early Viking flyby missions of the 1960s, pictures of the Martian surface have shown evidence of the movement of water. There is evidence of permafrost activity on the surface, indicating a "youthful" phase of water-related activity. In addition, integrated networks of channels and tributaries (Figure 5) cross the surface in what appear to be complex drainage and erosion patterns formed by surface water activity (Baker 2001).

Recent photographs and high-resolution laser topography data of Mars sent back by the *Mars Global Surveyor* show large plains in the northern hemisphere that appear at one time to have been oceans (Figure 6). What could be at least two concurrent northern ocean beds have been determined, and supporting characteristics such as shoreline erosion, meteorite ejecta properties, and the interaction of outflow channels and the ocean bed have been identified near the surface of the northern lowlands (Head et al. 1999). In addition, topographic and free-air gravity models have shown that these large, flat northern lowlands contain large buried channels that are consistent with the northward transport of water and sediment (Zuber et al. 2000). Some scientists have recently described evidence that the ridged areas that appear to be shorelines may have been caused by tectonic activity (Withers and Neumann 2001,

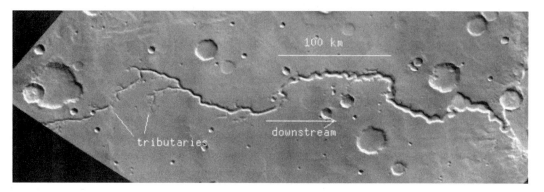

FIGURE WB. 5 Nirgal Vallis.
Source: Malin Space Science Systems

Lloyd 2001). Since the evidence is contradictory, the debate continues as to what caused these features.

These possible northern oceans are not the only features on Mars that may indicate water. Findings from the *Mars Orbital Camera* (MOC) have revealed pictures of what look like flood gullies very near the surface, where liquid water may burst forth from underground for a brief flash flood before evaporating. Since the temperature and atmospheric pressure conditions on the Martian surface make the existence of liquid fresh water almost impossible, scientists have hypothesized that any liquid water underground must be saltwater. Since saltwater freezes at temperatures significantly lower than that of fresh water, this expands the regions on Mars where ice melting could occur and increases the total time such conditions might exist for liquid water (NASA 2000). However, an alternative explanation posed by Australian geoscientist Nick Hoffman for these features involves a much colder Mars, where its carbon dioxide atmosphere could have condensed into layers of CO_2 ice covering the planet. Pressure from these ice layers would have allowed a layer of liquid CO_2 to form underneath, as liquid water forms under glaciers here on Earth. This liquid CO_2 would bubble to the surface and vaporize explosively in "pyroclastic flows," and the resulting gas, mixed with dust and rocky debris, could mimic the movement of water and carve the same sorts of channels and gullies. Hoffman came to this conclusion when he noticed a lack of calcium carbonate in the rocks on the Martian surface, a geologic byproduct formed when liquid water absorbs carbon dioxide out of the atmosphere (Davidson 2001).

Water may exist on Mars to this day. Research by Nadine Barlow, director of the University of Central Florida's Robinson Observatory, has uncovered what appears to be an Arizona-sized underground ice reservoir. Located in the Solis Planum region, the ice could

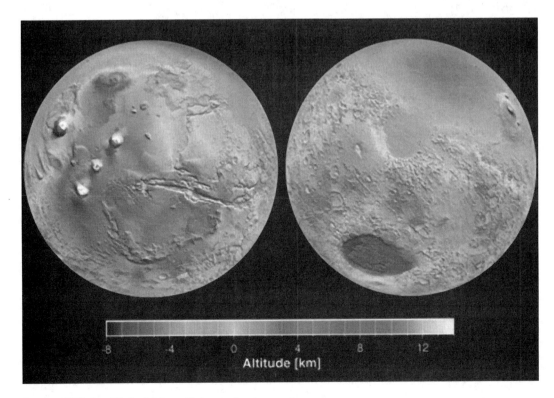

FIGURE WB.6   Global Mars (false color image).

Source: NASA

start just 200 meters from the Martian surface, making it an ideal water-mining area for future manned landings. Barlow hypothesizes that water has gathered in that spot because of volcanic activity tilting the water table into the low-lying region (David 2000).

While all this physical evidence shows signs of past (and possibly present) oceans on Mars, a recent analysis of the interior of a 1.2-billion-year-old Martian meteorite may give researchers a glimpse into the chemical content of these oceans. The Nahkla meteorite, which landed in El-Nahkla, Egypt, in 1911, showed residue from what is believed to be salts from brine. Researchers at Arizona State University examined the interior of the meteorite and found ionic chloride and sodium. Since these elements are predominant in Earth's oceans, scientists have interpreted a chemical similarity between oceans on Earth and an early Martian ocean (Hathaway 2000).

Being There

All the evidence described above has been gathered long-distance, via satellites and cameras. In order to confirm the existence of water on Mars, we must go there, in person or robotically. NASA has planned missions to search for evidence of water on the Martian surface. In April of 2001, NASA launched *Odyssey,* the latest Mars mission designed to search for water, with the goal of determining if life ever existed there. *Odyssey* successfully arrived in Mars orbit in fall 2001 equipped with an array of infrared, visible-light, and gamma-ray spectrometers, and in March 2002, NASA scientists announced findings of a significant hydrogen signature near the Martian equator (Britt 2002).

FIGURE WB.7   Mars Water-Sniffing Rover.
Source: NASA artist's rendition

NASA's Mars exploration plans also include the 2003 launch of a water-sniffing rover (Figure 7) similar to the extremely successful *Pathfinder*. The rover would carry among its six instruments a color camera and a detector to search for evidence of liquid water

Finding water in any form on Mars could provide the same benefits as finding water on the Moon: oxygen and hydrogen could be produced for breathing and fuel. In addition, water itself will be necessary for human colonization and possible terraforming of Mars, though the latter may take hundreds if not thousands of years (Bonsor n.d.).

Jupiter's Moons

Early radar observations of Jupiter's moon Europa detected a layer of water up to 160 kilometers deep covering the surface, but it was unknown if the water was frozen or liquid (Perlman 1999). In 1999, however, detailed photographs from NASA's *Galileo* spacecraft, orbiting Jupiter since 1995, clearly showed an ice surface with a network of unusual cracks and fissures (Figure 8) (Wilford 1999). Gregory Hoppa and other space scientists from the University of Arizona have theorized that these scalloped cracks were caused when the ever-changing tidal stress from Jupiter exceeded the tensile strength of the ice. As Europa moves around Jupiter in its irregular orbit, Jupiter's gravitational pull causes subsurface tides on Europa to flex and crack the ice shell. Hoppa and his colleagues modeled the stress fractures and concluded that for these sorts of cracks to appear the ice crust must sit on top of a liquid layer deep enough to account for significant tidal amplitude (Hoppa et al. 1999). However, the photographic evidence was not sufficient to determine if liquid water currently exists on Europa, only that it likely existed at some point in the past.

Further physical evidence for a liquid-water layer existing on Europa at present was detected in the late 1990s by *Galileo*'s magnetometer. Magnetometers function by detecting changes in magnetic fields over time. *Galileo* made close passes in 1996 and again in 1998, and from these magnetic readings, Margaret Kivelson and fellow researchers from UCLA have presented strong evidence that Europa currently has a global, spherical conducting layer just under the surface. Although this layer might be some other conductive material like graphite (Weinstock 2000), salty water is the most likely explanation, in light of the geologic and photographic evidence previously found

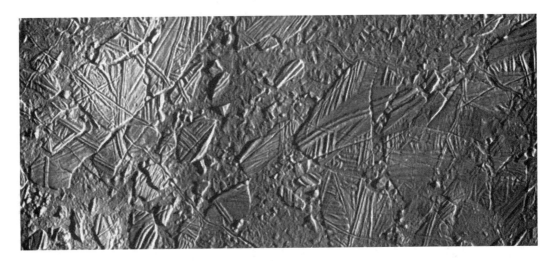

FIGURE WB.8   The surface of Europa.
Source: NASA JPL

(Kivelson et al. 2000). NASA hopes to answer this question definitively when it launches the *Europa Orbiter* in 2003, a spacecraft with ice-penetrating radar. Liquid water on Europa would be a major find and would give rise to the same speculations surrounding human use of the Moon and Mars (see Sidebar 2).

Two other Jovian satellites, Ganymede and Callisto, may also have liquid water, though not as near the surface as Europa. Infrared measurements from *Galileo's* near-infrared mapping spectrometer identified hydrated minerals on the moon's surface, and researchers hypothesize that Ganymede may have a subsurface layer of briny fluid like Europa's (McCord et al. 2001). In addition, magnetometer readings from *Galileo* flybys of Callisto indicated that this moon may also have a layer of melted salty water deep below its surface. However, unlike Europa, Callisto's surface is heavily cratered, indicating that the watery layer is much farther below (NASA 2001).

The existence of water on Europa or other Jovian moons raises again the question of extraterrestrial life. We know that Earth's least hospitable environments can support life in various forms. The liquid-water ocean on Europa could also not support life if other conditions permit. Researchers have considered the biochemical conditions necessary for life, or at least life as we know it, and most concluded that it is not the lack of water that makes Europan life problematic, but rather the lack of a chemically appropriate energy source. Some scientists have concluded that conditions on Europa change too quickly to sustain evolution of life beyond single-celled bacteria (Gaidos et al. 1999), though even this would, of course, be a remarkable finding.

Water Beyond Our Solar System

Water has also been found to exist in space beyond our own solar system. Hydrogen, originally produced in the Big Bang, is found everywhere in the universe. Oxygen is made in stars and dispersed out into the universe in events such as supernova explo-

---

Box WB.2   The Artemis Society

The relative closeness of the Europan subterranean ocean to the surface has given rise to some peculiar plans. The Artemis Society, a private venture dedicated to developing a permanent, self-supporting human colony on Earth's moon, has also formulated a plan to put human colonists on Europa. No such plans are possible without the presence of water. However, a colony poses more than logistical problems. Europa is right in the middle of Jupiter's deadly radiation belt, and humans on the surface would not survive more than ten minutes. However, Peter Kokh, the author of the Artemis plan, assumes that the engineering challenges inherent in radiation shielding will have been solved by the time humans are ready to mount a manned expedition to Europa. A manned mission would erect a surface hangar made of ice and would then deploy a submersible vehicle to tunnel under the surface ice and into the global ocean below (Kohk et al. 1997). NASA is skeptical, although several technical details envisioned by the Artemis Society have later shown up in NASA plans (Lipper 2001).

sions. The two elements mix in star-forming clouds and form large amounts of water (Figure 9). The molecules of water leave the clouds and end up in many different places: comets, planets, and the centers of galaxies. When stars age, more oxygen is made available to the cosmic water factory. Various efforts have been made to search for water outside of our solar system.

Interstellar Clouds

The Submillimeter Wave Astronomy Satellite (SWAS), operated by the Harvard-Smithsonian Center for Astrophysics, is a NASA Small Explorer project designed to study the chemical composition of interstellar gas clouds. SWAS was launched into low Earth orbit on December 5, 1998. Its primary objective is to look for the spectral emission lines of water, isotopic water, molecular oxygen, neutral carbon, and isotopic carbon monoxide in a variety of galactic star-forming regions. The results from SWAS observations will help determine if these elements and compounds are the dominant coolants of molecular clouds during their collapse to form stars and planets (Melnick et al. 2000). If true, this raises the interesting possibility that water is or was part of every star and planet in the galaxy.

On July 11, 2001, SWAS scientists announced their findings of water around a distant star, IRC+10216. More commonly known as CW Leonis, this star is located 500 light-years from Earth in the constellation Leo. The SWAS team found that it is a red giant star, vaporizing a belt of water ice comets surrounding its system and depositing large

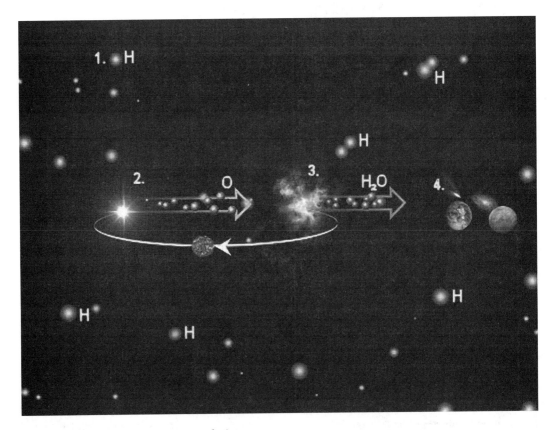

FIGURE WB.9   How water is made in space.
Source: European Space Agency

amounts of water vapor into the surrounding space (Vedantam 2001, Melnick et al. 2001). Our solar system has a similar set of comets, called the Kuiper Belt. Located beyond the orbit of Neptune, the Kuiper Belt contains many billions of icy objects, of which Pluto is the largest currently known. Most Kuiper Belt objects orbit the Sun quietly and unobserved, except very occasionally when one of them is deflected onto an elliptical orbit. As a comet comes close to the Sun, it heats up and starts to vaporize, yielding the fireball and tail that we recognize. Giant stars like CW Leonis have grown roughly 10,000 times more luminous than the Sun and are now so powerful that they vaporize comets even at the distance of the Kuiper Belt (see Sidebar 3).

SWAS is not the only low-orbit craft designed to look for extraterrestrial water in our galaxy. Using the European Space Agency's Infrared Space Observatory (ISO), Spanish and Italian astronomers have found water in the cold, dark areas of the galaxy: stellar clouds and the interstellar medium (Moneti et al. 2001). Previously, scientists assumed that the water-forming process would most likely be sustained only in warm stellar clouds such as the Orion Nebula, where astronomers found enough water vapor being produced to fill the earth's oceans every 24 minutes (Harwit et al. 1998). In this process, however, the water is heated to 200 degrees Fahrenheit or more (Roylance 1998). ISO's observations found substantial water vapor and ice in quiescent regions of the galaxy where the mean temperature is just 10 degrees above absolute zero. Astronomers estimate that there are millions of these cold clouds in the galaxy, containing as much water as the active, star-forming regions, and that by colliding and sticking together, they may form planets and comets (Recer 2001).

Water on the Other Side of the Universe

Water has even been detected outside of our own galaxy. In 1994, radio astronomers from the University of Maryland and the Max Planck Institut für Radioastronomie announced the discovery of water signatures in five active galaxies, the farthest of which, Markarian 1, is 200 million light years away (Braatz et al. 1994, Dawson, pers. comm. 2001). Specifically, these water signatures are called H2O "megamasers," or photon-emitting water-vapor clouds at the center of active galaxies. A maser is the microwave analogue of a laser; naturally occurring masers are frequently found in

---

Box WB.3   Could This Happen Here?

SWAS scientists point out that the process going on in IRC+10216 will be our fate eventually. In several billion years, our Sun will expand into a red giant star, and its power output will increase 5,000-fold. A wave of water vaporization will spread outward through the solar system, starting with Earth's oceans and extending into the Kuiper Belt, well beyond the orbit of Neptune. Icy bodies as large as Pluto will be vaporized, leaving nothing but a hot cinder of rock. From Earth, we are able to detect water vapor from similar events elsewhere in the galaxy.

Source: SWAS Press Release

dense clouds of gas in our own galaxy, where molecules such as water amplify the radiation from stars. Very powerful masers (megamasers) are observed in other galaxies, where dense clouds of molecular material orbit the black hole at the galactic center and amplify the intense radiation from its vicinity. Since 1994, at least 15 other H_2O megamasers have been discovered using a technique called Very Long Baseline Interferometry (VLBI; see Sidebar 4).

Water-vapor megamasers are useful to radio astronomers in two ways. By examining the orbital motions of megamasers, astronomers can calculate an absolute geometric distance between galaxies, making extragalactic measurements more accurate than they have ever been (Herrnstein et al. 1999). NASA and the National Radio Astronomy Observatory are currently planning a project called ARISE (Advanced Radio Interferometry between Space and Earth) to measure the distances to galactic centers containing these water-vapor megamasers. Also, emissions from water-vapor megamasers give astronomers detailed evidence of the structure and dynamics of active galaxies, furthering the study of black holes thought to be at the center of these galaxies (Ishihara et al. 2001).

Conclusion

Why do we care if there is water in deep space, or even on Europa? It's not likely that humans will have access to this water for hundreds of years, if ever. Its existence will certainly not provide relief to drought-stricken areas, supplement human consumption, or increase aquatic streamflows here on Earth, and huge physical and engineering challenges must be overcome before such water can be of use for human space exploration.

We care because the presence of accessible water greatly improves the chances of successful human exploration of space. Water is heavy to transport, but needed for the health and physical well-being of the explorers. If future manned NASA missions had to carry all the water the crew would need even to Mars and back, the cost of such a journey would be prohibitive. If such a mission had to carry its own water to Europa and back, current size limitations of spacecraft might make such a mission physically impossible. However, if water could be found and utilized *in situ*, suddenly the whole possibility of sending humans into space enters the tantalizing realm of possibility. The presence of water also offers a source of fuel for future missions, further reducing the resources that must be provided from Earth.

Box WB.4   Very Long Baseline Interferometry (VLBI)

Interferometry is a type of data gathering in which one image is made from the results of several different telescopes. The farther apart the telescopes are (the longer the baseline), the more detailed the picture. Current VLBI baselines are limited by the size of the earth, but an orbiting radio telescope, such as the ARISE project, combined with Earth-bound telescopes can increase the size of the baseline to four times the size of Earth!

Source: NASA JPL

Finally, we also care because where there is water, in any form, there *might* be life. Theoretical considerations suggest that prebiotic chemical evolution could lead to the development of self-replicating life, and we have already seen that water and other organic chemical molecules, like formaldehyde and benzene, exist in space. Because the known conditions under which life has actually arisen are limited to Earth, most exobiologists think that liquid water is necessary for life (Carr et al. 1995), or at least where water exists we know that a fundamental precursor to life is present. Thus, the past or present existence of liquid water anywhere in the universe gives us a kind of roadmap to discovering extraterrestrial organisms, whether one-celled or complex. Water has certainly defined our human past and now may point the way toward our human future.

REFERENCES

Baker, V.R. 2001. "Water and the Martian landscape." *Nature,* Vol. 412, July 12, pp. 228–36.

Binder, A.B, W.C. Feldman, G.S. Hubbard, A.S. Konopliv, R.P. Lin, M. Acuna, and L.L. Hood. 1998. "Lunar Prospector searches for polar ice, a metallic core, gas release events, and the Moon's origin." *EOS Transactions,* Vol. 79, No. 8, February 24, pp. 97, 108.

Bonsor, K. n.d. "How terraforming Mars will work." *Marshall Brain's How Stuff Works.* As found at www.howstuffworks.com/terraforming.htm.

Braatz, J.A., A.S. Wilson, and C. Henkel. 1994. "The discovery of five new H_2O megamasers in active galaxies." *Astrophysical Journal,* Vol. 437, No. 1, December 20, pp. L99–L102.

Britt, R.R. 2002. "Odyssey discovers abundant water ice on Mars." Space.com. 1 March 2002. As found at www.space.com/scienceastronomy/solarsystem/odyssey_update_020301.htm.

Broad, W.J. 1997a. "Nourishing space snowballs striking Earth's atmosphere." *New York Times,* May 29, pp. A1, A13.

Broad, W.J. 1997b. "Spotlight on comets in shaping of Earth." *New York Times,* June 3, pp. B7, B11.

Carlowitz, M. 1997. "New 'small comet' images challenges researchers." *EOS, Transactions,* Vol. 78, No. 25, June 24, pp. 257–58.

Carr, M.H., B. Clark, D.J. DesMarais, D.L. DeVincenzi, J.D. Farmer, J.M. Hayes, H. Holland, B. Jakosky, G.F. Joyce, J.F. Kerridge, H.P. Klein, A.H. Knoll, G.D. McDonald, C.P. McKay, M.A. Meyer, K.H. Nealson, E.L. Shock, and D.M. Ward. 1995. "An exobiological strategy for Mars exploration." Exobiology Program Office, Ames Research Center, National Aeronautics and Space Administration. 59 pp.

Cowen, R. 1999. "Found: primordial water." *Science News,* October 30. As found at www.findarticles.com/cf_0/m1200/18_156/57799553/print.jhtml.

Davidson, K. 2001. "Mars enigma: One scientist's contentious theory about planet may hold water." *San Francisco Chronicle,* August 13, p. A4.

Dawson, S. 2001. UC Berkeley Radio Astronomy Laboratory Public Affairs Officer. Personal communication, 17 July.

Frank, L., J.B. Sigwarth, and J.D. Craven. 1986. "On the influx of small comets into the Earth's upper atmosphere. I. Observations. II. Interpretation." *Geophysical Resource Letters,* Vol. 13, No. 303, pp. 303–7.

Frank, L., and J.B. Sigwarth. 2001. "Detections of small comets with a ground-based telescope." *Journal of Geophysical Research-Space Physics,* Vol. 106, No. 3, March 1. As found at http://smallcomets.physics.uiowa.edu/iro/iro_ground.html.

Friebele, E. 1997. "In brief: Dehydration." *EOS, Transactions,* Vol. 78, No. 31, August 5.

Gaidos, E.J., K.H. Nealson, and J.L. Kirschvink. 1999. "Life in ice-covered oceans." *Science,* Vol. 284, No. 5,420, 4 June 1999, pp. 1,631–33.

Hall, A. 1998. "Moon ice." *Scientific American,* March 9. As found at www.sciam.com/explorations/1998/0309moon.

Harwit, M., D.A. Neufeld, G.J. Melnick, and M.J. Kaufman. 1998. "Thermal water vapor emission from shocked regions in Orion." *Astrophysical Journal,* Vol. 497, No. 2, April 20, p. L105.

Hathaway, J. 2000. "Meteorite research indicates Mars had Earth-like oceans." ASU Press Release, June 23. As found at www.asu.edu/asunews/Releases/Meteorite0600.htm.

Herrnstein, J.R., J.M. Moran, L.J. Greenhill, P.J. Diamond, M. Inoue, N. Nakai, M. Miyoshi, C. Henkel, and A. Riess. 1999. "A geometric distance to the Galaxy NGC4258 from orbital motions in a nuclear gas disk." *Nature*, Vol. 400, August 5, pp. 539–41.

Hoppa, G.V., B.R. Tufts, R. Greenberg, and P.E. Geissler. 1999. "Formation of cycloidal features on Europa." *Science*, Vol. 285, September 17, pp. 1,899–1,902.

"In search of elusive little comets." 1988. *Science*, Vol. 240, June 10, pp. 1,403–4.

Ishihara, Y., N. Nakai, N. Iyomoto, K. Makishima, P. Diamond, and P. Hall. 2001. "Water-vapor maser emission from the Seyfert-2 Galaxy IC-2560: Evidence for a super-massive black hole." *Publications of the Astronomical Society of Japan*, Vol. 53, No. 2, April 25, pp. 215–25.

Kerr, R.A. 1997. "Spots confirmed, tiny comets spurned." *Science*, Vol. 276, No. 5,317, May 30, p. 1,333.

Kivelson, M.G., K.K. Khurana, C.T. Russell, M. Volwerk, R.J. Walker, and C. Zimmer. 2000. "Galileo magnetometer measurements: A stronger case for a subsurface ocean at Europa." *Science*, Vol. 289, August 25, pp. 1,340–43.

Kluger, J. 1997. "The gentle cosmic rain." *Time*, June 9, p. 52.

Kohk, P., M. Kaehny, D. Armstrong, and K. Burnside. 1997. "Europa II workshop report." *Moon Miners' Manifesto. Newsletter of the Artemis Society*, Vol. 110, November 1997. 7 pp. As found at www.asi.org/adb/06/09/03/02/110/europa2-wkshp.html.

Lipper, D. 2001. "Humans on Europa: A plan for colonies on the icy moon." Space.com. June 6. As found at www.space.com/missionlaunches/missions/europa_colonies_010606-1.html.

Lloyd, R. 1999. "Space water discoveries enhance odds for early life." CNN.com. September 8. As found at www.cnn.com/TECH/space/9909/08/space.water.folo/index.

Lloyd, R. 2001. "Mars' missing ocean: A new look at the Northern Plains." Space.com. April 5. As found at www.space.com/scienceastronomy/solarsystem/flat_mars_010405.html.

MacDonald, M. 1999. "Water trapped in meteorite gives hints to early solar system." Explorezone.com. August 30. As found at www.explorezone.com/archives/99_08/30_meteorite_water.htm.

McCord, T.B., G.B. Hansen, and C.A. Hibbitts. 2001. "Hydrated salt minerals on Ganymede's surface: Evidence of an ocean below." *Science*, Vol. 292, May 25, pp. 1,523–25.

Melnick, G.J., D.A. Neufeld, K.E. Saavik Ford, D.J. Hollenbach, and M.L.N. Ashby. 2001. "Discovery of water vapour around IRC+10216 as evidence for comets orbiting another star." *Nature*, Vol. 412, July 12, pp. 160–63.

Melnick, G.J., J.R. Stauffer, M.L.N. Ashby, E.A. Bergin, G. Chin, N.R. Erickson, P.F. Goldsmith, M. Harwit, J.E. Howe, S.C. Kleiner, D.G. Koch, D.A. Neufeld, B. Patten, R. Plume, R. Schieder, R.L. Snell, V. Tolls, Z. Wang, G. Winnewisser, and Y.F. Zhang. 2000. "The *Submillimeter Wave Astronomy Satellite*: Science objectives and instrument description." *Astrophysical Journal*, Vol. 539, No. 2, August 20, pp. L77–L85.

Moneti, A., J. Cernicharo, and J.R. Pardo. 2001. "Cold H2O and CO ice and gas toward the galactic center." *Astrophysical Journal*, Vol. 549, No. 2, March 10, pp. L203–L207.

NASA. 1999. "No water ice detected from Lunar Prospector impact." National Aeronautics and Space Administration Press Release 99-119, October 13. As found at ftp://ftp.hq.nasa.gov/pub/pao/pressrel/1999/99-119.txt.

NASA. 2000. "Making a splash on Mars." Science@NASA. June 29. As found at http://science.nasa.gov/headlines/y2000/ast29jun_1m.htm.

NASA. 2001. "Galileo succeeds in its closest flyby of a Jovian moon." 2001. NASA Press Release, May 25. As found at www.jpl.nasa.gov/galileo/news/release/press010525.html.

NASA Astrobiology Institute. 2001. "The case of the missing Mars water." Science@NASA. January 5. As found at http://science.nasa.gov/headlines/y2001/ast05jan_1.htm.

Recer, P. 2001. "Water found in deep space." Space.com. February 20. As found at www.space.com/scienceastronomy/astronomy/water_space_010220.html.

Roylance, F.D. 1998. "Astronomers observe water forming in the Orion Nebula." *Baltimore Sun*, April 14, p. 3.

"Small comets/Big flap." 1992. *Science*, Vol. 227, July 31, pp. 622–23.

Sorid, D. 2000. "Salt in meteorites may hold clues to how life formed." Space.com. June 8. As found at www.space.com/scienceastronomy/astronomy/purple_salt_000607.html.

Stevenson, D. 2000. "Europa's ocean-The case strengthens." *Science*, Vol. 289, No. 5483, August 25, pp. 1,305–7.

SWAS Press Release. 2001. "Stellar apocalypse yields first evidence of water-bearing worlds beyond our solar system." July 11, 2001. As found at www.harvard.edu/swas/pr010711/pr010711text.html.

Vedantam, S. 2001. "Water found in comets in distant solar systems." *Washington Post,* July 11, p. A2. As found at www.washingtonpost.com/ac2/wp-dyn/A46732-2001Jul11.

"Water droplets in meteorite may date to beginning of solar system." 1999. *U.S. Water News Online,* October. As found at www.uswaternews.com/archives/arcglobal/9watdro10.html.

Whitby, J., R. Burgess, G. Turner, J. Gilmour, and J. Bridges. 2000. "Extinct [129]I in halite from a primitive meteorite: Evidence for evaporite formation in the early solar system." *Science,* Vol. 288, No. 5472, June 9, pp. 1,819–21.

Withers, P., and G.A. Neumann. 2001. "Enigmatic Northern Plains of Mars." *Nature,* Vol. 410, April 5, p. 651.

Zolensky, M.E., R.J. Bodnar, E.K. Gibson Jr., L.E. Nyquist, Y. Reese, C.-Y. Shih, and H. Wiesmann. 1999. "Asteroidal water within fluid inclusion-bearing halite in an H5 chondrite, Monahans (1998)." *Science,* Vol. 285, No. 5432, August 27, pp. 1,377–79.

Zuber, M.T., S.C. Solomon, R.J. Phillips, D.E. Smith, G.L. Tyler, O. Aharonson, G. Balmino, W.B. Banerdt, J.W. Head, C.L. Johnson, F.G. Lemoine, P.J. McGovern, G.A. Neumann, D.D. Rowlands, and S. Zhong. 2000. "Internal structure and early thermal evolution of Mars from Mars Global Surveyor topography and gravity." *Science,* Vol. 287, March 10, pp. 1,788–93.

Water-Related Web Sites

The use of the Internet continues to mushroom for entertainment, commerce, research, and education. For those interested in any aspect of water resources, the Internet has become an increasingly crucial tool for finding references, data, and information. In the first two volumes of *The World's Water*, the Water Briefs sections contained lists of water-related web sites. We continue that policy with this updated list of both simple and sophisticated sites addressing every aspect of water resources. The following list is roughly broken into five categories:

- Governmental and Inter-Governmental
- Non-Governmental Organizations and Associations
- University and Educational
- Commercial
- Water Publications

We also, however, continue with our traditional warning:

Much that is useful is not on the Internet.
And much that is on the Internet is not useful.

Despite this caveat, more and more information is being made available electronically—a trend that we believe will only accelerate in the future. Beginning in 2002, the Pacific Institute will begin a comprehensive effort to digitize a wide range of water-related data (see the Water Data section) and make it available for free. We will also continue to expand our web links section.

By the time this book is published, some of the following Internet addresses may have changed, and new resources will certainly have become available. While this Web Directory will continue to be a part of future editions of *The World's Water*, the most up-to-date information will be found on the Internet itself. For that reason, the reader is encouraged to look at www.worldwater.org for current information and web addresses.

Water-Related Web Sites

Governmental and Inter-Governmental

UN FAO Water Information System	http://www.fao.org/waicent/faoinfo/agricult/agl/aglw/aquastat/aquastat.htm
Arizona Department of Water Resources	http://www.water.az.gov
Association of State Dam Safety Officials	http://www.damsafety.org
Australia Centre for Groundwater Studies	http://www.dwr.csiro.au/CGS
Australian CSIRO Division of Water Resources	http://www.dwr.csiro.au/
British Hydrological Society	http://www.salford.ac.uk/civils/BHS
California Department of Water Resources	http://www.dwr.water.ca.gov
California Environmental Resources Evaluation System (CERES): Water	http://ceres.ca.gov/cgi-bin/theme?keyword=Water%20resources
Center for Groundwater Studies	http://www.clw.csiro.au/CGS/home.htm
Co-operative Research Centre for Catchment Hydrology	http://www.catchment.crc.org.au
Colorado River Basin Forecast Center (NOAA)	http://www.cbrfc.noaa.gov
CSIRO Land and Water (Australia)	http://www.clw.csiro.au
Resources Links	http://ceres.ca.gov/cgi-bin/theme?keyword=Water%20resources
Environment Canada: Water	http://www.ec.gc.ca/water/index.htm
EPA Center for Environmental Statistics (surface water quality study U.S-Mexico border)	http://www.epa.gov/ceis
European-Mediterranean Information System on the Know-How in the Water	http://www.emwis.org
European Water Association	http://www.ewpca.de
FIVAS Association for International Water and Forest Studies (Norway)	http://www.solidaritetshuset.org/fivas
Global Applied Research Network (GARNET)	http://info.lut.ac.uk/departments/cv/wedc/garnet/grntback.html
Global Environment Monitoring System (GEMS), Freshwater Quality Programme, UNEP	http://www.cciw.ca/gems
Global International Waters Assessment (GIWA)	http://www.giwa.net
Global International Waters Assessment (GIWA) Links	http://www.giwa.net/links/links.phtml
Global Runoff Data Centre	http://www.bafg.de/grdc.htm
Global Water Partnership	http://www.gwp.sida.se
Global Water Partnership Forum	http://www.gwpforum.org
Groundwater Atlas of the United States, U.S. Geological Survey	http://sr6capp.er.usgs.gov/gwa/index.html
H2O-China	http://www.h2o-china.com
Hydrology and Water Resources Programme, World Meteorological Organization	http://www.wmo.ch/web/homs/hwrphome.html
Hydrologic Information Center: Current Hydrologic Conditions (NOAA)	http://www.nws.noaa.gov/oh/hic/current
International Boundary and Water Commission (IBWC): U.S. Section	http://www.ibwc.state.gov
International Geosphere Biosphere Programme	http://www.igbp.kva.se
International Hydrological Programme (UNESCO)	http://www.pangea.org/orgs/unesco
International Lake Environment Committee	http://www.ilec.or.jp
International Network of Basin Organizations	http://www.oieau.fr/riob/anglais/list_org.htm

Water-Related Web Sites *Continued*

Governmental and Inter-Governmental

International Office for Water (France)	http://www.oieau.fr/anglais/index.htm
Les Enjeux Internationaux De L'eau	http://www.mri.gouv.qc.ca/la_bibliotheque/eau/index.html
Mexican Institute of Water Technology (Spanish)	http://www.imta.mx
Ministerial Declaration of the Hague on Water Security in the 21st Century	http://www.worldwaterforum.net/Ministerial/declaration.html
Minnesota Ground Water Association	http://www.mgwa.org
NASA Laboratory for Hydrospheric Processes	http://hydro4.gsfc.nasa.gov/hydrological.html
National Water Commission of Mexico	http://www.cna.gob.mx
National Water Research Institute (Environment Canada)	http://www.cciw.ca/nwri/intro.html
Netherlands Water Partnership	http://www.nwp.nl
Nile Basin Initiative	http://www.nilebasin.org
Nile Basin Water Resources Digital Library	http://www.hydrosult.com/niledata
Norwegian Water Resources and Energy Directorate	http://webben.nve.no/english
Pacific Water Association	http://www.pwa.org.fj
Portuguese Water Institute	http://www.inag.pt
Programa Hidrologico Internacional (PHI): UNESCO	http://www.unesco.org.uy/phi
Reseau National des Donnes sur L'eau (French Water Data Network)	http://www.rnde.tm.fr/anglais/rnde.htm
Secretaria de Medio Ambiente, Recursos Naturales y Pesca (SEMARNAP)	http://www.semarnap.gob.mx
South African Department of Water Affairs and Forestry	http://www-dwaf.pwv.gov.za
South Africa Water Research Commission	http://www.wrc.org.za
Southern Africa Water Partnership	http://www.gwpsatac.org.zw
Texas Alliance of Groundwater Districts	http://www.texasgroundwater.org
Transboundary Resource Inventory Program (TRIP)	http://www.bic.state.tx.us/trip
U.S. Bureau of Water Reclamation	http://www.usbr.gov/water
U.S. Bureau of Reclamation's Water Conservation Page	http://www.usbr.gov/wrrl/rwc/rwc.html
U.S. Department of Agriculture: Water Quality Information Center: Water and Agriculture	http://www.nal.usda.gov/wqic
U.S. Environmental Protection Agency (USEPA)	http://www.epa.gov/watrhome
U.S. Environmental Protection Agency Surf Your Watershed	http://www.epa.gov/surf
U.S. Geological Survey: San Francisco Bay/Delta	http://sfbay.wr.usgs.gov
U.S. Geological Survey: National Water Information System (NWISWeb)	http://water.usgs.gov/nwis
U.S. Geological Survey: U.S. Water Data	http://water.usgs.gov/data.html
U.S. Geological Survey: U.S. Water Use Data	http://water.usgs.gov/watuse
U.S. Geological Survey: Water Resources of California	http://water.wr.usgs.gov
U.S. Geological Survey: Water Resources of the United States	http://water.usgs.gov

Water-Related Web Sites *Continued*

Governmental and Inter-Governmental

U.S. National Agricultural Library: Water Quality Information Center	http://www.nal.usda.gov/wqic
UNDP-World Bank Water and Sanitation Program	http://www.wsp.org/English/index.html
United Nations Development Programme (UNDP)	http://www.undp.org
United Nations Educational, Scientific, and Cultural Organization (UNESCO)	http://www.unesco.org
United Nations Environment Programme (UNEP)	http://www.unep.org
United Nations Environment Programme (UNEP) Freshwater Unit	http://www.unep.org/unep/program/ natres/water/fwu/home.htm
United Nations Environment Programme (UNEP) Global Environmental Monitoring System (GEMS)	http://www.cciw.ca/gems http://www.cciw.ca/gems
United Nations Food and Agriculture Organization (UNFAO)	http://www.fao.org
Vision21: Water for People	http://www.wsscc.org/vision21/wwf/ index.html
Water and the Forest Service (U.S. Forest Service)	http://www.fs.fed.us
Water Resources (India)	http://sdnp.delhi.nic.in/resources/ waterharvesting/water-frame.html
Water Research Commission, South Africa	http://www.wrc.org.za/default.htm
Water Supply and Sanitation Collaborative Council	http://www.wsscc.org/index.html
Water Quality Home Pages	http://hermes.ecn.purdue.edu/water
Western Water Policy Review Advisory Commission	http://www.den.doi.gov/wwprac
WHO Water, Sanitation and Health	http://www.who.int/ water_sanitation_health/index.htm
World Bank	http://www.worldbank.org
World Bank Group: Global Water Unit	http://www-esd.worldbank.org\water
World Bank Water Policy Reform Program	http://www.worldbank.org/html/ edi/edien.html
World Health Organization (WHO)	http://www.who.int/ctd
World Hydrological Cycle Observing System (WHYCOS)	http://www.wmo.ch/web/homs/ whycos.html
World Water Forum (The Hague)	http://www.worldwaterforum.org

Non-Governmental Organizations and Associations

African Water Page	http://www.africanwater.org
American Desalting Association	http://www.desalting-ada.org
American Institute of Hydrology	http://www.aihydro.org
American Rivers	http://www.amrivers.org
American Water Resources Association (United States)	http://www.awra.org
American Water Works Association (United States)	http://www.awwa.org

Water-Related Web Sites *Continued*

Non-Governmental Organizations and Associations

American Water Works Association
   Research Foundation (United States)   http://www.awwarf.com

Amigos Bravos: Friends of the Wild Rivers   http://www.newmex.com/amigosbravos

AquaRAP   http://www.conservation.org/aquarap

Boulder Area Sustainability
   Information Network   http://bcn.boulder.co.us/basin

British Hydrological Association   http://www.hydrology.org.uk

Canadian Water Resources Association   http://www.cwra.org

Centro del Agua del Trópico Humedo
   para America Latina y el Caribe   http://www.cathalac.org

Centre for Ecology and Hydrology   http://www.nwl.ac.uk/ih

Center for the Humid Tropics of
   Latin America and the Caribbean   http://www.cathalac.org

Coalition Eau Secours (France)   http://www.eausecours.org

Consejo para la Conservacion
   de Humedales de Norteamerica   http://uib.gym.itesm.mx/ine-nawcc

European Desalination Society   http://www.edsoc.com

European Rivers Network   http://www.rivernet.org

Foundation for Water Research   http://www.fwr.org

Freshwater Society   http://www.freshwater.org

Glen Canyon Institute   http://www.glencanyon.org

Global Water   http://www.globalwater.org

Green Cross International, Water Program   http://www.gci.ch/GreenCrossPrograms/
waterres/waterresource.html

Groundwater Foundation   http://www.groundwater.org

International Association of Hydraulic
   Engineering and Research (IAHR)   http://www.iahr.nl

International Association of Hydrogeologists   http://www.iah.org

International Association
   of Hydrological Sciences   http://www.wlu.ca/~wwwiahs/index.html

International Association on Water Quality   http://www.iawq.org.uk

International Commission on Irrigation
   and Drainage   http://www.icid.org

International Commission on Large Dams   http://genepi.louis-jean.com/cigb

International Water Commissions (various)   http://www.wlu.ca/~wwwiahs/
news-1295/comms.html#ICWQ

International Desalination Association   http://www.ida.bm

International Hydrological Programme   http://www.pangea.org/org/unesco

International Rivers Network   http://www.irn.org

International Water Academy   http://www.thewateracademy.org

International Water and Sanitation Centre   http://www.irc.nl

International Water Association   http://www.iawq.org.uk

International Water Management Institute (IWMI)   http://www.cgiar.org/iwmi

International Water Resources Association   http://www.iwra.siu.edu

The Irrigation Association   http://www.irrigation.org

Islamic Relief   http://www.irw.org/water.htm

LakeNet   http://www.worldlakes.org

Lifewater Canada   http://www.lifewater.ca

Middle East Desalination Research Center   http://www.medrc.org.om

Water-Related Web Sites *Continued*

Non-Governmental Organizations and Associations

National Ground Water Association (United States)	http://www.ngwa.org
New Zealand Hydrological Society	http://www.landcare.cri.nz/hosted/hydrosoc
New Zealand Water and Wastes Association	http://www.nzwwa.org.nz
NGOs and Freshwater Home Page (Earth Summit 2002)	http://www.earthsummit2002.org/freshwater
Nile Basin Initiative	http://www.nilebasin.org
Norway Institute for Water Research	http://www.niva.no/engelsk/niva/niva_hth.htm
Pacific Institute for Studies in Development, Environment, and Security	http://www.pacinst.org
River Network	http://www.rivernetwork.org
Solidarity Water Europe	http://www.s-w-e.org
Stockholm Environment Institute (SEI)	http://www.sei.se
Stockholm International Water Institute	http://www.siwi.org/menu/menu.html
Surfers Against Sewage	http://www.sas.org.uk
Terrene Institute	http://www.terrene.org
The Hydrogeologists Home Page	http://www.thehydrogeologist.com
United Nations University, International Network on Water, Environment, and Health	http://www.inweh.unu.edu
United States Society on Dams (USSD)	http://www.ussdams.org
WaterPartners International	http://www.water.org
Watershed Management Council	http://www.watershed.org
Water Aid	http://www.wateraid.org.uk/index.html
Water Education Foundation (United States)	http://www.water-ed.org
Water Engineering Development Centre	http://info.lut.ac.uk/departments/cv/wedc/index.html
Water Environment Federation	http://www.wef.org
Water for the People	http://www.water4people.org
Water Observatory	http://www.waterobservatory.org
Water Quality Association	http://www.wqa.org
West Bengal and Bangladesh Arsenic Crisis Information Centre	http://bicn.com/acic
World Commission on Dams (WCD)	http://www.dams.org
World Conservation Union (IUCN)	http://www.iucn.org
World Resources Institute, Water Resources and Freshwater Ecosystems	http://earthtrends.wri.org
Water Systems Council	http://www.watersystemscouncil.org/site/index.html
The World's Water	http://www.worldwater.org
World Water and Climate Atlas	http://www.iwmi.org/iwmihome.html
World Water Assessment Programme (WWAP)	http://www.unesco.org/water/wwap
World Water Council (WWC)	http://www.worldwatercouncil.org
World Water Vision for the 21st Century	http://www.watervision.org
World Wildlife Fund European Freshwater Programme	http://www.panda.org/europe/freshwater
World Wildlife Fund Living Waters Campaign	http://www.panda.org/livingwaters
WWF and the EU Water Framework Directive	http://www.wwffreshwater.org/initiatives/wfd.html

Water-Related Web Sites *Continued*

University and Educational

Arizona Water Resources Research Center, University of Arizona	http://ag.arizona.edu/azwater
Canberra Cooperative Research Centre for Freshwater Ecology (Australia)	http://enterprise.canberra.edu.au/WWW/www-crcfe.nsf
Centre for Water in Urban Areas (FSP-WIB)	http://www.FSP-WIB.TU-Berlin.de
Climate Change and U.S. Water Bibliography	http://www.pacinst.org/CCBib.html
Department of Water and Environmental Studies, Tema Institute at Linköping University (Sweden)	http://www.tema.liu.se/tema-v
Desalination Directory	http://www.desline.com
Drinking Water Activities for Teachers and Students, U.S. EPA	http://www.epa.gov/OGWDW/kids
Flood Information, Water Resources Engineering Website, University of Alberta	http://www.civil.ualberta.ca/water/MISC/OUTLINKS/floodlinks.html
Global Energy and Water Cycle Experiment (GEWEX)	http://www.gewex.com
Global Runoff Data Centre, Federal Institute of Hydrology (Germany)	http://www.bafg.de/grdc.htm
Great Lakes Information Network	http://www.great-lakes.net
Global Rivers Environmental Network (youth education)	http://www.earthforce.org/green
Hydrological Modeling Resources	http://hydromodel.com/duan/hydrology
Hydrology, Education Planet	http://www.educationplanet.com/search/Science_and_Engineering/Earth_Sciences/Hydrology
Hydrology Web	http://terrassa.pnl.gov:2080/EESC/resourcelist/hydrology.html
Hydrology Web, Links for Kids	http://terrassa.pnl.gov:2080/hydrology/kids.html
Info-Agua (Spanish)	http://www.infoagua.org
Institute for Groundwater Studies, the University of the Free State, South Africa	http://www.uovs.ac.za/igs
Institute of Hydrology (IH)	http://www.nwl.ac.uk/ih
Institute of Water Research, Michigan State University	http://www.iwr.msu.edu/index.html
Inter-American Water Resources Network	http://iwrn.net
International Ground Water Modeling Center	http://www.mines.edu/igwmc
International Institute for Infrastructural, Hydraulic and Environmental Engineering	http://www.ihe.nl
International Water History Association	http://www.iwha.net
International Water Law Project	http://www.internationalwaterlaw.org
Institute of Water and Environment, Cranfield University, Silsoe	http://www.silsoe.cranfield.ac.uk/iwe
IRC International Water and Sanitation Centre	http://www.irc.nl
Island Press Publisher, Environment	http://www.islandpress.org
KeyWATER	http://keywater.vub.ac.be
Learning to Be Water Wise and Efficient (fourth to eighth grade)	http://www.getwise.org/wwise/index.html
Middle East Water Information Network (MEWIN)	http://www.ssc.upenn.edu/~mewin
Montana Water	http://water.montana.edu/default.htm
Nature, Society and Water Programme, University of Bergen, Norway	http://www.getwise.org/wwise/index.html

Water-Related Web Sites *Continued*

University and Educational

New Mexico Water Resources Research Institute	http://wrri.nmsu.edu
Nile Basin Water Resources Digital Library	http://www.hydrosult.com/niledata/biblio.htm
OneWorld Water Think Tank	http://www.oneworld.org
Powell Consortium: Water Resource Research Institutes	http://wrri.nmsu.edu/powell
Research Centre for Sustainability in Ecological Engineering and Water Resources Technology	http://www.uws.edu.au/seewrt
Texas Water Resources Institute; Texas Waternet	http://twri.tamu.edu
The Global Water Sampling Projects (for students)	http://k12science.stevens-tech.edu/curriculum/waterproj
The Hydrologic Cycle, University of Illinois	http://ww2010.atmos.uiuc.edu/(Gh)/guides/mtr/hyd/home.rxml
The Water Institute, University of South Florida	http://water.grad.usf.edu
Transboundary Freshwater Dispute Database	http://terra.geo.orst.edu/users/tfdd
U.C. Berkley Center for Sustainable Resource Development	http://www.cnr.berkeley.edu/csrd/html/projects.html
U.S. Water News Online	http://www.uswaternews.com
Waternet	http://waternet.rug.ac.be
Universities Council on Water Resources (UCOWR)	http://www.uwin.siu.edu/ucowr/index.html
Universities Water Information Network	http://www.uwin.siu.edu
University College Dublin Centre for Water Resources Research	http://www.ucd.ie/~civileng/cwrr.html
University of London: Water Issues Group	http://endjinn.soas.ac.uk/geography/waterissues
University of South Australia: Water Policy and Law Group	http://business.unisa.edu.au/waterpolicylaw
Useful Internet Sites for Water Resources and GIS	http://civil.ce.utexas.edu/prof/maidment/gishyd97/library/websites/othr_web.htm
Virtual Irrigation Library	http://www.wiz.uni-kassel.de/kww/projekte/irrig/irrig_i.html
Water Education Resources, EcoIQ	http://www.ecoiq.com/water/#2
Water, Engineering & Development Center, Loughborough University (UK)	http://www.lboro.ac.uk/departments/cv/wedc

Water Information Web Sites

Water Information Organization, Colorado	http://www.waterinfo.org/interest.html
Waters Issues Group, School of Oriental and African Studies, University of London	http://www.soas.ac.uk/Geography/WaterIssues/Home.html
Water Librarian's Home Page	http://www.wco.com/~rteeter/waterlib.html
Water Magazine	http://www.watermagazine.com
Water Page	http://www.thewaterpage.com
Water Resources Databases	http://www.nal.usda.gov/wqic/dbases.html
Water Resources Worldwide Database	http://turboguide.com/cdprod1/cdhrec/010/283.shtml

Water-Related Web Sites *Continued*

Water Information Web Sites

Water Science for Schools, U.S. Geological Survey	http://wwwga.usgs.gov/edu
Water Resources Center Archives: University of California, Berkeley	http://www.lib.berkeley.edu/ WRCA/index.html
WateReuse Association of California (United States)	http://www.watereuse.org
Watershed Management Professional Program of Portland State University	http://www.eli.pdx.edu/Watershed
WaterWeb Consortium	http://www.waterweb.org
WaterWise: Water Efficiency Clearinghouse	http://www.waterwiser.org
Wetlist, Universities Water Information Network	http://www.uwin.siu.edu/WaterSites/ index.html
WhyFiles	http://whyfiles.org

Commercial

Anglian Water	http://www.anglianwater.co.uk
Atlantic Geoscience Corporation	http://www.geo-science.com
Azurix	http://www.azurix.com
Berliner Wasser Betriebe	http://www.bwb.de
BOSS International	http://www.bossintl.com
Bouygues	http://www.bouygues.fr
Brooke Water Services	http://www.brookwater.co.uk/index.html
Denver Water Company (United Statess)	http://www.water.denver.co.gov
Dragados	http://www.dragados.com
Groupe Suez Lyonnaise des Eaux	http://www.suez.fr/metiers/ english/index.htm
Hydrology Web Directory	http://www.webdirectory.com/ Science/Hydrology
Integrated Resource Management Research (IRMR)	http://www.catchment.com
Safe Water Systems	http://www.safewatersystems.com
Spragg Waterbag Homepage	http://www.waterbag.com
Stockholm, Vatten	http://www.stockholmvatten.se/ indexEng.htm
Thames Water	http://www.thames-water.com
Water Bank	http://www.waterbank.com
Water Strategist Community (WaterChat)	http://www.waterchat.com
Water Center.com	http://www.watercenter.com
Water Online	http://www.wateronline.com/content/ homepage/default.asp
Water Resources Publications, LLC	http://www.wrpllc.com
Water and Sanitation in Latin America	http://www.latinsectors.com/ sectors/water.htm
Water Utility Homepages: List at AWWA	http://www.awwa.org/asp/utility.asp
WL	Delft Hydraulics	http://www.wldelft.nl
World Business Council on Sustainable Development, Water Project	http://www.wbcsd.org/projects/ pr_water.htm

Water-Related Web Sites *Continued*

Water Publications

Advances in Water Resources	http://www.elsevier.nl/inca/publications/store/4/2/2/9/1/3
Agricultural Water Management	http://www.elsevier.nl/inca/publications/store/5/0/3/2/9/7/index.htt
Catch Water Newsletter (India), Center for Science and Environment	http://www.cseindia.org
Desalination	http://www.elsevier.nl/inca/publications/store/5/0/2/6/8/3/index.htt
European Water Management News	http://www.riza.nl/ewa_news/fr_ewa.html
Ground Water	http://www.ngwa.org/publication/grwtr.html
Ground Water Monitoring & Remediation	http://www.ngwa.org/publication/gwmrinfo.html
Hydrogeology Journal	http://link.springer.de/link/service/journals/10040/index.htm
Hydrological Processes	http://www.interscience.wiley.com/jpages/0885-6087
Hydrological Sciences Journal	http://www.wlu.ca/~wwwiahs/hsj/hsjindex.htm
International Journal of Water Resources Development	http://www.tandf.co.uk/journals/frameloader.html?http://www.tandf.co.uk/journals/carfax/07900627.html
Journal of Contaminant Hydrology	http://www.elsevier.nl/inca/publications/store/5/0/3/3/4/1/index.htt
Journal of Environmental Hydrology	http://www.hydroweb.com/jeh.html
Journal of Hydrology	http://www.elsevier.nl/inca/publications/store/5/0/3/3/4/3
Journal of the American Water Resources Association	http://www.awra.org/jawra/jawratoc_35_6.html
International News, International Office for Water (France)	http://www.oieau.fr/anglais/oie/news/flatest.htm
Safedrinkingwater.com	http://www.safedrinkingwater.com
September 2000 Clean Water News	http://www.amsa-cleanwater.org/pubs/cleanwater/sept00/september00.htm
Source: Water and Sanitation News	http://www.wsscc.org/source
Stormwater	http://www.stormh2o.com
The Aquifer	http://www.groundwater.org/Foundation/aquifer.htm
University of Denver Water Law Review	http://www.law.du.edu/waterlaw
Urban Water	http://www.urbanwater.net
U.S. Water News Online	http://www.uswaternews.com/homepage.html
WaterCom Engineering	http://www.watercom.ca
Water Law	http://www3.interscience.wiley.com/cgi-bin/jtoc?ID=60501795
Water Log Legal Reporter	http://www.olemiss.edu/orgs/masglp/waterlog.html
Water Policy	http://www.elsevier.nl/inca/publications/store/6/0/0/6/5/6/index.htt

Water-Related Web Sites *Continued*

Water Publications

Water Research	http://www.elsevier.nl/inca/publications/store/3/0/9/index.htt
Water Resources Research	http://www.agu.org/wrr
Water Science and Technology	http://www.elsevier.nl/inca/publications/store/4/6/4/index.htt
Water Strategist	http://www.waterstrategist.com
Water Supply	http://www.iwap.co.uk/journals/ws
Water Technology Online	http://www.waternet.com
Water Well Journal	http://www.ngwa.org/publication/wwjinfo.html
Water World	http://www.pennnet.com/home/home.cfm

Data Section

Table 1 Total Renewable Freshwater Supply, by Country (2002 Update)

Description

Average annual renewable freshwater resources are listed by country. All quantities are in cubic kilometers per year (km³/yr). These data represent average freshwater resources in a country—actual annual renewable supply will vary from year to year. The data typically include both renewable surface water and groundwater supplies, including surface inflows from neighboring countries. The United Nations Food and Agriculture Organization (FAO) refers to this as "total natural renewable water resources." Flows to other countries are not subtracted from these numbers—they thus represent the water made available by the natural hydrologic cycle, unconstrained by political, institutional, or economic factors. Updated since the last volume of *The World's Water* are most of the countries of Latin America and the Caribbean, Germany, and China. Interestingly, some of the countries most in need of updating are in Europe, such as Ireland and Denmark. Data are not available, for example, on the countries created by the breakup of Yugoslavia.

Limitations

As described in previous volumes of *The World's Water*, these detailed country data should be viewed, and used, with caution. The data come from different sources and were estimated over different periods. Many countries do not directly measure or report internal water-resources data, so some of these entries were produced using indirect methods. In the past few years, new assessments have begun to standardize definitions and assumptions, particularly the work of the UN FAO.

Not all of the annual renewable water supply is available for use by the countries to which they are credited here; some flows are committed to downstream users. For example, the Sudan is listed as having 154 cubic kilometers per year, but treaty commitments require it to pass significant flows downstream to Egypt. Other countries, such as Turkey, Syria, and France, to name only a few, also pass significant amounts of water to other users. The average annual figures also hide large seasonal, interannual, and long-term variations.

Sources (Compiled by P.H. Gleick)

nd: No data

a: Total natural renewable surface and groundwater. Typically includes flows from other countries. (FAO: "Natural total renewable water resources.")

b: Belyaev, 1987, Institute of Geography, National Academy of Sciences, Moscow, USSR.

c: UN FAO, 1995, *Water Resources of the African Countries: A Review*, Food and Agriculture Organization, United Nations, Rome.

d: World Resources Institute, 1994, *World Resources 1994–95*, in collaboration with the United Nations Environment Programme and the United Nations Development Programme, Oxford University Press, New York.

e: Margat, J., 1989, "The sharing of common water resources in the European Community (EEC)," *Water International* 14:59–91, as cited in P. Gleick (ed.), *Water in Crisis*, Oxford University Press, New York, Table A11.

f: Shahin, M., 1989, "Review and assessment of water resources in the Arab region," *Water International* 14 (4): 206–19, as cited in P. Gleick (ed.), *Water in Crisis*, Oxford University Press, New York, Table A17.

g: Goscomstat, USSR, 1989, *Protection of the Environment and Rational Utilization of Natural Resources in the USSR, Statistical Handbook*, Government Committee on Statistics, Moscow (in Russian), as cited in P. Gleick (ed.), *Water in Crisis*, Oxford University Press, New York, Table A16.

h: World Resources Institute, 1996, *World Resources 1996–97*, A Joint Publication of the World Resources Institute, United Nations Environment Programme, United Nations Development Programme, and the World Bank, Oxford University Press, New York.

i: Economic Commission for Europe, 1992, *Environmental Statistical Database*, the Environment in Europe and North America, United Nations, New York.

j: UN FAO, 1997, *Water Resources of the Near East Region: A Review*, Food and Agriculture Organization, United Nations, Rome, Italy.

k: UN FAO, 1997, *Irrigation in the Countries of the Former Soviet Union in Figures*, Rome, Italy.

l: UN FAO, 1999, *Irrigation in Asia in Figures*, Food and Agriculture Organization, United Nations, Rome, Italy.

m: Nix, H., 1995, *Water/Land/Life: The Eternal Triangle*, Water Research Foundation of Australia, Canberra, Australia.

n: UN FAO, 2000, *Irrigation in Latin America and the Caribbean*, Food and Agricultural Organization, United Nations, Rome, Italy.

p: AQUASTAT web site as of January 2002, www.fao.org.

q: Bundesministerium fur Umwelt, Naturschutz, und Reaktorsicherheit, 2001, *Water Resources Management in Germany, Part 1: Fundamentals*, Bonn, Germany (October).

TABLE 1   Total Renewable Freshwater Supply, by Country (2002 Update)

Region and Country	Annual Renewable Water Resources[a] (km³/yr)	Year of Estimate	Source of Estimate
AFRICA			
Algeria	14.3	1997	c,j
Angola	184.0	1987	b
Benin	25.8	1994	c
Botswana	14.7	1992	c
Burkina Faso	17.5	1992	c
Burundi	3.6	1987	b
Cameroon	268.0	1987	b
Cape Verde	0.3	1990	c
Central African Republic	141.0	1987	b
Chad	43.0	1987	b
Comoros	1.0	1987	b
Congo	832.0	1987	b
Congo, Democratic Republic (formerly Zaire)	1,019.0	1990	c
Cote D'Ivoire	77.7	1987	b
Djibouti	0.3	1997	j
Egypt	86.8	1997	j
Equatorial Guinea	30.0	1987	b
Eritrea	8.8	1990	c
Ethiopia	110.0	1987	b
Gabon	164.0	1987	b
Gambia	8.0	1982	c
Ghana	53.0	1970	c
Guinea	226.0	1987	b
Guinea-Bissau	27.0	1991	c
Kenya	30.2	1990	c
Lesotho	5.2	1987	b
Liberia	232.0	1987	b
Libya	0.6	1997	c,j
Madagascar	337.0	1984	c
Malawi	18.7	1994	c
Mali	100.0	2001	p
Mauritania	11.4	1997	c,j
Mauritius	2.2	2001	p
Morocco	30.0	1997	c,j
Mozambique	216.0	1992	c
Namibia	45.5	1991	c
Niger	32.5	1988	c
Nigeria	280.0	1987	b
Reunion	nd	nd	
Rwanda	6.3	1993	c
Senegal	39.4	1987	b
Sierra Leone	160.0	1987	b
Somalia	15.7	1997	j
South Africa	50.0	1990	c
Sudan	154.0	1997	c,j

TABLE 1 *Continued*

Region and Country	Annual Renewable Water Resources[a] (km³/yr)	Year of Estimate	Source of Estimate
AFRICA *(continued)*			
Swaziland	4.5	1987	b
Tanzania	89.0	1994	c
Togo	12.0	2001	p
Tunisia	4.1	1997	j
Uganda	66.0	1970	c
Zambia	116.0	1994	c
Zimbabwe	20.0	1987	b
NORTH AND CENTRAL AMERICA			
Antigua and Barbuda	0.1	2000	n
Bahamas	nd	nd	
Barbados	0.1	2000	n
Belize	18.6	2000	n
Canada	2901.0	1980	d
Costa Rica	112.4	2000	n
Cuba	38.1	2000	n
Dominica	nd	nd	
Dominican Republic	21.0	2000	n
El Salvador	25.3	2000	n
Grenada	nd	nd	
Guatemala	111.3	2000	n
Haiti	14.0	2000	n
Honduras	95.9	2000	n
Jamaica	9.4	2000	n
Mexico	457.2	2000	n
Nicaragua	196.7	2000	n
Panama	148.0	2000	n
St. Kitts and Nevis	0.02	2000	n
Trinidad and Tobago	3.8	2000	n
United States of America	2478.0	1985	d
SOUTH AMERICA			
Argentina	814.0	2000	n
Bolivia	622.5	2000	n
Brazil	8233.0	2000	n
Chile	922.0	2000	n
Colombia	2132.0	2000	n
Ecuador	432.0	2000	n
Guyana	241.0	2000	n
Paraguay	336.0	2000	n
Peru	1913.0	2000	n
Suriname	128.0	2000	n
Uruguay	139.0	2000	n
Venezuela	1233.2	2000	n
ASIA			
Afghanistan	65.0	1997	j
Bahrain	0.1	1997	j

TABLE 1 *Continued*

Region and Country	Annual Renewable Water Resources[a] (km³/yr)	Year of Estimate	Source of Estimate
Bangladesh	1210.6	1999	l
Bhutan	95.0	1987	b
Brunei	8.5	1999	l
Cambodia	476.1	1999	l
China	2829.6	1999	l
Cyprus	0.9	1997	d,j
India	1907.8	1999	l
Indonesia	2838.0	1999	l
Iran	137.5	1997	j
Iraq	96.4	1997	j
Israel	2.2	1986	d
Japan	430.0	1999	l
Jordan	0.9	1997	j
Korea DPR	77.1	1999	l
Korea Rep	69.7	1999	l
Kuwait	0.02	1997	j
Laos	331.6	1999	l
Lebanon	4.8	1997	j
Malaysia	580.0	1999	l
Maldives	0.03	1999	l
Mongolia	34.8	1999	l
Myanmar	1045.6	1999	l
Nepal	210.2	1999	l
Oman	1.0	1997	j
Pakistan	429.4	1997	j
Philippines	479.0	1999	l
Qatar	0.1	1997	j
Saudi Arabia	2.4	1997	j
Singapore	0.6	1975	d
Sri Lanka	50.0	1999	l
Syria	46.1	1997	j
Thailand	409.9	1999	l
Turkey	200.7	1997	j
United Arab Emirates	0.2	1997	j
Vietnam	891.2	1999	l
Yemen	4.1	1997	j
EUROPE			
Albania	21.3	1970	d
Austria	90.3	1980	d
Belgium	12.5	1980	e
Bosnia and Herzegovina	nd	nd	
Bulgaria	205.0	1980	d
Croatia	nd	nd	
Czech Republic	58.2	1990	h
Denmark	13.0	1977	e
Finland	113.0	1980	d

TABLE 1   *Continued*

Region and Country	Annual Renewable Water Resources[a] (km³/yr)	Year of Estimate	Source of Estimate
EUROPE *(continued)*			
France	198.0	1990	i
Germany	182.0	2001	q
Greece	58.7	1980	e
Hungary	120.0	1991	i
Iceland	170.0	1987	d
Ireland	50.0	1972	e
Italy	167.0	1990	h
Luxembourg	5.0	1976	e
Macedonia	nd	nd	
Malta	0.02	1997	j
Netherlands	90.0	1980	e
Norway	392.0	1991	i
Poland	56.2	1980	e
Portugal	69.6	1990	h
Romania	208.0	1980	d
Slovak Republic	30.8	1990	h
Slovenia	nd	nd	
Spain	111.3	1985	e
Sweden	180.0	1980	d
Switzerland	50.0	1985	d
United Kingdom	120.0	1980	e
Yugoslavia	265.0	1980	d
Russia	4498.0	1997	g,k
Armenia	10.5	1997	k
Azerbaidzhan	30.3	1997	k
Belarus	58.0	1997	k
Estonia	12.8	1997	k
Georgia	63.3	1997	k
Kazakhstan	109.6	1997	k
Kyrgyzstan	20.6	1997	k
Latvia	35.4	1997	k
Lithuania	24.9	1997	k
Moldavia	11.7	1997	k
Tadjikistan	16.0	1997	k
Turkmenistan	24.7	1997	k
Ukraine	139.5	1997	k
Uzbekistan	50.4	1997	k
OCEANIA			
Australia	398.0	1995	m
Fiji	28.6	1987	b
New Zealand	397.0	1995	m
Papua New Guinea	801.0	1987	b
Solomon Islands	44.7	1987	b

Table 2   Fresh Water Withdrawals, by Country and Sector (2002 Update)

Description

Data on water use by regions and by different economic sectors are still among the most sought-after and the most unreliable in the water resources area. The following table presents the most up-to-date data on total freshwater withdrawals by country in cubic kilometers per year and cubic meters per person per year, using estimated water withdrawal for the year noted and the United Nations population estimates (medium variant) by country for the year 2000. The table also gives the breakdown of that water use for the domestic, agricultural, and industrial sectors, in both percentage of total water use and cubic meters per person per year. The data sources are identified in the final column. Updated from the last volume of *The World's Water* are countries in Latin America and the Caribbean, China, and some countries in Africa and Europe.

The use of water varies greatly from country to country and from region to region. "Withdrawal" refers to water taken from a water source for use; it does not refer to water "consumed" in that use. The domestic sector typically includes household and municipal uses as well as commercial and governmental water use. The industrial sector includes water used for power plant cooling and industrial production. The agricultural sector includes water for irrigation and livestock.

Limitations

Extreme care should be used when applying these data—they are often the least reliable and most inconsistent of all water-resources information. They come from a wide variety of sources and are collected using different approaches, with few formal standards. As a result, this table includes data that are actually measured, are estimated, are modeled using different assumptions, or are derived from other data. The data also come from different years, making direct comparisons difficult. For example, some water-use data are over 20 years old. Separate data are now provided for the independent states of the former Soviet Union, but not for the former states of Yugoslavia. Industrial withdrawals for Panama, St. Lucia, St. Vincent, and the Grenadines are included in the domestic category.

Another major limitation of these data is that they do not include the use of rainfall in agriculture. Many countries use a significant fraction of the rain falling on their territory for agricultural production, but this water use is neither accurately measured nor reported. In the past several years, the United Nations Food and Agriculture Organization (UN FAO) has begun a systematic reassessment of water-use data, and there is reason to hope that over the next several years a more accurate picture of global, regional, and sectoral water use will emerge.

Sources

a: UN FAO, 1995, *Water Resources of African Countries,* Food and Agriculture Organization, United Nations, Rome, Italy.

b: UN FAO, 1997, *Water Resources of the Near East Region: A Review,* Food and Agriculture Organization, United Nations, Rome, Italy.

c: World Resources Institute, 1994, *World Resources 1994–95,* in collaboration with the United Nations Environment Programme and the United Nations Development Programme, Oxford University Press, New York.

d: UN FAO, 1997, *Irrigation in the Countries of the Former Soviet Union in Figures,* Rome, Italy.

e: U.S. Geological Survey, 1997, "Estimated use of water in the United States in 1995," U.S. Department of the Interior, U.S. Geological Survey.

f: Eurostat Yearbook, 1997, *Statistics of the European Union,* EC/C/6/Ser.26GT, Luxembourg.

g: World Resources Institute, 1990, *World Resources 1990–1991,* Oxford University Press, New York.

h: UN FAO, 1999, *Irrigation in Asia in Figures,* Food and Agriculture Organization, United Nations, Rome, Italy.

i: Nix, H., 1995, *Water/Land/Life,* Water Research Foundation of Australia, Canberra.

j: UN FAO, 2000, *Irrigation in Latin America and the Caribbean,* Food and Agricultural Organization, United Nations, Rome, Italy.

k: AQUASTAT web site, January 2002, www.fao.org.

l: Ministry of Water Resources, China, 2001, *Water Resources Bulletin of China,* 2000, People's Republic of China, Beijing, China (September).

m: Geological Survey of Denmark and Greenland, 1997, *National Water Resources Model of Denmark,* available at http://www.geus.dk/publications/geo-nyt-geus/Gi97-2-uk.pdf.

TABLE 2   Freshwater Withdrawals, by Country and Sector (2002 Update)

Region and Country	Year of Withdrawal Data	Total Freshwater Withdrawal (km³/yr)	Estimated Year 2000 Per Capita Withdrawal (m³/p/yr)	USE Domestic (%)	Industrial (%)	Agricultural (%)	Domestic m³/p/yr	Industrial m³/p/yr	Agricultural m³/p/yr	Source for Water Use Data	Estimated Population Year 2000 (millions)
AFRICA											
Algeria	1990	4.50	142	25	15	60	35	22	85	a	31.60
Angola	1987	0.48	38	14	10	76	5	4	29	a	12.80
Benin	1994	0.15	23	23	10	67	5	2	16	a	6.20
Botswana	1992	0.11	70	32	20	48	22	14	33	a	1.62
Burkina Faso	1992	0.38	31	19	0	81	6	0	25	a	12.06
Burundi	1987	0.10	14	36	0	64	5	0	9	a	6.97
Cameroon	1987	0.40	26	46	19	35	12	5	9	a	15.13
Cape Verde	1990	0.03	59	10	19	88	6	11	51	a	0.44
Central African Republic	1987	0.07	19	21	5	74	4	1	14	a	3.64
Chad	1987	0.18	25	16	2	82	4	0	20	a	7.27
Comoros	1987	0.01	14	48	5	47	7	1	7	g	0.71
Congo	1987	0.04	13	62	27	11	8	4	1	a	2.98
Congo, Democratic Republic (formerly Zaire)	1990	0.36	7	61	16	23	4	1	2	a	51.75
Cote D'Ivoire	1987	0.71	47	22	11	67	10	5	31	a	15.14
Djibouti	1985	0.01	11	13	0	87	1	0	9	a	0.69
Egypt	1993	55.10	809	6	8	86	49	65	696	k	68.12
Equatorial Guinea	1987	0.01	22	81	13	6	18	3	1	a	0.45
Ethiopia (and Eritrea)	1987	2.20	31	11	3	86	3	1	27	a	69.99
Gabon	1987	0.06	49	72	22	6	35	11	3	a	1.24
Gambia	1982	0.02	16	7	2	91	1	0	15	a	1.24
Ghana	1970	0.30	15	35	13	52	5	2	8	a	19.93
Guinea	1987	0.74	94	10	3	87	9	3	82	a	7.86
Guinea-Bissau	1991	0.02	14	60	4	36	8	1	5	a	1.18

245

TABLE 2 *Continued*

Region and Country	Year of Withdrawal Data	Total Freshwater Withdrawal (km³/yr)	Estimated Year 2000 Per Capita Withdrawal (m³/p/yr)	USE Domestic (%)	Industrial (%)	Agricultural (%)	Domestic m³/p/yr	Industrial m³/p/yr	Agricultural m³/p/yr	Source for Water Use Data	Estimated Population Year 2000 (millions)
AFRICA *(continued)*											
Kenya	1990	2.05	68	20	4	76	13	3	52	a	30.34
Lesotho	1987	0.05	22	22	22	56	5	5	12	a	2.29
Liberia	1987	0.13	40	27	13	60	11	5	24	a	3.26
Libya	1994	4.60	720	11	2	87	78	16	626	a	6.39
Madagascar	1984	16.30	937	1	0	99	9	0	928	a	17.40
Malawi	1994	0.94	85	10	3	86	9	3	74	a	10.98
Mali	1987	1.36	108	2	1	97	2	1	105	a	12.56
Mauritania	1985	1.63	632	6	2	92	39	11	581	a	2.58
Mauritius	1995	0.62	522	18	8	75	94	42	391	k	1.18
Morocco	1991	11.05	381	5	3	92	19	11	351	a	28.98
Mozambique	1992	0.61	31	9	2	89	3	1	28	a	19.56
Namibia	1991	0.249	144	29	3	68	41	5	98	a	1.733
Niger	1988	0.50	46	16	2	82	7	1	38	a	10.81
Nigeria	1987	3.63	28	31	15	54	9	4	15	a	128.79
Rwanda	1993	0.77	100	5	2	94	5	2	94	a	7.67
Senegal	1987	1.36	143	5	3	92	7	4	132	a	9.50
Sierra Leone	1987	0.37	76	7	4	89	5	3	68	a	4.87
Somalia	1987	0.81	70	3	0	97	2	0	68	a	11.53
South Africa	1990	13.31	288	17	11	72	49	31	207	a	46.26
Sudan	1995	17.80	597	4	1	94	27	7	563	a	29.82
Swaziland	1980	0.66	667	2	2	96	11	16	640	a	0.98
Tanzania	1994	1.17	35	9	2	89	3	1	31	a	33.69
Togo	1987	0.09	19	62	13	26	12	3	5	a	4.68
Tunisia	1990	3.08	313	9	3	89	27	9	277	a	9.84
Uganda	1970	0.20	9	32	8	60	3	1	5	a	22.46

	Year										
Zambia	1994	1.71	187	16	7	77	30	13	144	a	9.13
Zimbabwe	1987	1.22	98	14	7	79	14	7	78	a	12.42
NORTH AND CENTRAL AMERICA											427.23
Antigua and Barbuda	1990	0.005	75	60	20	20	45	15	15	j	0.07
Barbados	1996	0.08	312	77	0	23	240	0	72	j	0.26
Belize	1993	0.095	396	12	88	0	48	348	0	j	0.24
Canada	1990	43.89	1,431	11	80	8	157	1144	114	f	30.68
Costa Rica	1997	5.77	1,520	13	7	80	198	106	1216	j	3.80
Cuba	1995	5.21	465	49	0	51	228	0	237	j	11.20
Dominica	1996	0.02	239	0	0	100	0	0	239	k	0.07
Dominican Republic	1994	8.34	982	11	0	89	108	0	874	j	8.50
El Salvador	1992	0.73	115	34	20	46	39	23	53	j	6.32
Guatemala	1992	1.16	95	9	17	74	9	16	70	j	12.22
Haiti	1991	0.98	125	5	1	94	6	1	118	j	7.82
Honduras	1992	1.52	234	4	5	91	9	12	213	j	6.49
Jamaica	1993	0.90	348	15	7	77	52	24	268	j	2.59
Mexico	1998	77.81	787	17	5	78	134	39	614	j	98.88
Nicaragua	1998	1.29	274	14	2	84	38	5	230	j	4.69
Panama	1990	1.64	575	28	2	70	161	12	403	j	2.86
St. Lucia	1997	0.01	89	100	0	0	89	0	0	j	0.15
St. Vincent and the Grenadines	1995	0.01	88	100	0	0	88	0	0	j	0.11
Trinidad and Tobago	1997	0.30	221	68	26	6	151	58	13	j	1.34
United States of America	1995	469.00	1,688	12	46	42	203	777	709	e	277.83
SOUTH AMERICA											296.72
Argentina	1995	28.58	772	16	9	75	123	69	579	c	37.03
Bolivia	1987	1.21	145	10	3	87	15	4	126	c	8.33
Brazil	1992	54.87	324	21	18	61	68	58	198	c	169.20
Chile	1987	20.29	1,334	5	11	84	67	147	1120	c	15.21
Colombia	1996	8.94	230	59	4	37	136	9	85	c	38.91
Ecuador	1997	16.99	1,343	12	6	82	161	81	1101	c	12.65

TABLE 2   Continued

Region and Country	Year of Withdrawal Data	Total Freshwater Withdrawal (km³/yr)	Estimated Year 2000 Per Capita Withdrawal (m³/p/yr)	USE Domestic (%)	Industrial (%)	Agricultural (%)	USE Domestic m³/p/yr	Industrial m³/p/yr	Agricultural m³/p/yr	Source for Water Use Data	Estimated Population Year 2000 (millions)
SOUTH AMERICA *(continued)*											
Guyana	1992	1.46	1,670	1	0	99	17	0	1654	c	0.87
Paraguay	1987	0.43	78	15	7	78	12	5	61	c	5.50
Peru	1992	18.97	739	7	7	86	52	52	636	c	25.66
Suriname	1987	0.46	1,018	6	5	89	61	51	906	c	0.45
Uruguay	1965	0.65	199	6	3	91	12	6	181	c	3.27
Venezuela	1970	4.10	170	44	10	46	75	17	78	c	24.17
ASIA											3,112.70
Afghanistan	1991	26.11	1,020	1	0	99	10	0	1010	b	25.59
Bahrain	1991	0.24	387	39	4	56	151	15	217	b	0.62
Bangladesh	1990	14.64	114	12	2	86	14	2	98	h	128.31
Bhutan	1987	0.02	10	36	10	54	4	1	5	h	2.03
Brunei	1994	0.92	2,788	nd	nd	nd	4	h		h	0.33
Cambodia	1987	0.52	46	5	1	94	2	0	44	h	11.21
China	2000	549.76	431	11	21	69	45	89	297	l	1,276.30
Cyprus	1993	0.21	267	7	2	91	19	5	243	b	0.79
India	1990	500.00	497	5	3	92	25	15	457	h	1,006.77
Indonesia	1990	74.35	350	6	1	93	21	3	325	h	212.57
Iran	1993	70.03	916	6	2	92	55	18	843	b	76.43
Iraq	1990	42.80	1,852	3	5	92	56	93	1704	b	23.11
Israel	1990	1.70	280	16	5	79	45	14	221	c	6.08
Japan	1992	91.40	723	19	17	64	137	123	463	h	126.43
Jordan	1993	0.98	155	22	3	75	34	5	117	b	6.33
Korea Democratic People's Republic	1987	14.16	592	11	16	73	65	95	432	h	23.91

Country	Year										
Korea Rep	1994	23.67	505	26	11	63	131	56	318	h	46.88
Kuwait	1994	0.54	274	37	2	60	101	5	164	b	1.97
Laos	1987	0.99	174	8	10	82	14	17	143	h	5.69
Lebanon	1994	1.29	393	28	4	68	110	16	267	b	3.29
Malaysia	1995	12.73	571	10	13	77	57	74	440	h	22.30
Maldives	1987	0.003	10	98	2	0	10	0	0	h	0.29
Mongolia	1993	0.43	157	20	27	53	31	42	83	h	2.74
Myanmar	1987	3.96	80	7	3	90	6	2	72	h	49.34
Nepal	1994	28.95	1,189	1	0	99	12	0	1177	h	24.35
Oman	1991	1.22	450	5	2	94	23	9	423	b	2.72
Pakistan	1991	155.60	997	2	2	97	20	20	967	b	156.01
Philippines	1995	55.42	739	8	4	88	59	30	650	h	75.04
Qatar	1994	0.28	476	23	3	74	109	14	352	b	0.60
Saudi Arabia	1992	17.02	786	9	1	90	71	8	707	b	21.66
Singapore	1975	0.19	53	45	51	4	24	27	2	c	3.59
Sri Lanka	1990	9.77	519	2	2	96	10	10	498	h	18.82
Syria	1993	14.41	894	4	2	94	36	18	840	b	16.13
Thailand	1990	33.13	548	5	4	91	27	22	498	h	60.50
Turkey	1992	31.60	481	16	11	72	77	53	346	b	65.73
United Arab Emirates	1995	2.11	863	24	9	67	207	78	578	b	2.44
Vietnam	1990	54.33	674	4	10	86	27	67	580	h	80.55
Yemen	1990	2.93	162	7	1	92	11	2	149	c	18.12
EUROPE											498.37
Albania	1970	0.20	57	6	18	76	3	10	44	c	3.49
Austria	1991	2.52	304	19	73	8	58	222	24	f	8.29
Belgium	1990	9.00	877	11	85	4	97	746	35	f	10.26
Bulgaria	1988	13.90	1,673	3	75	22	50	1255	368	k	8.31
Czech Republic	1991	2.74	269	23	68	9	62	183	24	c	10.20
Denmark	1995	1.00	190	30	27	43	57	51	82	m	5.27
Finland	1994	2.43	469	12	85	3	56	399	14	f	5.18

TABLE 2 Continued

Region and Country	Year of Withdrawal Data	Total Freshwater Withdrawal (km³/yr)	Estimated Year 2000 Per Capita Withdrawal (m³/p/yr)	USE Domestic (%)	Industrial (%)	Agricultural (%)	Domestic m³/p/yr	Industrial m³/p/yr	Agricultural m³/p/yr	Source for Water Use Data	Estimated Population Year 2000 (millions)
EUROPE (continued)											
France	1994	34.88	591	16	69	15	94	407	89	f	59.06
Germany	1990	58.85	712	14	68	18	100	484	128	f	82.69
Greece	1990	6.00	566	8	29	63	45	164	357	f	10.60
Hungary	1991	6.81	694	9	55	36	62	382	250	c	9.81
Iceland	1994	0.16	567	31	63	6	176	357	34	f	0.28
Ireland	1990	1.20	336	16	74	10	54	248	34	f	3.57
Italy	1990	56.20	983	14	27	59	138	265	580	f	57.19
Luxembourg	1994	0.06	133	42	45	13	56	60	17	f	0.43
Malta	1995	0.06	147	87	1	12	128	1	18	b	0.38
Netherlands	1991	7.80	491	5	61	34	25	300	167	f	15.87
Norway	1985	2.03	461	20	72	8	92	332	37	f	4.41
Poland	1991	12.28	317	16	60	24	51	190	76	f	38.73
Portugal	1990	7.29	745	15	37	48	112	276	357	f	9.79
Romania	1994	26.00	1,155	8	33	59	92	381	682	c	22.51
Slovak Republic	1991	1.78	331								5.37
Spain	1994	33.30	837	12	26	62	100	218	519	f	39.80
Sweden	1994	2.96	333	36	55	9	120	183	30	f	8.90
Switzerland	1994	2.60	351	23	73	4	81	256	14	f	7.41
United Kingdom	1994	11.75	201	20	77	3	40	155	6	f	58.34
Yugoslavia	1980	8.77	368	16	72	12	59	265	44	c	23.81
FORMER SOVIET UNION											
Armenia	1994	2.93	800	30	4	66	240	32	528	d	3.66
Azerbaijan	1995	16.53	2,112	5	25	70	106	528	1478	d	7.83
Belarus	1990	2.73	265	22	43	35	58	114	93	d	10.28

Estonia	1995	0.16	113	56	39	5	63	44	6	d	1.42
Georgia	1990	3.47	640	21	20	59	134	128	378	d	5.42
Kazakhstan	1993	33.67	1,989	2	17	81	40	338	1611	d	16.93
Kyrgyz Republic	1994	10.09	2,221	3	3	94	67	67	2088	d	4.54
Latvia	1994	0.29	121	55	32	13	67	39	16	d	2.40
Lithuania	1995	0.25	68	81	16	3	55	11	2	d	3.69
Moldova	1992	2.96	664	9	65	26	60	432	173	d	4.46
Russian Federation	1994	77.10	527	19	62	20	100	327	105	d	146.20
Tajikistan	1994	11.87	1,855	3	4	92	56	74	1707	d	6.40
Turkmenistan	1994	23.78	5,309	1	1	98	53	53	5203	d	4.48
Ukraine	1992	25.99	512	18	52	30	92	266	153	d	50.80
Uzbekistan	1994	58.05	2,320	4	2	94	93	46	2181	d	25.02
OCEANIA											26.48
Australia	1995	17.80	945	15	10	75	142	94	709	I	18.84
Fiji	1987	0.03	35	20	20	60	7	7	21	c	0.85
New Zealand	1991	2.00	532	46	10	44	245	53	234	c	3.76
Papua New Guinea	1987	0.10	21	29	22	49	6	5	10	c	4.81
Solomon Islands	1987		40	20	40	0	0	0	c	0.44	

Table 3   Access to Safe Drinking Water by Country, 1970 to 2000

Description

Safe drinking water is one of the most basic human requirements. As a result, estimates of access to safe drinking water are a cornerstone of most international assessments of progress—or lack thereof—toward solving global and regional water problems. The provision of safe drinking water for all was the goal of the United Nations' International Drinking Water Supply and Sanitation Decade (the Decade) from 1981 to 1990. In 2000, the international water community set a new goal of reducing by half by 2015 the proportion of people unable to reach or afford safe drinking water.

Data are given here for the percentage of urban, rural, and total populations, by country, with access to safe drinking water for 1970, 1975, 1980, 1985, 1990, 1994, and 2000—the most recent year for which data are available. The World Health Organization (WHO) collected the data presented here over various periods. Most of the data presented were drawn from responses by national governments to WHO questionnaires. Participants in data collection include the Joint Monitoring Programme (JMP) of WHO, the United Nations Children's Fund, and the Water Supply and Sanitation Collaborative Council, which has continued sector monitoring and aims to support and strengthen the monitoring efforts of individual countries.

The most recent data (WHO 2000) reflect a significant change in definition. Data are now reported for populations without access to "improved" water supply. The 40 largest countries in the developing world account for 90 percent of population in these regions. As a result, WHO spent extra effort to collect comprehensive data for these countries. According to WHO, the following technologies were included in the assessment as representing "improved" water supply:

  Household connection

  Public standpipe

  Borehole

  Protected dug well

  Protected spring

  Rainwater collection

Limitations

A review of water and sanitation coverage data from the 1980s and 1990s shows that the definition of safe, or improved, water supply and sanitation facilities differs from one country to another and for a given country over time. Indeed, some of the data from individual countries often showed rapid and implausible changes in the level of coverage from one assessment to the next. This indicates that some of the data are also unreliable, irrespective of the definition used. Countries used their own definitions of "rural" and "urban."

For the 1996 data, two-thirds of the countries reporting indicated how they defined "access." At the time, the definition most commonly centered on walking distance or time from household to water source, such as a public standpipe, which varied from 50 to 2,000 meters and 5 to 30 minutes. Definitions sometimes included considerations of quantity, with the acceptable limit ranging from 15 to 50 liters per capita per day. The WHO considers safe drinking water to be treated surface water or untreated water from protected springs, boreholes, and wells.

The 2000 WHO Assessment attempts to shift from gathering information from water providers only to include consumer-based information. The current approach uses household surveys in an effort to assess the actual use of facilities. "Reasonable access" was broadly defined as the availability of at least 20 liters per person per day from a source within one kilometer of the user's dwelling. A drawback of this approach is that household surveys are not conducted regularly in many countries. Thus, direct comparisons between countries, and across time within the same country, are difficult. Direct comparisons are additionally complicated by the fact that these data hide disparities between regions and socioeconomic classes.

Access to water, as reported by WHO 2000, does not imply that the level of service or quality of water is "adequate" or "safe." The assessment questionnaire did not include any methodology for discounting coverage figures to allow for intermittence or poor quality of the water supplies. However, the instructions stated that piped systems should not be considered "functioning" unless they were operating at over 50 percent capacity on a daily basis; and that handpumps should not be considered "functioning" unless they were operating for at least 70 percent of the time with a lag between breakdown and repair not exceeding two weeks. These aspects were taken into consideration when estimating coverage for countries for which national surveys had not been conducted. More details of the methods used, and their limitations, can be found at http://www.who.int/water_sanitation_health/Globassessment/globalAnnex.htm.

Sources

United Nations Environment Programme, 1989, *Environmental Data Report*, GEMS Monitoring and Assessment Research Centre, Basil Blackwell, Oxford.

United Nations Environment Programme, 1993–94, *Environmental Data Report*, GEMS Monitoring and Assessment Research Centre in cooperation with the World Resources Institute and the UK Department of the Environment, Basil Blackwell, Oxford.

WHO 1996, *Water Supply and Sanitation Sector Monitoring Report: 1996 (Sector Status as of 1994)*, in collaboration with the Water Supply and Sanitation Collaborative Council and the United Nations Children's Fund, UNICEF, New York.

WHO 2000, *Global Water Supply and Sanitation Assessment 2000 Report*, available in full at http://www.who.int/water_sanitation_health/Globassessment/GlobalTOC.htm.

WRI 1988, World Health Organization data, cited by the World Resources Institute, 1988, *World Resources 1988–89*, World Resources Institute and the International Institute for Environment and Development in collaboration with the United Nations Environment Programme, Basic Books, New York.

TABLE 3 Access to Safe Drinking Water by Country, 1970 to 2000

Region and Country	Fraction of Population With Access to Drinking Water																				
	Urban							Rural							Total						
	1970	1975	1980	1985	1990	1994	2000	1970	1975	1980	1985	1990	1994	2000	1970	1975	1980	1985	1990	1994	2000
AFRICA																					
Algeria	66	68	69	77			88		61			55		94		77		68			94
Angola	84	100	85	85	73	69	34			10	15	20	15	40			26	33	35	32	38
Benin	83		26	87	73	41	74	20	20	15	34	43	53	55	29	34	18	50	54	50	63
Botswana	71	95		80	100		100	26	39	15	46	88			29	45		53	91		
Burkina Faso	35	50	27	84			84	10	23	31	69	70	49		12	25	31	67		78	
Burundi	77		90	43	92	92	96			20	21	43	49	42			23	25	45	52	
Cameroon	77			98			82	21		21	24	45		42	32			32	44	52	62
Cape Verde			100	83	42	70	64				50		34	89			25	52		51	74
Central African Republic				13	19	18	80					26	18	43					23	18	60
Chad	47	43				48	31	24	23				17	26	27	26				24	27
Comoros							98							95							96
Congo	63	81	42	52	68	37	71	6	9	7		24	23	17	27	38	20	32	36	27	51
Congo, Democratic Rep.	33	38			57	59	89	4	12		21			26	11	19					45
Cote D'Ivoire	98		50	50		77	90	29		20		80	81	65	44					72	77
Djibouti						82	100						100	100			43	45		90	100
Egypt	94		88		95	82	96	93		64		86	50	94	93		84		90	64	95
Equatorial Guinea			47		65	88	45					18	100	42				32		95	43
Eritrea							63							42							46
Ethiopia	61	58		69		67	77	3	1		9			13	6	8		16			24
Gabon						14	73					48		55							70
Gambia	97		85	97	100		80				50	48		53	12			59	60	76	62
Ghana	86	86	72	93	63	70	87		14	33	39	37	49	49	35	35	45	56	21	56	64
Guinea	68	69	69	41	100	61	72			2	12		62	36		14	15	18	53	62	48
Guinea-Bissau			18	17		38	29			8	22		57	55			10	21		53	49
Kenya	100	100	85			67	87	2	4	15			49	31	15	17	26			53	49
Lesotho	100	65	37	65		14	98	1		11	30		64	88	3	17	15	36		52	91
Liberia	100			100		58		6	14		23		8		15		15	53		30	
Libya	100	100	100				72	42	82	90				68	58	87	96				72

Madagascar	67	80	81	85	83		17		31	10	31	11	25	21	31		29	47		
Malawi	76	77	97	95	52		50	37	44	44		41			56		45	57		
Mali	29	37	46	74	36	41	10			38	61			84	16	13	37	65		
Mauritania	98	80	73	34	84		100	85		69	40	17		99	100	100	76	37		
Mauritius	100	100	100	100	95	100	29	98	100	100	100	61	60			59	98	100		
Morocco	92	100	100	86	98	18	28		25	14	58	51			15	56	52	82		
Mozambique			38		17				9	40	43						32	60		
Namibia	37		90	100	87	37			37	42	67				52	52	57	77		
Niger	36	41	35	98	46	45	19	32	49	55	56	20	27	33	47	56	53	59		
Nigeria	100	100	100	70	63	22			20	26	39				38	49	39	57		
Reunion																68				
Rwanda	81	48	79	60		84	66	55	48		40	67	68	55	50	68		41		
Sao Tome and Principe									45						45					
Senegal	87	56	79	92	82	65		25	38	28	65			43	53	42	50	78		
Seychelles		77							95						95					
Sierra Leone	75	50	68	23	58	80	1	2	7	21	31	12	38	14	24	39	34	28		
Somalia	17	58		92			14	22	22		80	15			34		70	86		
South Africa							13	31		45	80	19	50	51				75		
Sudan	61		96	86	66				43	45	69		37		31		50			
Swaziland	83	100	100		41		9		7	44					31		43			
Tanzania	61	88	90	80		90	5	42	42	58	42	13	39	38	53	53	63	54		
Togo	100	49	100	85	74	70		31	41		38	17	16	60	54	54		54		
Tunisia	92	93	100	100	100		17	17	31	89		49	35		70		99			
Uganda	88	100	37	72	47	60	17	29	18	32	46	22			20	33	34	50		
Zambia	70	86	76	88	64		22	16	41	27	48	37	42		58		43	64		
Zimbabwe				100		95			32		77					84		85		

NORTH & CENTRAL
AMERICA & CARIBBEAN

   Anguilla

   Antigua and Barbuda

   Aruba

TABLE 3 *Continued*

Region and Country	Urban							Rural							Total						
	1970	1975	1980	1985	1990	1994	2000	1970	1975	1980	1985	1990	1994	2000	1970	1975	1980	1985	1990	1994	2000
NORTH & CENTRAL																					
AMERICA & CARIBBEAN *(continued)*																					
Bahamas	100	100	100	100	98		99	12	13			75		86	65	65	100	100	90		96
Barbados	95	100	99	100	100		100	100	100	98	99	100		100	98	100	99	99	100		100
Belize			99	100	95	96	83			36	26	53	82	69			68	64	74	89	76
British Virgin Islands					100																
Canada							100					100		99							100
Cayman Islands			100	98																	
Costa Rica	98	100	100	100		85	98	59	56	82	83		99	98	74	72	90	91		92	98
Cuba	82	96			100	96	99	15				91	85	82	56				98	93	95
Dominican Republic	72	88	85	85	82	74	83	14	27	34	33	45	67	70	37	55	60	62	67	71	79
Dominica							100							100							100
El Salvador	71	89	67	68	87	78	88	20	28	40	40	15	37	61	40	53	50	51	47	55	74
Grenada	100	100					97	47	77					93							94
Guadeloupe							94							94							94
Guatemala	88	85	90	72	92		97	12	14	18	14	43		88	38	39	46	37	62		92
Haiti		46	51	59	56	37	46		3	8	30	35	23	45		12	19	38	41	28	46
Honduras	99	99	93	56	85	81	97	10	13	40	45	48	53	82	34	41	59	49	64	65	90
Jamaica	100	100	55	99			81	48	79	46	93			59	62	86	51	96			71
Martinique																					
Mexico	71	70	90	99	94	91	94	29	49	40	47		62	63	54	62	73	83	69	83	86
Montserrat																					
Netherlands Antilles																					
Nicaragua	58	100	67	76		81	95	16	14	6	11		27	59	35	56	39	48		61	79
Panama	100	100	100	100			88	41	54	62	64			86	69	77	81	82		83	87
Puerto Rico																					
St Kitts																					

	Values (reading left to right)
St Lucia	100
St Vincent	79, 100
Trinidad/Tobago	100, 100, 100, 96, 93, 97, 98, 96, 86
Turks/Caicos Islands	87, 68, 77, 100
United States of America	100, 100, 100, 100, 100, 100
United States Virgin Islands	
SOUTH AMERICA	
Argentina	69, 76, 61, 63, 85, 95, 12, 17, 17, 26, 30, 56, 66, 54, 56, 53, 79
Bolivia	92, 81, 69, 75, 78, 93, 2, 10, 13, 6, 22, 30, 55, 33, 34, 36, 43, 53, 55, 79
Brazil	78, 87, 83, 95, 85, 76, 95, 28, 51, 56, 61, 54, 55, 70, 72, 77, 87, 72, 87
Chile	67, 78, 100, 98, 94, 99, 13, 17, 29, 37, 66, 56, 84, 87, 85, 94
Colombia	88, 86, 93, 100, 87, 88, 98, 28, 73, 76, 82, 48, 73, 63, 64, 86, 86, 76, 91
Ecuador	76, 67, 79, 81, 63, 82, 81, 7, 20, 31, 44, 55, 51, 34, 36, 50, 57, 55, 70, 71
Falkland Islands (Malvinas)	
French Guiana	88, 71, 84
Guyana	100, 100, 100, 90, 98, 63, 75, 60, 65, 71, 45, 91, 75, 84, 72, 76, 81, 61, 94
Paraguay	22, 25, 39, 53, 61, 95, 5, 5, 9, 8, 9, 58, 11, 13, 21, 28, 34, 79
Peru	58, 72, 68, 73, 68, 74, 87, 8, 15, 18, 17, 24, 24, 51, 35, 47, 50, 55, 55, 60, 77
Suriname	100, 71, 100, 94, 79, 94, 96, 92, 98, 88, 83, 95
Uruguay	100, 100, 96, 95, 100, 98, 59, 2, 27, 93, 92, 81, 85, 89, 98
Venezuela	92, 93, 93, 80, 88, 38, 53, 65, 36, 75, 58, 75, 86, 89, 79, 84
ASIA	
Afghanistan	18, 40, 28, 38, 40, 39, 19, 1, 8, 17, 19, 5, 11, 3, 9, 8, 17, 23, 12, 13
Armenia	
Azerbaijan	100, 100, 100, 100, 100, 94, 40, 0, 100, 99, 100, 100, 100
Bahrain	100, 100, 100, 99, 47, 61, 40, 49, 89, 97, 97, 45, 56, 39, 46, 81, 97, 97
Bangladesh	13, 22, 26, 24, 39, 86, 5, 19, 30, 54, 60, 32, 64, 62
Bhutan	50, 60, 75, 19, 7
Brunei Darus	100, 100, 95, 5
Cambodia	53, 25, 73, 30
China	87, 93, 94, 68, 89, 66, 73, 90, 75

TABLE 3   Continued

258

Fraction of Population With Access to Drinking Water

Region and Country	Urban							Rural							Total						
	1970	1975	1980	1985	1990	1994	2000	1970	1975	1980	1985	1990	1994	2000	1970	1975	1980	1985	1990	1994	2000
ASIA (*continued*)																					
Cyprus	100	94		100	100		100	92	96		100	100		100	95	95		100	100		100
East Timor																					
Gaza Strip																					
Georgia																					
Hong Kong			100		100					95		96							100		
India	60	80	77	76	86	85	92	6	18	31	50	69	79	86	17	31	42	56	73	81	88
Indonesia	10	41	35	43	35	78	91	1	4	19	36	33	54	65	3	11	23	38	34	62	76
Iran	68	76	82		100	89	99	11	30	50		75	77	89	35	51	66		89	83	95
Iraq	83	100		100	93		96	7	11	54	54	41		48	51	66		86	78	44	85
Israel																					
Japan																					
Jordan	98		100	100	100		100	59		65	88	97		84	77		86	96	99	89	96
Kazakhstan							98							82							91
Korea DPR							100							100							100
Korea Rep	84	95	86	90	100		97	38	33	61	48	76		71	58	66	75	75	93		92
Kuwait	60	100	86	97						100					51	89	87				
Kyrgyzstan							98							66							77
Laos	97	100	28		47	40	59	39	32	20		25	39	100	48	41	21		29	39	90
Lebanon						100	100						100	100						100	100
Macau																					
Malaysia	100	100	90	96	96			1	6	49	76	66		94	29	34	63	84	79		94
Maldives			11	58	77	98	100			3	12	68	86	100			2	21		89	100
Mongolia					100		77					58		30					82		60
Myanmar (Burma)	35	31	38	36	79	36	88	13	14	15	24	72	39	60	18	17	21	27	74	38	68
Nepal	53	85	83	70	66	66	85		5	7	25	34	41	80	2	8	11	28	38	44	81
Oman		100		90			41	48	48		49			30		52		53		63	39

Pakistan	88	60	55	44	35	25	21		84	52	42	27	20	5	4		96	77	82	83	72
	75	77																			
Philippines	87	85	81	52	45	50	36		80	77	72	54	43	31	20		92	93	93	49	49
	82	67																			
Qatar			100	94	71	97	95		64			88	43	83	75		100		100	100	76
	100	100																			
Saudi Arabia	95		60	100	90	64	49		80	47	55	29	87	56	37		100	97	100	100	92
Singapore	100			40	100	19	21						18	13	14		91		80	82	100
Sri Lanka	83	46			28		71		64	78			54		50		94	43			65
	36	46																		98	
Syria	80	85		64	74						85	66	63	16	10		89	92			98
		60																			
Tajikistan																			56		65
	69	60																			
Thailand	80		63			25	17		77			39					89			70	95
Turkey	83		76						84			25	62				82	53	47	100	95
Turkmenistan			92						78				81				96				
United Arab Emirates	85		92	45		69			50	32	33	39	32				81		70	100	
Uzbekistan	56	36		40	31		4		64			25	18		2		85		100		
Vietnam	69				52		57		64				25		43		85				
Yemen A R	45																				
Yemen Dem	88																				
OCEANIA																					
American Samoa	100								100								100				
Australia	100								100								100				
Cook Islands	100			92					100		100	88					100		100	99	100
Fiji	47				77	69	37		51		69		66				43	100	96	94	
French Polyneisa	100								100		18						100		100	100	
Guam									25		63		25				82		91	93	
Kiribati	47										45								100		
Marshall Islands			100								38										
Micronesia										100								100			
Nauru																					
New Caledonia																					
New Zealand																	100	100			
Niue	100	100	100						100	100	100	100	25	56	15			100		0	100
Northern Mariana Islands	100	100							100		0							100		100	

259

TABLE 3 *Continued*

Fraction of Population With Access to Drinking Water

Region and Country	Urban 1970	1975	1980	1985	1990	1994	2000	Rural 1970	1975	1980	1985	1990	1994	2000	Total 1970	1975	1980	1985	1990	1994	2000
OCEANIA *(continued)*																					
Palau					100		100					97		20							79
Papua New Guinea		30	55	95	94	84	88	72	19	10	15	20	17	32	70	20	16	26	32	28	42
Pitcairn																					
Samoa	86		97		100		95		23	94		77		100	17	43					99
Solomon Islands			96		82		94			45		58		65					62		71
Tokelau						100					100		100							100	
Tonga	100	100	86	99	92	100	100	53	71	70	99	98	100	100	63	83	17	99	96	100	100
Tuvalu				100		100	100				100		95	100						98	100
Vanuatu			65	95			63			53	54			94				64			88
Wallis and Futuna Islands																					
Western Samoa			97	75						94	67							69			
Sources:	UNEP 1989, WRI 1988	UNEP 1989, WRI 1988	UNEP 1989, WRI 1988	UNEP 1989, WRI 1988	UNEP 1993	WHO 1996	WHO 2000	UNEP 1989, WRI 1988	UNEP 1989, WRI 1988	UNEP 1989, WRI 1988	UNEP 1989, WRI 1988	UNEP 1993	WHO 1996	WHO 2000	UNEP 1989, WRI 1988	UNEP 1989, WRI 1988	UNEP 1989, WRI 1988	UNEP 1989, WRI 1988	calculated from UNEP 1993	WHO 1996	WHO 2000

Table 4 Access to Sanitation by Country, 1970 to 2000

Description

Adequate sanitation is also a fundamental requirement for basic human well-being. Data are given here for the percentage of urban, rural, and total populations, by country, with access to sanitation services for 1970, 1975, 1980, 1985, 1990, 1994, and 2000—the most recent year for which data are available. The World Health Organization (WHO) collected these data over various periods. Most of the data presented were drawn from responses by national governments to WHO questionnaires. Participants in data collection include the Joint Monitoring Programme (JMP), the United Nations Children's Fund, and the Water Supply and Sanitation Collaborative Council, which has continued sector monitoring and aims to support and strengthen the monitoring efforts of individual countries. Countries used their own definitions of "rural" and "urban."

For the 2000 WHO assessment, new definitions were provided of "improved" sanitation with allowance for acceptable local technologies. The 40 largest countries in the developing world account for 90 percent of population. As a result, WHO spent extra effort to collect comprehensive data for these countries. The excreta disposal system was considered adequate if it was private or shared (but not public) and if it hygienically separated human excreta from human contact. The following technologies were included in the 2000 assessment as representing "improved" sanitation:

Connection to a public sewer

Connection to septic system

Pour-flush latrine

Simple pit latrine

Ventilated improved pit latrine

Limitations

As is the case with drinking water data, definitions for access to sanitation vary from country to country, and from year to year within the same country. Countries generally regard sanitation facilities that break the fecal-oral transmission route as adequate. In urban areas, adequate sanitation may be provided by connections to public sewers or by household systems such as pit privies, flush latrines, septic tanks, and communal toilets. In rural areas, pit privies, pour-flush latrines, septic tanks, and communal toilets are considered adequate. Direct comparisons between countries, and across time within the same country, are difficult and are additionally complicated by the fact that these data hide disparities between regions and socioeconomic classes.

The 2000 WHO Assessment attempts to shift from gathering information from water providers only to include consumer-based information. The current approach uses household surveys in an effort to assess the actual use of facilities. Access to sanitation services, as reported by WHO 2000, does not imply that the level of service is "adequate" or "safe." The assessment questionnaire did not include any methodology for discounting coverage figures to allow for intermittence or poor quality of the service provided. More details of the methods used, and their limitations, can be found at http://www.who.int/water_sanitation_health/Globassessment/globalAnnex.htm.

Sources

United Nations Environment Programme, 1989, *Environmental Data Report,* GEMS Monitoring and Assessment Research Centre, Basil Blackwell, Oxford.

United Nations Environment Programme, 1993–94, *Environmental Data Report,* GEMS Monitoring and Assessment Research Centre in cooperation with the World Resources Institute and the UK Department of the Environment, Basil Blackwell, Oxford.

WHO 1996, World Health Organization, 1996, *Water Supply and Sanitation Sector Monitoring Report: 1996 (Sector Status as of 1994),* in collaboration with the Water Supply and Sanitation Collaborative Council and the United Nations Children's Fund, UNICEF, New York.

WHO 2000, World Health Organization, 2000, *Global Water Supply and Sanitation Assessment 2000 Report,* available in full at http://www.who.int/water_sanitation_health/ Globassessment/GlobalTOC.htm.

WRI 1988, World Health Organization data, cited by the World Resources Institute, 1988, *World Resources 1988–89,* World Resources Institute and the International Institute for Environment and Development in collaboration with the United Nations Environment Programme, Basic Books, New York.

TABLE 4 Access to Sanitation by Country, 1970 to 2000

Fraction of Population With Access to Sanitation

	Urban							Rural							Total						
Region and Country	1970	1975	1980	1985	1990	1994	2000	1970	1975	1980	1985	1990	1994	2000	1970	1975	1980	1985	1990	1994	2000
AFRICA																					
Algeria	13	100	80	75		90	6	50		40		20	47	9	47	75	57	75		73	44
Angola			40	29	25	34	70			15	16	20	8	30			20	19	21	16	23
Benin	83		48	58	60	54	46	1		4	20	35	6	6	14		16	33	45	20	
Botswana				93	100		88				28	85						40	89		29
Burkina Faso	49	47	38	44		42				5	6		11	16	4	4	7	9	18	18	29
Burundi	96		40	84	64	60	79			35	56	16	50	16			35	58		51	
Cameroon				100			99				1			85				43			92
Cape Verde			34	32		40	95			10	9		10	32			11	10		24	71
Central African Republic	64	100			45		43	96	100			46		23	72	100			46	46	31
Chad	7	9				73	81		1				7	13	1	1				21	29
Comoros							98							98							98
Congo	8	10					14	6	9					6	6	9				9	20
Congo, Democratic Republic	5	65			46	23	53	5	6		9	11	4		5	22		21	21	54	20
Cote D'Ivoire	23		43		81	59						100	51		5				92	54	
Djibouti				78		77	99			20	17		100	50			39	64		90	91
Egypt					80	20	98			10		26	5	91					50	11	94
Equatorial Guinea					54	61	60					24	48	46					33	54	53
Eritrea							66							1						15	13
Ethiopia	67	56		96			58	8	8		96			6	14	14					15
Gabon							25							4						21	21
Gambia					100	83	41	40	40	17		27	23	35	55		26		44	37	37
Ghana	92	95	47	51	63	53	62	2		1	16	60	36	64	13	56	11	30	61	42	63
Guinea	70		54				94	2				0		41						70	58
Guinea-Bissau			21	29		32	88			13	18		17	34			15	21		20	47
Kenya	85	98	89			69	96	45	48	19			81	81	50	55	30			77	86
Lesotho	44	51	13	22		1	93	10	12	14	14		7	96	11	13	14	15		6	92

263

TABLE 4 *Continued*

Fraction of Population With Access to Sanitation

Region and Country	Urban							Rural							Total						
	1970	1975	1980	1985	1990	1994	2000	1970	1975	1980	1985	1990	1994	2000	1970	1975	1980	1985	1990	1994	2000
AFRICA (*continued*)																					
Liberia	100			6		38		9			2		2		19					18	
Libya	100	100	100				97	54	69	72	2			30	67	79	88				97
Madagascar	88		9	55		50	70		9				3	70						15	42
Malawi			100			70	96			81			51	98			83			53	77
Mali	63		79	90	81	58	93	8			3	10	21	100	8			19	27	31	69
Mauritania	100		5	8			44							19	7						33
Mauritius	51	63	100	100	100	100	100	99	100	90	86	100	100	99	77	82	94	92	100	100	99
Morocco	75			62	100	69	100	4			16		18	42	29					40	75
Mozambique				53		70	69				12		70	26				20			43
Namibia					24		96					11		17				15			41
Niger	10	30	36		71	71	79		1	3	5	4	4	5	1	3			17	15	20
Nigeria					80	61	85				5	11	21	45					35	36	63
Reunion																					
Rwanda	83	87	60	77	88		12	52	56	50	55	17		8	53	57	51	56	21		8
Sao Tome and Principe											15							15			
Senegal			100	87	57	83	94			2		38	40	48			36		46	58	70
Seychelles																					
Sierra Leone			31	60	55	17	23			6	10	31	8	31			12	24	39	11	28
Somalia		77		44					35		5					47		18			
South Africa						79	99						12	73						46	86
Sudan	100	100	73	73		79	87	4	10		25		4	48	16	22				22	62
Swaziland				100		36			25		25		37			36		45		36	
Tanzania		88		93			98		14	66	58			86		17	66				90
Togo	4	36	24	31		57	69		12	10	9	4	13	17	1	15	13	14		26	34
Tunisia	100		100	84		100		34			16		85		62			55	57	96	
Uganda	84	82		32	32	75	96	76	95		30	60	55	72	76	94		30	57	57	75

Zambia	78 23 55 42 16 64 10 34 18 16 99 40 76 12
Zimbabwe	68 43 22 15 51 99 95 87
NORTH & CENTRAL AMERICA & CARIBBEAN	
Anguilla	
Antigua and Barbuda	
Aruba	
Bahamas	93 92 63 100 88 66 65 2 94 13 13 93 88 100 98 100 100
Barbados	100 100 100 66 100 100 68 100 16 44 100 100
Belize	42 57 50 66 69 87 21 45 75 54 4 10 76 41 23 59 76 62 25 87
British Virgin Islands	57 75 100 100
Canada	100 99 94 96
Cayman Islands	96 92 95 91 52 37 94 89 43 26 17 43 45 41 82 48 94 99
Costa Rica	95 66 93 91 15 35 59 38 11 20 16 18 41 72 41 82 85 48 99
Cuba	71 78 87 23 58 51 83 64 10 1 13 16 54 44 41 24 57 100 74
Dominican Republic	59 50 63 87 69 42 15 39 84 4 26 17 18 24 89 25 41 75 95
Dominica	100 100 100
El Salvador	83 68 59 35 39 37 78 59 38 43 17 43 48 82 88 85 66 71
Grenada	97 97 61 96
Guadeloupe	61 61 61
Guatemala	85 60 24 30 76 52 12 20 16 11 14 45 72 98 41
Haiti	28 25 21 19 16 17 13 1 10 43 42 44 42 50
Honduras	77 65 63 30 35 26 57 42 34 26 13 9 49 24 94
Jamaica	84 91 7 94 66 90 2 91 92 12 92 98
Martinique	
Mexico	73 66 58 55 32 26 13 12 14 13 77 77 85 87
Montserrat	
Netherlands Antilles	
Nicaragua	84 31 27 27 68 16 24 8 34 35 96
Panama	99 86 81 71 94 61 59 76 69 83 99 87
Puerto Rico	78 77 94 87

265

TABLE 4 *Continued*

Fraction of Population With Access to Sanitation

Region and Country	Urban							Rural							Total						
	1970	1975	1980	1985	1990	1994	2000	1970	1975	1980	1985	1990	1994	2000	1970	1975	1980	1985	1990	1994	2000
NORTH & CENTRAL																					
AMERICA & CARIBBEAN *(continued)*																					
St Kitts																					
St Lucia																					
St Vincent																					
Trinidad/Tobago					100							92							97		88
Turks/Caicos Islands	51	83	96	100				96	97	88	95				81	92	93	98			
United States of America							100							100							100
United States Virgin Islands																					
SOUTH AMERICA																					
Argentina	87	100	80	75		58	89	79	83	35	35			48	85	97		69			85
Bolivia	25		37	33	38	55	82	4	9	4	10	14	16	38	12		18	21	26	41	66
Brazil	85			86	84		85	24		1	1	32	3	40	58			63	71	44	77
Chile	33	36	100	100		82	98	10	11	10	4			93	29	32	83	84			97
Colombia	75	73	93	96	84	76	97	8	13	4	13	18	33	51	47	48	61		64	63	85
Ecuador			73	98	56	87	70		7	17	29	38	34	37			43	65	48	64	59
Falkland Islands (Malvinas)																					
French Guiana							85							57							79
Guyana	95	99	73	100	97		97	92	94	80	79	81		81	93	96	78	86	86		87
Paraguay	16	28	95	89	31		95			80	83	60		95	6	10	86	85	46		95
Peru	52		57	67	76	62	90	16		0	12	20	10	40	36		36	49	59	44	76
Suriname			100	78			100			79	48			34			88	62			83
Uruguay	97	97	59	59			96	13	17	6	59			89	82	83	51	59			95
Venezuela			60	57		64	86	45		12	5	72	30	69			52	50		58	74
ASIA																					
Afghanistan	69	63		5	13	38	25	16	15				1	8	21	21				8	12
Armenia																					

Country																		
Azerbaijan																		
Bahrain	87	40	21	100	100	77	82		1	0	30	44	6	5	3	100	10	35	53
Bangladesh		24	40	80	66	65			3	18	70				7	41	69
Bhutan											18
Brunei Darus	100	100			58			76	7	10		86	18
Cambodia			100	58	58	68		81	7	24	21	38
China	94	100	100	96	100		100	100	95	95	100	98	100
Cyprus								92	95
East Timor																		
Gaza Strip																		
Georgia			90			50			88
Hong Kong	85	87	27	31	44	70	73	1	1	2	3	14	14	18	20	7	9	14	29	31
India	50	60	29	33	79	73	87	4	21	5	38	30	40	52	12	15	23	37	44	51	66
Indonesia	100	100	96	100	100	89	86	48	43	59	35	37	74	70	78	69	72	67	81
Iran	82	75	100	96	96	93		1	11	31	47	47	74	36	79
Iraq																		
Israel																		
Japan			94	92	100		34	100	98	70	95	99
Jordan				100			98	99
Kazakhstan				99			100	99
Korea DPR	59	80	100	100	67	76		50	100	12	4	25	64	100	52	63
Korea Rep		100	100					100	100	100
Kuwait					100		100	100	100
Kyrgyzstan		10	13	30	70	84		2	8	13	34	3	5	12	24	46
Laos			100	100	92		4	100	100	5	100	99
Lebanon																		
Macau	100	100	100	94	95	41	43	55	94	59	60	70	75	94
Malaysia	21	60	100	95	58	1	2	4	26	58	3	13	22	44	56
Maldives		100	100	46		47	2	78	30
Mongolia	45	38	38	33	50	42	65	13	40	39	35	33	20	24	22	41	46
Myanmar (Burma)																		

TABLE 4 *Continued*

Fraction of Population With Access to Sanitation

Region and Country	Urban							Rural							Total						
	1970	1975	1980	1985	1990	1994	2000	1970	1975	1980	1985	1990	1994	2000	1970	1975	1980	1985	1990	1994	2000
ASIA *(continued)*																					
Nepal	14	14	16	17	34	51	75			1	1	3	16	20	1	1	1		6	20	27
Oman	100	100		88			98		5		25		19	61		12		31		76	92
Pakistan	12	21	42	51	53	53	94			2	6	12	19	42	3	6	13	19	25	30	61
Philippines	90	76	81	83	79		92	40	44	67	56	63		71	57	56	75	67	70		83
Qatar	100	100		100	100			16	100			85			83	100					
Saudi Arabia	67	91	81	100			100	11	35	50	33	85		100	21	47	70	82			100
Singapore			80	99	99		100										80	99			100
Sri Lanka	76	68	80	65	68	33	91	61	55	63	39	45	58	80	64	59	67	44	50	52	83
Syria			74			77	98			28			35	81			50			56	90
Tajikistan																					
Thailand	65	58	64	78			97	8	36	41	46	86		96	17	40	45	52			96
Turkey			56				98							70							91
Turkmenistan																					
United Arab Emirates			93							22							80				
Uzbekistan							100							100							100
Vietnam	100				23	43	87	26	2	55		10	15	70	26				13	21	73
Yemen A R			60	83			99							31							45
Yemen Dem			70				99			15				31			35				45
OCEANIA																					
American Samoa																					
Australia							100							100							100
Cook Islands			100	100	100		100			76	99	100		100				99			100
Fiji	100		85		91	100	75	87	93	60		65	85	12	91	96	70		75	92	43
French Polyneisa					98		99					95		97							98
Guam		100				100							100								
Kiribati					91	100	54					49		44						100	48

Marshall Islands					100							45									
Micronesia					99	100						46	100							100	
Nauru																					
New Caledonia																					
New Zealand																					
Niue					0	100	100				100	100	100	100						100	100
Northern Mariana Islands					100							71		92							
Palau					95		100					100		100							100
Papua New Guinea	100	100	96	99	57	82	92	5	5	3	35		11	80	14	18	15	44		22	82
Pitcairn																					
Samoa	100	100	86		100		95	80	99	83		92	17	100	84	99					99
Solomon Islands			80		73		98			21		2		18					13		34
Tokelau						100				41			100							100	
Tonga	100	100	97	99	88	100		100	100	94	40	78	100		100	100	19	52	82	100	
Tuvalu			100	81		90	100			80	73		85	100						87	100
Vanuatu			95	86			100			68	25			100				40			100
Wallis and Futuna Islands																					
Western Samoa			86	88						83	83							84			
Sources:	UNEP 1989, WRI 1988	UNEP 1989, WRI 1988	UNEP 1989, WRI 1988	UNEP 1989, WRI 1988	UNEP 1993	WHO 1996	WHO 2000	UNEP 1989, WRI 1988	UNEP 1989, WRI 1988	UNEP 1989, WRI 1988	UNEP 1989, WRI 1988	UNEP 1993	WHO 1996	WHO 2000	UNEP 1989, WRI 1988	UNEP 1989, WRI 1988	UNEP 1989, WRI 1988	UNEP 1989, WRI 1988	calculated from UNEP 1993	WHO 1996	WHO 2000

Table 5   Access to Water Supply and Sanitation by Region, 1990 and 2000

Description

Total population and the populations without access to improved water supply or sanitation services ("unserved") are shown here by World Health Organization (WHO) regions and globally, for 1990 and 2000, the most recent year for which data are available. The corresponding information for urban and rural areas is also given. Overall, global water supply coverage for the year 2000 is estimated to be 82 percent and global sanitation coverage is estimated at 60 percent. Data were available for 76 percent of the global population for 1990, while 89 percent were represented in the 2000 figures.

These data form the basis for all major international policy statements on lack of access to water. A previous data set for 1980, 1990, and 1994 was provided in *The World's Water 1998–1999*. The data here are more recent and have been revised to account for differences in definitions and data collection methods. Since 1980 countries have adopted increasingly stringent definitions of access, and WHO utilized these definitions for their 2000 survey.

Limitations

These data give a good picture of the current lack of access to improved water and sanitation services, but comparisons from different assessments should be done with extreme care, or not at all, because of changing definitions.

Country-reported data may reflect national definitions of "improved," unlike survey data, which were standardized as far as possible. For example, in many African countries the population "without access" to improved sanitation means people with no access to any sanitary facility. In Latin America and the Caribbean, however, it is more likely that those "without access" in fact have a sanitary facility, but the facility is deemed unsatisfactory by the local or national authorities. Low coverage figures for Latin America and the Caribbean may in part be a reflection of the comparatively narrow definitions used within that region.

The 2000 data come from both household surveys and from past information made available to WHO by providers of services (usually government agencies). Both sources are considered to be reasonably reliable concerning data on household connections, but we have chosen not to reprint the summary of the 1980 or earlier 1990 and 1994 data here because of differences in how those data were collected. These data can be found in the two previous tables at the country level.

Source

World Health Organization, 2000, *Global Water Supply and Sanitation Assessment 2000 Report*, available in full at http://www.who.int/water_sanitation_health/Globassessment/tab2-2.gif.

TABLE 5   Access to Water Supply and Sanitation by Region, 1990 and 2000

Region	1990 Population (millions)				2000 Population (millions)			
	Total Population	Population Served	Population Unserved	Percent Served	Total Population	Population Served	Population Unserved	Percent Served
GLOBAL								
Urban water supply	2,292	2,178	114	95%	2,845	2,671	174	94%
Rural water supply	2,974	1,961	1,013	66%	3,210	2,284	926	71%
Total water supply	5,266	4,139	1,127	79%	6,055	4,955	1,100	82%
Urban sanitation	2,292	1,877	415	82%	2,845	2,443	402	86%
Rural sanitation	2,974	1,027	1,947	35%	3,210	1,209	2,001	38%
Total sanitation	5,266	2,904	2,362	55%	6,055	3,652	2,403	60%
AFRICA								
Urban water supply	197	166	31	84%	297	253	44	85%
Rural water supply	418	183	235	44%	487	231	256	47%
Total water supply	615	349	266	57%	784	484	300	62%
Urban sanitation	197	167	30	85%	297	251	46	85%
Rural sanitation	418	206	212	49%	487	220	267	45%
Total sanitation	615	373	242	61%	784	471	313	60%
ASIA								
Urban water supply	1,029	972	57	94%	1,352	1,254	98	93%
Rural water supply	2,151	1,433	718	67%	2,331	1,736	595	74%
Total water supply	3,180	2,405	775	76%	3,683	2,990	693	81%
Urban sanitation	1,029	690	339	67%	1,352	1,055	297	78%
Rural sanitation	2,151	496	1,655	23%	2,331	712	1,619	31%
Total sanitation	3,180	1,186	1,994	37%	3,683	1,767	1,916	48%
LATIN AMERICA AND THE CARIBBEAN								
Urban water supply	313	287	26	92%	391	362	29	93%
Rural water supply	128	72	56	56%	128	79	49	62%
Total water supply	441	359	82	81%	519	441	78	85%
Urban sanitation	313	267	46	85%	391	340	51	87%
Rural sanitation	128	50	78	39%	128	62	66	48%
Total sanitation	441	317	124	72%	519	402	117	77%

TABLE 5 *Continued*

Region	1990 Population (millions)				2000 Population (millions)			
	Total Population	Population Served	Population Unserved	Percent Served	Total Population	Population Served	Population Unserved	Percent Served
OCEANIA								
Urban water supply	18	18	0	100%	21	21	0	100%
Rural water supply	8	5	3	63%	9	6	3	67%
Total water supply	26	23	3	88%	30	27	3	90%
Urban sanitation	18	18	0	100%	21	21	0	100%
Rural sanitation	8	7	1	88%	9	7	2	78%
Total sanitation	26	25	1	96%	30	28	2	93%
EUROPE								
Urban water supply	522	522	0	100%	545	542	3	99%
Rural water supply	200	199	1	100%	184	161	23	88%
Total water supply	722	721	1	100%	729	703	26	96%
Urban sanitation	522	522	0	100%	545	537	8	99%
Rural sanitation	200	199	1	100%	184	137	47	74%
Total sanitation	722	721	1	100%	729	674	55	92%
NORTH AMERICA								
Urban water supply	213	213	0	100%	239	239	0	100%
Rural water supply	69	69	0	100%	71	71	0	100%
Total water supply	282	282	0	100%	310	310	0	100%
Urban sanitation	213	213	0	100%	239	239	0	100%
Rural sanitation	69	69	0	100%	71	71	0	100%
Total sanitation	282	282	0	100%	310	310	0	100%

Due to rounding, coverage figures might not total 100%, even if the population unserved is shown as 0.
The global totals are the sums of the regional figures given here.
Definitions have changed over time. See the description of this table.

Table 6 Reported Cases of Dracunculiasis by Country, 1972 to 2000

Description

Dracunculiasis, or guinea worm disease, is directly related to drinking unclean water. A global campaign is underway to eradicate guinea worm. Although the original goal of elimination by the year 2000 has failed, substantial progress continues to be made. We first reported on this disease in *The World's Water 1998–1999*. This update adds the last four years of data and provides some insight into the remaining challenges.

Guinea worm cases are reported here by country. The accompanying figure shows reported cases from 1972 to 2000. Comprehensive monitoring began in the late 1980s. In 2000, the number of cases declined to just over 75,000 in 16 countries, with the vast majority of cases remaining in the Sudan, which has been racked by civil war for the past decade, greatly limiting the ability of health workers to implement a comprehensive guinea worm detection and prevention program. The total number of cases reported from 1990 to 1996 may not equal the totals obtained by adding the country numbers. In these cases, the World Health Organization has adjusted totals to reflect estimates for countries not reporting at the time or reporting by a different system than that used by WHO.

Limitations

Dracunculiasis cases occur primarily in remote, rural areas. Reporting has improved tremendously in the past decade through implementation of comprehensive monitoring programs, but some cases may still be missed or underreported. Data for early years should be viewed with care, since many national reporting programs were not established before the late 1980s or early 1990s. For some countries, the totals reflect imported cases only. See Table 7 for further information.

Sources

UNICEF Statistics, 2001, Http://www.childinfo.org/eddb/gw/countdata.

World Health Organization, "Dracunculiasis surveillance," *Weekly Epidemiological Record* 57 (9): 65–67, 5 March 1982; 60 (9): 61–63, 1 March 1985; 61 (5): 31–36, 31 January 1986; 66 (31): 225–30, 2 August 1991; 68 (18): 125–26, 30 April 1993; 69 (17): 121–28, 29 April 1994; 70 (18): 125–26, 5 May 1995; 71 (19): 141–47, 10 May 1996; 72 (19): 133–39, 9 May 1997.

World Health Organization, 2001, "Dracunculaisis global surveillance summary, 2000," *Weekly Epidemiological Record* 76 (18): 133–40, 4 May 2001.

TABLE 6   Reported Dracunculiasis Cases by Country, 1972 to 2000

Region and Country	1972	1973	1974	1975	1976	1977	1978	1979	1980	1981	1982	1983	1984	1985
AFRICA														
Benin	1,480		820											
Burkina Faso	5,822	4,404	4,008	6,277	1,557		2,885	2,694	2,565			4,362	1,739	458
Cameroon				251									0	168
Central African Republic														
Chad							172						1,472	9
Cote d'Ivoire	4,891	4,654	6,283	4,971	4,656	5,207	6,993		6,712	7,978		2,259	2,573	1,889
Ethiopia													2,882	1,467
Ghana	693	1,606	1,226	4,052	1,421	1,617	1,676		2,703	853	3,413	3,040	4,244	4,501
Kenya														
Mali	668	786	737	542	760	1,084			816	777	401	428	5,008	4,072
Mauritania						127			651	663	903	1,612	1,241	1,291
Niger				2,600	3,000	5,560			1,906	2,113	1,530			1,373
Nigeria	98			1,007					1,693				8,777	5,234
Senegal		334	208	65	137				161					62
Sudan														
Togo				3,261	1,648		2,617	2,673	1,748	951	2,592		1,839	1,456
Uganda													6,230	4,070
ASIA AND THE MIDDLE EAST														
India							7,052	6,827	2,846	2,729	5,406	42,926	44,818	39,792	30,950
Pakistan								250		14,155				
Yemen		25												
Number of countries reporting	6	6	6	8	7	6	8	3	11	7	6	6	12	14
Number of Cases	13,652	11,809	13,282	20,426	12,779	18,087	26,980	8,213	35,839	18,741	51,765	56,519	75,797	57,000

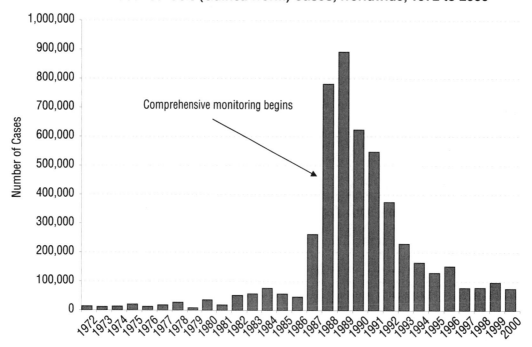

Dracunculiasis (Guinea Worm) Cases, Worldwide, 1972 to 2000

Comprehensive monitoring begins

1986	1987	1988	1989	1990	1991	1992	1993	1994	1995	1996	1997	1998	1999	2000
	400	33,962	7,172	37,414	4,006	4,315	13,887	4,302	2,273	1,472	855	695	492	186
2,558	1,957	1,266	45,004	42,187		11,784	8,281	6,861	6,281	3,241	2,477	2,227	2,184	1,956
86		752	871	742	393	127	72	30	15	17	19	23	8	5
									18	9	5	34	26	35
314						156	1,231	640	149	127	25	3	1	3
1,177	1,272	1,370	1,555	1,360	12,690		8,034	5,061	3,801	2,794	1,254	1,414	476	297
3,385	2,302	1,487	3,565	2,333		303	1,120	1,252	514	371	451	366	249	60
4,717	18,398	71,767	179,556	123,793	66,697	33,464	17,918	8,432	8,894	4,877	8,921	5,473	9,027	7,402
			5	6			35	53	23	0	6	7	1	4
5,640	435	564	1,111	884	16,024		12,011	5,581	4,218	2,402	1,099	650	410	290
	227	608	447	8,036		1,557	3,533	5,029	1,762	562	388	379	255	136
	699		288		32,829	500	21,564	18,562	13,821	2,956	3,030	2,700	1,920	1,166
2,821	216,484	653,492	640,008	394,082	281,937	183,169	75,752	39,774	16,374	12,282	12,590	13,420	13,237	7,869
128	132	138		38	1,341	728	630	195	76	19	4	0	0	0
822	399	542				2,447	2,984	53,271	64,608	118,578	43,596	47,977	66,097	54,890
1,325		178	2,749	3,042	5,118	8,179	10,349	5,044	2,073	1,626	1,762	2,128	1,589	828
		1,960	1,309	4,704		126,369	42,852	10,425	4,810	1,455	1,374	1,061	321	96
23,070	17,031	12,023	7,881	4,798	2,185	1,081	755	371	60	9	0	0	0	0
	2,400	1,110	534	160	106	23	2	0	0	0	0	0	0	0
								94	82	62	7	0	0	0
12	13	15	15	15	11	15	18	19	20	20	20	20	20	20
46,043	262,136	781,219	892,055	623,844	547,575	374,202	229,773	164,973	129,852	152,814	77,863	78,557	96,293	75,223

Table 7   Reported Dracunculiasis Cases, Eradication Progress, 2000

Description

As described in the previous table, a global campaign is underway to eradicate guinea worm, a disease directly related to drinking unclean water. Although the original goal of elimination by the year 2000 has failed, substantial progress continues to be made. This table reports on the current status of eradication efforts in various countries and over the past decade.

Guinea worm cases are reported here by country for the year 2000 and for a "comparison" year between 1990 and 1995, when comprehensive monitoring had begun. The drop in cases over the decade can be seen in the fourth column, "Cases in 2000 as Percentage of Comparison Year." For almost all countries, the number of reported cases has dropped by more than 90 percent. The last column shows the fraction of cases "contained." Containment entails identifying cases promptly, keeping the lesion bandaged for up to three weeks, and educating the sufferer and community how to prevent reinfection. The hope for eradication requires that all cases be contained. About half of all cases in 2000 were reported contained.

Limitations

The comparison year varies from 1990 to 1995. The comparison year was chosen based on when a country had established a consistent national reporting program. Some of the reported cases are imported. For example, all five of the cases reported in Cameroon in 2000 were imported from Nigeria. All four cases reported in Kenya in 2000 were imported from the Sudan.

Source

UNICEF Statistics, 2001, Guinea worm disease, UNICEF End Decade Databases, available at http://childinfo.org/eddb/gw/currrent.htm.

TABLE 7 Reported Dracunculiasis Cases, Eradication Progress, 2000

Country	Guinea Worm Cases Reported Comparison Year	2000	Cases in 2000 as Percent of Comparison Year	Percent of Cases Contained
Eradication Certified (no cases in three years)				
India (1990)	4,798	–	0	–
Pakistan (1990)	160	–	0	–
Countries under Precertification Surveillance				
Senegal (1991)	1,341	–	0	–
Cameroon (1997)	742	5	1	100
Yemen (1994)	94	–	0	–
Kenya (1994)	53	4	8	100
Active Cases being Reported				
Sudan (1995)	64,608	54,890	85	42
Nigeria (1990)	394,082	7,869	2	58
Ghana (1990)	123,793	7,402	6	80
Burkina Faso (1990)	42,187	1,956	5	71
Niger (1991)	32,829	1,166	4	63
Togo (1993)	10,394	828	8	72
Cote d'Ivoire (1991)	12,690	297	2	70
Mali (1991)	16,024	290	2	57
Benin (1990)	37,414	186	0	81
Mauritania (1990)	8,036	136	2	57
Uganda (1992)	126,369	96	0	79
Ethiopia (1990)	2,333	60	3	95
Central African Republic (1995)	18	35	Increase	0
Chad (1993)	1,231	3	0	100

Table 8   National Standards for Arsenic in Drinking Water

Description

Growing concern over the presence of arsenic in drinking water, especially in Bangladesh, India, and the United States, has revived debate over the establishment of national standards for arsenic. This table presents information on such standards for a number of countries. In 1993, the World Health Organization (WHO) set new provisional guidelines (Guidelines for Drinking-water Quality) of 0.01 milligrams per liter (mg/L), replacing an earlier WHO guideline of 0.05 mg/L. Several countries have retained the earlier standard as their national standard or as an interim target. A number of European countries have now adopted the WHO provisional guideline of 0.01 mg/L as their standard. The national standard for arsenic in drinking water remains at 0.05 mg/L in Bangladesh, China, and India.

In the United States of America, the Safe Drinking Water Act (SDWA) directs the U.S. Environmental Protection Agency (EPA) to establish national standards for public drinking-water supplies. EPA's interim maximum contaminant level (MCL) for arsenic in drinking water is 0.05 mg/L. Under the 1996 SDWA amendments, EPA proposed a new standard (an MCL) for arsenic in drinking water in June 2000, but it has yet to be finalized.

Limitations

Drinking arsenic-rich water over a long period is unsafe. New information from some countries around the world, such as Bangladesh and India, has revived the debate about appropriate standards. Different definitions of "safe," and uncertainty about the relationship between arsenic concentration in drinking water and human health effects, make determining a single standard difficult. Delayed effects from arsenic poisoning and poor reporting further complicate developing standards. The official WHO position is that "in developing national drinking water standards based on the [WHO] guideline values, it will be necessary to take account of a variety of geographical, socioeconomic, dietary and other conditions affecting potential exposure."

Source

United Nations Synthesis Report on Arsenic in Drinking Water, 2001, available at http://www.who.int/water_sanitation_health/Arsenic/ArsenicUNReptoc.htm.

TABLE 8   National Standards for Arsenic in Drinking Water

Countries whose standard is lower than 0.01 mg/L
Australia (0.007 mg/L, 1996)

Countries whose standard is 0.01 mg/L
European Union (1998), Japan (1993), Jordan (1991), Laos (1999), Laos, Mongolia (1998), Namibia, Syria (1994)

Countries whose standard is lower than 0.05 mg/L but higher than 0.01 mg/L
Canada (1999) 0.025 mg/L

Countries considering lowering the standard from 0.05 mg/L
United States (1986*), Mexico (1994)

Countries whose standard is 0.05 mg/lL
Bahrain (unknown), Bangladesh (unknown), Bolivia (1997), China (unknown), Egypt (1995), India (unknown), Indonesia (1990), Oman, Philippines (1978), Saudi Arabia (unknown), Sri Lanka (1983), Viet Nam (1989), Zimbabwe (unknown)

( ) shows the year standard was established
* new standard value is being proposed

Table 9   United States National Primary Drinking Water Regulations

Description

The Safe Drinking Water Act (SDWA) of the United States, passed in 1974 and amended in 1986 and 1996, gives the U.S. Environmental Protection Agency (EPA) the authority to set drinking water standards. Drinking water standards are regulations that EPA sets to control the level of contaminants in drinking water. These standards are part of the SDWA's "multiple barrier" approach to drinking water protection, which includes assessing and protecting drinking water sources; protecting wells and collection systems; making sure water is treated by qualified operators; ensuring the integrity of distribution systems; and making information available to the public on the quality of their drinking water. In most cases, EPA delegates responsibility for implementing drinking water standards to states and tribes.

This table summarizes the U.S. National Primary Drinking Water Regulation (NPDWR, or primary standard). The NPDWR is a legally enforceable standard that applies to public water systems. Primary standards protect drinking water quality by limiting the levels of specific contaminants that can adversely affect public health and are known or anticipated to occur in water.

Limitations

Different standards may be appropriate for different places—the example below offers background on the comprehensive U.S. standards but is not meant to prevent different countries or regions from choosing different levels as appropriate.

Source

United States Environmental Protection Agency, no date, National Primary Drinking Water Standards, available at http://www.epa.gov/safewater/mcl.html.

TABLE 9   National Primary Drinking Water Standards

Contaminant	MCLG[1] (mg/L)[2]	MCL or TT[1] (mg/L)[2]	Potential health effects from exposure above the MCL	Common sources of contaminant in drinking water
MICROORGANISMS				
Cryptosporidium	as of 01/01/02: zero	as of 01/01/02: TT[3]	Gastrointestinal illness (e.g., diarrhea, vomiting, cramps)	Human and fecal animal waste
Giardia lamblia	zero	TT[3]	Gastrointestinal illness (e.g., diarrhea, vomiting, cramps)	Human and animal fecal waste
Heterotrophic plate count (HPC)	n/a	TT[3]	HPC has no health effects; it is an analytic method used to measure the variety of bacteria that are common in water. The lower the concentration of bacteria in drinking water, the better maintained the water system is.	HPC measures a range of bacteria that are naturally present in the environment
Legionella	zero	TT[3]	Legionnaire's Disease, a type of pneumonia	Found naturally in water; multiplies in heating systems
Total Coliforms (including fecal coliform and *E. coli*)	zero	5.0%[4]	Not a health threat in itself; it is used to indicate whether other potentially harmful bacteria may be present[5]	Total coliforms are naturally present in the environment; fecal coliforms and *E. coli* come from human and animal fecal waste.
Turbidity	n/a	TT[3]	Turbidity is a measure of the cloudiness of water. It is used to indicate water quality and filtration effectiveness (e.g., whether disease-causing organisms are present). Higher turbidity levels are often associated with higher levels of disease-causing micro-organisms such as viruses, parasites, and some bacteria. These organisms can cause symptoms such as nausea, cramps, diarrhea, and associated headaches.	Soil runoff
Viruses (enteric)	zero	TT[3]	Gastrointestinal illness (e.g., diarrhea, vomiting, cramps)	Human and animal fecal waste
DISINFECTANTS AND DISINFECTION BYPRODUCTS				
Bromate	as of 01/01/02: zero	as of 01/01/02: 0.010	Increased risk of cancer	Byproduct of drinking water disinfection
Chloramines (as Cl_2)	as of 01/01/02: MRDLG=4[1]	as of 01/01/02: MRDL=4.0[1]	Eye/nose irritation; stomach discomfort; anemia	Water additive used to control microbes
Chlorine (as Cl_2)	as of 01/01/02: MRDLG=4[1]	as of 01/01/02: MRDL=4.0[1]	Eye/nose irritation; stomach discomfort	Water additive used to control microbes
Chlorine dioxide (as ClO_2)	as of 01/01/02: MRDLG=0.8[1]	as of 01/01/02: MRDL=0.8[1]	Anemia; infants & young children: nervous system effects	Water additive used to control microbes

TABLE 9   *Continued*

Contaminant	MCLG[1] (mg/L)[2]	MCL or TT[1] (mg/L)[2]	Potential health effects from exposure above the MCL	Common sources of contaminant in drinking water
DISINFECTANTS AND DISINFECTION BYPRODUCTS *(continued)*				
Chlorite	as of 01/01/02: 0.8	as of 01/01/02: 1.0	Anemia; infants & young children: nervous system effects	Byproduct of drinking water disinfection
Haloacetic acids (HAA5)	as of 01/01/02: n/a[6]	as of 01/01/02: 0.060	Increased risk of cancer	Byproduct of drinking water disinfection
Total Trihalomethanes (TTHMs)	none[7] as of 01/01/02: n/a[6]	0.10 as of 01/01/02: 0.080	Liver, kidney or central nervous system problems; increased risk of cancer	Byproduct of drinking water disinfection
INORGANIC CHEMICALS				
Antimony	0.006	0.006	Increase in blood cholesterol; decrease in blood sugar	Discharge from petroleum refineries; fire retardants; ceramics; electronics; solder
Arsenic	none[7]	0.05	Skin damage; circulatory system problems; increased risk of cancer	Erosion of natural deposits; runoff from orchards; runoff from glass and electronics production wastes
Asbestos (fibers >10 micrometers)	7 million fibers per Liter (MFL)	7 MFL	Increased risk of developing benign intestinal polyps	Decay of asbestos cement in water mains; erosion of natural deposits
Barium	2	2	Increase in blood pressure	Discharge of drilling wastes; discharge from metal refineries; erosion of natural deposits
Beryllium	0.004	0.004	Intestinal lesions	Discharge from metal refineries and coal-burning factories; discharge from electrical, aerospace, and defense industries
Cadmium	0.005	0.005	Kidney damage	Corrosion of galvanized pipes; erosion of natural deposits; discharge from metal refineries; runoff from waste batteries and paints
Chromium (total)	0.1	0.1	Allergic dermatitis	Discharge from steel and pulp mills; erosion of natural deposits
Copper	1.3	TT[8]; Action Level= 1.3	Short term exposure: Gastrointestinal distress Long term exposure: Liver or kidney damage People with Wilson's Disease should consult their personal doctor if the amount of copper in their water exceeds the action level	Corrosion of household plumbing systems; erosion of natural deposits

TABLE 9   *Continued*

Contaminant	MCLG[1] (mg/L)[2]	MCL or TT[1] (mg/L)[2]	Potential health effects from exposure above the MCL	Common sources of contaminant in drinking water
INORGANIC CHEMICALS (*continued*)				
Cyanide (as free cyanide)	0.2	0.2	Nerve damage or thyroid problems	Discharge from steel/metal factories; discharge from plastic and fertilizer factories
Fluoride	4.0	4.0	Bone disease (pain and tenderness of the bones); Children may get mottled teeth	Water additive which promotes strong teeth; erosion of natural deposits; discharge from fertilizer and aluminum factories
Lead	zero	TT[8]; Action Level= 0.015	Infants and children: Delays in physical or mental development; children could show slight deficits in attention span and learning abilities. Adults: Kidney problems; high blood pressure	Corrosion of household plumbing systems; erosion of natural deposits
Mercury (inorganic)	0.002	0.002	Kidney damage	Erosion of natural deposits; discharge from refineries and factories; runoff from landfills and croplands
Nitrate (measured as Nitrogen)	10	10	Infants below the age of six months who drink water containing nitrate in excess of the MCL could become seriously ill and, if untreated, may die. Symptoms include shortness of breath and blue-baby syndrome.	Runoff from fertilizer use; leaching from septic tanks, sewage; erosion of natural deposits
Nitrite (measured as Nitrogen)	1	1	Infants below the age of six months who drink water containing nitrite in excess of the MCL could become seriously ill and, if untreated, may die. Symptoms include shortness of breath and blue-baby syndrome.	Runoff from fertilizer use; leaching from septic tanks, sewage; erosion of natural deposits
Selenium	0.05	0.05	Hair or fingernail loss; numbness in fingers or toes; circulatory problems	Discharge from petroleum refineries; erosion of natural deposits; discharge from mines
Thallium	0.0005	0.002	Hair loss; changes in blood; kidney, intestine, or liver problems	Leaching from ore-processing sites; discharge from electronics, glass, and drug factories
ORGANIC CHEMICALS				
Acrylamide	zero	TT[9]	Nervous system or blood problems; increased risk of cancer	Added to water during sewage/wastewater treatment

TABLE 9   *Continued*

Contaminant	MCLG[1] (mg/L)[2]	MCL or TT[1] (mg/L)[2]	Potential health effects from exposure above the MCL	Common sources of contaminant in drinking water
ORGANIC CHEMICALS *(continued)*				
Alachlor	zero	0.002	Eye, liver, kidney, or spleen problems; anemia; increased risk of cancer	Runoff from herbicide used on row crops
Atrazine	0.003	0.003	Cardiovascular system or reproductive problems	Runoff from herbicide used on row crops
Benzene	zero	0.005	Anemia; decrease in blood platelets; increased risk of cancer	Discharge from factories; leaching from gas storage tanks and landfills
Benzo(a)pyrene (PAHs)	zero	0.0002	Reproductive difficulties; increased risk of cancer	Leaching from linings of water storage tanks and distribution lines
Carbofuran	0.04	0.04	Problems with blood, nervous system, or reproductive system	Leaching of soil fumigant used on rice and alfalfa
Carbon tetrachloride	zero	0.005	Liver problems; increased risk of cancer	Discharge from chemical plants and other industrial activities
Chlordane	zero	0.002	Liver or nervous system problems; increased risk of cancer	Residue of banned termiticide
Chlorobenzene	0.1	0.1	Liver or kidney problems	Discharge from chemical and agricultural chemical factories
2,4-D	0.07	0.07	Kidney, liver, or adrenal gland problems	Runoff from herbicide used on row crops
Dalapon	0.2	0.2	Minor kidney changes	Runoff from herbicide used on rights of way
1,2-Dibromo-3-chloropropane (DBCP)	zero	0.0002	Reproductive difficulties; increased risk of cancer	Runoff/leaching from soil fumigant used on soybeans, cotton, pineapples, and orchards
o-Dichlorobenzene	0.6	0.6	Liver, kidney, or circulatory system problems	Discharge from industrial chemical factories
p-Dichlorobenzene	0.075	0.075	Anemia; liver, kidney or spleen damage; changes in blood	Discharge from industrial chemical factories
1,2-Dichloroethane	zero	0.005	Increased risk of cancer	Discharge from industrial chemical factories
1,1-Dichloroethylene	0.007	0.007	Liver problems	Discharge from industrial chemical factories
cis-1,2-Dichloroethylene	0.07	0.07	Liver problems	Discharge from industrial chemical factories

TABLE 9   *Continued*

Contaminant	MCLG[1] (mg/L)[2]	MCL or TT[1] (mg/L)[2]	Potential health effects from exposure above the MCL	Common sources of contaminant in drinking water
ORGANIC CHEMICALS *(continued)*				
trans-1,2-Dichloroethylene	0.1	0.1	Liver problems	Discharge from industrial chemical factories
Dichloromethane	zero	0.005	Liver problems; increased risk of cancer	Discharge from drug and chemical factories
1,2-Dichloropropane	zero	0.005	Increased risk of cancer	Discharge from industrial chemical factories
Di(2-ethylhexyl) adipate	0.4	0.4	General toxic effects or reproductive difficulties	Discharge from chemical factories
Di(2-ethylhexyl) phthalate	zero	0.006	Reproductive difficulties; liver problems; increased risk of cancer	Discharge from rubber and chemical factories
Dinoseb	0.007	0.007	Reproductive difficulties	Runoff from herbicide used on soybeans and vegetables
Dioxin (2,3,7,8-TCDD)	zero	0.00000003	Reproductive difficulties; increased risk of cancer	Emissions from waste incineration and other combustion; discharge from chemical factories
Diquat	0.02	0.02	Cataracts	Runoff from herbicide use
Endothall	0.1	0.1	Stomach and intestinal problems	Runoff from herbicide use
Endrin	0.002	0.002	Liver problems	Residue of banned insecticide
Epichlorohydrin	zero	TT [9]	Increased cancer risk, and over a long period of time, stomach problems	Discharge from industrial chemical factories; an impurity of some water treatment chemicals
Ethylbenzene	0.7	0.7	Liver or kidney problems	Discharge from petroleum refineries
Ethylene dibromide	zero	0.00005	Problems with liver, stomach, reproductive system, or kidneys; increased risk of cancer	Discharge from petroleum refineries
Glyphosate	0.7	0.7	Kidney problems; reproductive difficulties	Runoff from herbicide use
Heptachlor	zero	0.0004	Liver damage; increased risk of cancer	Residue of banned termiticide
Heptachlor epoxide	zero	0.0002	Liver damage; increased risk of cancer	Breakdown of heptachlor
Hexachlorobenzene	zero	0.001	Liver or kidney problems; reproductive difficulties; increased risk of cancer	Discharge from metal refineries and agricultural chemical factories
Hexachloro-cyclopentadiene	0.05	0.05	Kidney or stomach problems	Discharge from chemical factories

TABLE 9 *Continued*

Contaminant	MCLG[1] (mg/L)[2]	MCL or TT[1] (mg/L)[2]	Potential health effects from exposure above the MCL	Common sources of contaminant in drinking water
ORGANIC CHEMICALS *(continued)*				
Lindane	0.0002	0.0002	Liver or kidney problems	Runoff/leaching from insecticide used on cattle, lumber, gardens
Methoxychlor	0.04	0.04	Reproductive difficulties	Runoff/leaching from insecticide used on fruits, vegetables, alfalfa, livestock
Oxamyl (Vydate)	0.2	0.2	Slight nervous system effects	Runoff/leaching from insecticide used on apples, potatoes, and tomatoes
Polychlorinated biphenyls (PCBs)	zero	0.0005	Skin changes; thymus gland problems; immune deficiencies; reproductive or nervous system difficulties; increased risk of cancer	Runoff from landfills; discharge of waste chemicals
Pentachlorophenol	zero	0.001	Liver or kidney problems; increased cancer risk	Discharge from wood-preserving factories
Picloram	0.5	0.5	Liver problems	Herbicide runoff
Simazine	0.004	0.004	Problems with blood	Herbicide runoff
Styrene	0.1	0.1	Liver, kidney, or circulatory system problems	Discharge from rubber and plastic factories; leaching from landfills
Tetrachloroethylene	zero	0.005	Liver problems; increased risk of cancer	Discharge from factories and dry cleaners
Toluene	1	1	Nervous system, kidney, or liver problems	Discharge from petroleum factories
Toxaphene	zero	0.003	Kidney, liver, or thyroid problems; increased risk of cancer	Runoff/leaching from insecticide used on cotton and cattle
2,4,5-TP (Silvex)	0.05	0.05	Liver problems	Residue of banned herbicide
1,2,4-Trichlorobenzene	0.07	0.07	Changes in adrenal glands	Discharge from textile finishing factories
1,1,1-Trichloroethane	0.20	0.2	Liver, nervous system, or circulatory problems	Discharge from metal degreasing sites and other factories
1,1,2-Trichloroethane	0.003	0.005	Liver, kidney, or immune system problems	Discharge from industrial chemical factories
Trichloroethylene	zero	0.005	Liver problems; increased risk of cancer	Discharge from metal degreasing sites and other factories
Vinyl chloride	zero	0.002	Increased risk of cancer	Leaching from PVC pipes; discharge from plastic factories
Xylenes (total)	10	10	Nervous system damage	Discharge from petroleum factories; discharge from chemical factories

TABLE 9   *Continued*

Contaminant	MCLG[1] (mg/L)[2]	MCL or TT[1] (mg/L)[2]	Potential health effects from exposure above the MCL	Common sources of contaminant in drinking water
RADIONUCLIDES				
Alpha particles	none[7]	15 picocuries per Liter (pCi/L)	Increased risk of cancer	Erosion of natural deposits of certain minerals that are radioactive and may emit a form of radiation known as alpha radiation
Beta particles and photon emitters	none[7]	4 millirems per year (mrem/yr)	Increased risk of cancer	Decay of natural and man-made deposits of certain minerals that are radioactive and may emit forms of radiation known as photons and beta radiation
Radium 226 and Radium 228 (combined)	none[7]	5 pCi/L	Increased risk of cancer	Erosion of natural deposits

NOTES

1 - Definitions
- Maximum Contaminant Level Goal (MCLG) - The level of a contaminant in drinking water below which there is no known or expected risk to health. MCLGs allow for a margin of safety and are non-enforceable public health goals.
- Maximum Contaminant Level (MCL) - The highest level of a contaminant that is allowed in drinking water. MCLs are set as close to MCLGs as feasible using the best available treatment technology and taking cost into consideration. MCLs are enforceable standards.
- Maximum Residual Disinfectant Level Goal (MRDLG) - The level of a drinking water disinfectant below which there is no known or expected risk to health. MRDLGs do not reflect the benefits of the use of disinfectants to control microbial contaminants.
- Maximum Residual Disinfectant Level (MRDL) - The highest level of a disinfectant allowed in drinking water. There is convincing evidence that addition of a disinfectant is necessary for control of microbial contaminants.
- Treatment Technique (TT) - A required process intended to reduce the level of a contaminant in drinking water.

2 - Units are in milligrams per liter (mg/L) unless otherwise noted. Milligrams per liter are equivalent to parts per million (ppm).

3 - EPA's surface water treatment rules require systems using surface water or ground water under the direct influence of surface water to (1) disinfect their water, and (2) filter their water or meet criteria for avoiding filtration so that the following contaminants are controlled at the following levels:
- *Cryptosporidium:* (as of January 1, 2002) 99% removal
- *Giardia lamblia:* 99.9% removal/inactivation
- Viruses: 99.99% removal/inactivation
- *Legionella:* No limit, but EPA believes that if *Giardia* and viruses are removed/inactivated, *Legionella* will also be controlled.
- Turbidity: At no time can turbidity (cloudiness of water) go above 5 nephelolometric turbidity units (NTU); systems that filter must ensure that the turbidity go no higher than 1 NTU (0.5 NTU for conventional or direct filtration) in at least 95% of the daily samples in any month. As of January 1, 2002, turbidity may never exceed 1 NTU, and must not exceed 0.3 NTU in 95% of daily samples in any month.
- HPC: No more than 500 bacterial colonies per milliliter

4 - No more than 5.0% of samples may be total coliform-positive in a month. (For water systems that collect fewer than 40 routine samples per month, no more than one sample may be total coliform-positive during a month). Every sample that has total coliforms must be analyzed for either *E. coli* or fecal coliforms to

determine whether human or animal fecal matter is present (fecal coliform and *E. coli* are part of the total coliform group). There may not be any fecal coliforms or *E. coli*.

5 -   Fecal coliform and *E. coli* are bacteria whose presence indicates that the water may be contaminated with human or animal wastes. Disease-causing microbes (pathogens) in these wastes can cause diarrhea, cramps, nausea, headaches, or other symptoms. These pathogens may pose a special health risk for infants, young children, and people with severely compromised immune systems.

6 -   Although there is no collective MCLG for this contaminant group, there are individual MCLGs for some of the individual contaminants:
   •   Haloacetic acids: dichloroacetic acid (zero); trichloroacetic acid (0.3 mg/L)
   •   Trihalomethanes: bromodichloromethane (zero); bromoform (zero); dibromochloromethane (0.06 mg/L)

7 -   MCLGs were not established before the 1986 Amendments to the Safe Drinking Water Act. The standard for this contaminant was set prior to 1986. Therefore, there is no MCLG for this contaminant.

8 -   Lead and copper are regulated by a Treatment Technique that requires systems to control the corrosiveness of their water. If more than 10% of tap water samples exceed the action level, water systems must take additional steps. For copper, the action level is 1.3 mg/L, and for lead is 0.015 mg/L.

9 -   Each water system must certify, in writing, to the state that when it uses acrylamide and/or epichlorohydrin to treat water, the combination (or product) of dose and monomer level does not exceed the levels specified, as follows: Acrylamide = 0.05% dosed at 1 mg/L (or equivalent); Epichlorohydrin = 0.01% dosed at 20 mg/L (or equivalent).

Table 10   Irrigated Area, by Region, 1961 to 1999

Description

Total irrigated areas by continent are listed here for 1961, 1965, 1970, 1975, 1980, 1985, 1990, 1995, and 1999—the latest year for which reliable data are available. Units are thousands of hectares. After 1990, all irrigated area in the former USSR is split between Europe and Asia. These data have been updated from *The World's Water 2000–2001* to correct misreporting to the UN Food and Agriculture Organization (UN FAO).

Limitations

These data depend on in-country surveys, national reports, and estimates by the UN FAO. In some regions, multiple cropping may increase the apparent area in production. These data are not reported here. No information is offered about the quality of the land in production. Recent changes in political borders and the independence of several countries make certain continental time-series comparisons misleading. Data for the Soviet Union, Yugoslavia, and Czechoslovakia are provided through 1990; thereafter the irrigated areas of the newly independent states are reported. When summing by continental area, however, trends will appear misleading because some of the newly independent states are now included in Asia, while others are in Europe. No meaningful time-series trends by continent can thus be seen for these areas. The time-series for Africa, North and Central America, South America, and Oceania do not suffer from this problem.

Source

UN FAO, 2001, web site at www.fao.org.

TABLE 10   **Irrigated Area, by Region, 1961 to 1999** Thousand hectares (2)

Region	1961	1965	1970	1975	1980	1985	1990	1995	1999
World	138,989	149,976	167,803	188,225	209,716	225,138	244,305	261,380	274,166
Africa	7,410	7,795	8,483	9,010	9,491	10,331	11,235	12,380	12,538
Asia	90,166	97,093	109,666	121,565	132,377	141,922	155,009	180,461	192,962
Europe	8,324	9,225	10,355	12,296	13,979	15,479	16,744	25,208	24,406
North & Central America	17,949	19,525	20,938	22,831	27,593	27,464	28,905	30,657	31,395
Oceania	1,079	1,368	1,588	1,620	1,684	1,957	2,113	2,688	2,539
South America	4,661	5,070	5,673	6,403	7,392	8,296	9,499	9,986	10,326
USSR	9,400	9,900	11,100	14,500	17,200	19,689	20,800	(1)	(1)
Total	**138,989**	**149,976**	**167,803**	**188,225**	**209,716**	**225,138**	**244,305**	**261,380**	**274,166**

Notes
1. After 1990, all irrigated area in the former USSR is split among Europe and Asia.
2. These data have been updated from the previous volume of *The World's Water (2000-2001)* to correct misreporting to the UN Food and Agriculture Organization.

Table 11   Irrigated Area, Developed and Developing Countries, 1960 to 1999

Description

Total irrigated areas by UN FAO's "developed" and "developing" country categories are listed here for 1960, 1965, 1970, 1975, 1980, 1985, 1990, 1995, and 1999—the latest year for which reliable data are available. Units are thousands of hectares. While total irrigated area worldwide continues to increase, the average annual rate of change has been dropping. In the developed world, total irrigated area actually dropped between 1995 and 1999. Overall, the rate of increase in irrigated area worldwide has fallen below 1 percent per year.

Limitations

These data depend on in-country surveys, national reports, and estimates by the UN Food and Agriculture Organization (UN FAO). In some regions, multiple cropping may increase the apparent area in production. These data are not reported here. No differentiation is made about the quality of the land in production. Recent changes in political borders and the independence of several countries make certain time-series comparisons very difficult. Data for the Soviet Union, Yugoslavia, and Czechoslovakia are provided here through 1990; thereafter the irrigated areas of the newly independent states are reported.

Source

UN FAO, 2001, web site at www.fao.org.

TABLE 11   Irrigated Area, Developed and Developing Countries, 1960 to 1999

	1960	1965	1970	1975	1980	1985	1990	1995	1999
Developing Countries	100,000	40,056	44,050	49,973	58,426	62,016	65,616	67,073	66,708
Developed Countries	36,000	109,920	123,753	138,252	151,290	163,122	178,689	194,307	207,458
World	136,000	149,976	167,803	188,225	209,716	225,138	244,305	261,380	274,166
Annual Rate of Change:									
Developing Countries			1.99	2.69	3.38	1.23	1.16	0.44	(0.14)
Developed Countries			2.52	2.34	1.89	1.56	1.91	1.75	1.69
World			2.38	2.43	2.28	1.47	1.70	1.40	0.98

FAOSTAT 2001

Notes: Updated from last version. FAO has recalculated slightly the area of irrigated land in developed countries.

Table 12 Number of Dams, by Continent and Country

Description

The total number of dams larger than 15 meters is presented here by continent and country, as of the late 1990s. More than 47,000 such dams are included. These data were developed by the International Commission on Large Dams and others for the World Commission on Dams.

Limitations

Nearly half of all of these dams are estimated to be in China, but no recent accurate estimate from China is available. No estimate is available of the total volume of water actually stored behind these reservoirs.

Source

World Commission on Dams, 2000, *Dams and Development: A New Framework for Decision-Making: Annex V (Dams, Water and Energy: A Statistical Profile)*, Capetown, South Africa (reproduced with permission).

TABLE 12 **Number of Dams, by Continent and Country**

Continent/Country	Number of Dams
AFRICA	
South Africa	539
Zimbabwe	213
Algeria	107
Morocco	92
Tunisia	72
Nigeria	45
Cote d'Ivoire	22
Angola	15
Democratic Republic of Congo	14
Kenya	14
Namibia	13
Libya	12
Madagascar	10
Cameroon	9
Mauritius	9
Burkina Faso	8
Ethiopia	8
Mozambique	8
Lesotho	7
Egypt	6
Swaziland	6
Ghana	5
Sudan	4
Zambia	4
Botswana	3
Malawi	3
Benin	2
Congo	2
Guinea	2
Mali	2
Senegal	2
Seychelles	2
Sierra Leone	2
Tanzania	2
Togo	2
Gabon	1
Liberia	1
Uganda	1
EUROPE	
Spain	1,196
France	569
Italy	524
United Kingdom	517
Norway	335
Germany	311
Sweden	190
Switzerland	156

TABLE 12   *Continued*

Continent/Country	Number of Dams
Austria	149
Portugal	103
Finland	55
Cyprus	52
Greece	46
Iceland	20
Ireland	16
Belgium	15
Denmark	10
Netherlands	10
Luxembourg	3
SOUTH AMERICA	
Brazil	594
Argentina	101
Chile	88
Venezuela	74
Colombia	49
Peru	43
Ecuador	11
Bolivia	6
Uruguay	6
Paraguay	4
Guyana	2
Suriname	1
CENTRAL/EASTERN EUROPE	
Albania	306
Romania	246
Bulgaria	180
Czech Republic	118
Poland	69
Yugoslavia	69
Slovakia	50
Slovenia	30
Croatia	29
Bosnia-Herzegovina	25
Ukraine	21
Lithuania	20
Macedonia	18
Hungary	15
Latvia	5
Moldova	2
NORTH/CENTRAL AMERICA AND THE CARIBBEAN	
United States	6,575
Canada	793
Mexico	537
Cuba	49
Dominican Republic	11

TABLE 12 *Continued*

Continent/Country	Number of Dams
NORTH/CENTRAL AMERICA AND THE CARIBBEAN	
continued	
Costa Rica	9
Honduras	9
Panama	6
El Salvador	5
Guatemala	4
Nicaragua	4
Trinidad & Tobago	4
Jamaica	2
Antigua	1
Haiti	1
ASIA AND THE MIDDLE EAST	
China	22,000
India	4,291
Japan	2,675
South Korea	765
Turkey	625
Thailand	204
Indonesia	96
Russia	91
Pakistan	71
North Korea	70
Iran	66
Malaysia	59
Taiwan	51
Sri Lanka	46
Syria	41
Saudi Arabia	38
Azerbaijan	17
Armenia	16
Philippines	15
Georgia	14
Uzbekistan	14
Iraq	13
Kazakhstan	12
Kyrgyzstan	11
Tajikistan	7
Jordan	5
Lebanon	5
Myanmar	5
Nepal	3
Vietnam	3
Singapore	3
Afghanistan	2
Brunei	2
Cambodia	2

TABLE 12   *Continued*

Continent/Country	Number of Dams
Bangladesh	1
Laos	1
OCEANIA	
Australia	486
New Zealand	86
Papua New Guinea	3
Fiji	2
TOTAL	47,655

Table 13 Number of Dams, by Country

Description

The total number of dams larger than 15 meters is presented here by country as of the late 1990s. More than 47,000 such dams are included. These data were developed by the International Commission on Large Dams and others for the World Commission on Dams.

Limitations

Nearly half of all of these dams are estimated to be in China, but no recent accurate estimate from China is available. No estimate is available of the total volume of water actually stored behind these reservoirs.

Source

World Commission on Dams, 2000, *Dams and Development: A New Framework for Decision-Making: Annex V (Dams, Water and Energy: A Statistical Profile)*, Capetown, South Africa (reproduced with permission).

TABLE 13   Number of Dams, by Country

Country	Number of Dams
China	22,000
United States	6,575
India	4,291
Japan	2,675
Spain	1,196
Canada	793
South Korea	765
Turkey	625
Brazil	594
France	569
South Africa	539
Mexico	537
Italy	524
United Kingdom	517
Australia	486
Norway	335
Germany	311
Albania	306
Romania	246
Zimbabwe	213
Thailand	204
Sweden	190
Bulgaria	180
Switzerland	156
Austria	149
Czech Republic	118
Algeria	107
Portugal	103
Argentina	101
Indonesia	96
Morocco	92
Russia	91
Chile	88
New Zealand	86
Venezuela	74
Tunisia	72
Pakistan	71
North Korea	70
Poland	69
Yugoslavia	69
Iran	66
Malaysia	59
Finland	55
Cyprus	52
Taiwan	51
Slovakia	50
Colombia	49
Cuba	49
Greece	46

TABLE 13   *Continued*

Country	Number of Dams
Sri Lanka	46
Nigeria	45
Peru	43
Syria	41
Saudi Arabia	38
Slovenia	30
Croatia	29
Bosnia-Herzegovina	25
Cote d'Ivoire	22
Ukraine	21
Iceland	20
Lithuania	20
Macedonia	18
Azerbaijan	17
Ireland	16
Armenia	16
Angola	15
Belgium	15
Hungary	15
Philippines	15
Democratic Republic of Congo	14
Kenya	14
Georgia	14
Uzbekistan	14
Namibia	13
Iraq	13
Libya	12
Kazakhstan	12
Ecuador	11
Dominican Republic	11
Kyrgyzstan	11
Madagascar	10
Denmark	10
Netherlands	10
Cameroon	9
Mauritius	9
Costa Rica	9
Honduras	9
Burkina Faso	8
Ethiopia	8
Mozambique	8
Lesotho	7
Tajikistan	7
Egypt	6
Swaziland	6
Bolivia	6
Uruguay	6
Panama	6
Ghana	5

TABLE 13   *Continued*

Country	Number of Dams
Latvia	5
El Salvador	5
Jordan	5
Lebanon	5
Myanmar	5
Sudan	4
Zambia	4
Paraguay	4
Guatemala	4
Nicaragua	4
Trinidad & Tobago	4
Botswana	3
Malawi	3
Luxembourg	3
Nepal	3
Vietnam	3
Singapore	3
Papua New Guinea	3
Benin	2
Congo	2
Guinea	2
Mali	2
Senegal	2
Seychelles	2
Sierra Leone	2
Tanzania	2
Togo	2
Guyana	2
Moldova	2
Jamaica	2
Afghanistan	2
Brunei	2
Cambodia	2
Fiji	2
Gabon	1
Liberia	1
Uganda	1
Suriname	1
Antigua	1
Haiti	1
Bangladesh	1
Laos	1
Total	**47,655**

This table includes dams larger than 15 meters.

Table 14 Regional Statistics on Large Dams

Description

Various statistics are reported here for a global set of large dams (over 15 meters in height), using data prepared for the World Commission on Dams report (see Chapter 7). These statistics include the total number of large dams by region, their average height (reported in meters), average reservoir volume (reported in million cubic meters), and technically feasible and actual hydroelectricity generation (reported in terawatt-hours per year). Technical feasibility for hydropower potential is based on the conversion of all river head and flow in major rivers in each region.

Limitations

Nearly half of all of these dams are estimated to be in China, but no recent accurate estimate from China is available. The estimate of "technically feasible" hydroelectric potential is a relatively meaningless number, since it assumes all theoretically available potential can be captured. This is not true for economical, ecological, and social reasons. The average height and average storage volume figures hide enormous variations among dams.

Source

World Commission on Dams, 2000, *Dams and Development: A New Framework for Decision-Making: Annex V (Dams, Water and Energy: A Statistical Profile)*, Capetown, South Africa (reproduced with permission).

TABLE 14   Regional Statistics on Large Dams

Category	World	Europe	Asia	North and Central America	South America	Africa	AustralAsia
Total Number of Large Dams	~48,000	5,480	31,340	8,010	979	1,269	577
Average Height (m)	31	33	33	28	37	28	33
Average Reservoir Capacity (million m³)	269	70	268	998	1,011	883	205
Technically Feasible Hydroelectric Potential (TWhr/yr)	14,370	1,225	6,800	1,660	2,665	1,750	270
Annual Hydroelectric Production (TWhr/yr)	2,643	552	753	700	534	62	42

Table 15 Commissioning of Large Dams in the 20th Century, by Decade

Description

The twentieth century was a century of major dam construction. Tens of thousands of large dams were built around the world to generate hydroelectricity, trap floods, provide water supply and recreation, and to irrigate crops. This table, and the accompanying figure, shows the commissioning of large dams (over 15 meters in height) by decade. The peak of dam construction occurred in the 1960s and 1970s and has slowed considerably since then for ecological, economical, and social reasons, and as local opposition to new dams has grown.

Limitations

This data set is limited to dams over 15 meters in height. A comparable data set, limited to reservoirs with storage volumes of 0.1 cubic kilometers or more, was presented in *The World's Water 2000–2001,* Tables 16 and 17. That data set showed the same trend. The current set does not include most large dams in China—a very substantial number. The International Commission on Large Dams (ICOLD), which compiled the information in this table, believes that the number of dams commissioned in the 1990s is underreported, though no evidence for that is available.

Source

World Commission on Dams, 2000, *Dams and Development: A New Framework for Decision-Making: Annex V (Dams, Water and Energy: A Statistical Profile),* Capetown, South Africa (reproduced with permission).

TABLE 15   Commissioning of Large Dams in the 20th Century, by Decade

	Number of Dams
Before 1900	630
1900s	353
1910s	601
1920s	809
1930s	964
1940s	913
1950s	2,735
1960s	4,788
1970s	5,418
1980s	4,431
1990s	2,069

Commissioning of Large Dams, by Decade, 20th Century

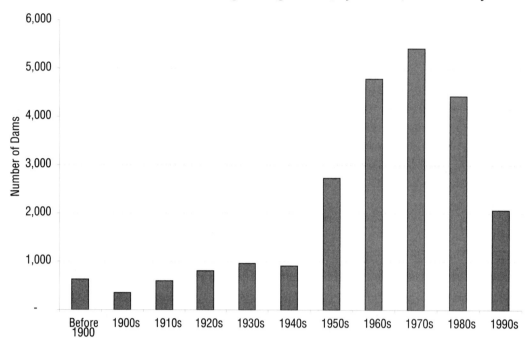

Table 16 Water System Rate Structures

Description

Every major water system sets rates in order to cover operation and maintenance costs, and typically, repayment of capital expenditures. The design of these rates has an influence on the rate at which investments are repaid, the demand for water, conservation and efficiency incentives, and equity. Little good information is available on the nature and prevalence of rate structures around the world. This short table lists the distribution of rate structures for a few nations, primarily in developed countries. Shown is the split, in percentage of utilities in a country, among four different broad categories:

> Flat Fee: A single fee for all users is charged for water service, regardless of the volume of water used.
> Volume Rate: A fee is charged per unit of water used.
> Increasing Block Rates: Blocks of water use are charged fees that increase with the volume of water used. Higher volumes of water are charged higher fees.
> Decreasing Block Rates: Blocks of water use are charged fees that decrease with volume. The more water used, the cheaper each block becomes.

Limitations

These data were collected from a wide range of sources and combined to produce national averages. No information was provided on the year the data were collected. Because water utilities periodically review and change rate designs, these data should be considered rough estimates. Each category hides variations in rate design.

Source

Ringskog, K., 2000, "International Trends in Water Pricing and Use," World Bank presentation, Riyadh, Saudi Arabia (June 5).

TABLE 16 Water System Rate Structures (as percent of utilities)

Country	Flat Fee	Volume Rate	Block Rates Increasing	Decreasing
Australia	–	68	27	5
Canada	56	27	4	13
France	2	98	–	–
Hungary	–	95	5	–
Japan	–	42	57	1
Netherlands	7	90	3	–
Norway	87	13	–	–
Spain	–	10	90	–
Sweden	–	100	–	–
Turkey	–	–	100	–
United Kingdom	90	10	–	–
United States	2	33	31	34

Table 17   Water Prices for Various Households

Description

Water prices, in US dollars per cubic meter of water, are shown for households in different cities or countries around the world. These data show nearly a factor of 50 between the prices paid by the cheapest and most expensive areas. The year of the estimate is also shown.

Limitations

These data were collected from a wide range of sources, over different time periods. Hence they are not comparable without several words of caution. Domestic water prices sometimes include costs for repaying fixed investments such as infrastructure; sometimes they only include operation and maintenance fees. The numbers in these tables are also affected by the different dates cited, and by different and changing currency rates. We hope that over time, more comprehensive and comparable data on water prices will be collected and made available.

Source

Ringskog, K., 2000, "International Trends in Water Pricing and Use," World Bank presentation, Riyadh, Saudi Arabia (June 5).

TABLE 17   **Water Prices for Various Households**

Location	Date	US$/m³
Cairo	1993	$0.04
Syria (countrywide)	no date	$0.07
Beheira (Egypt)	1993	$0.07
Gaza City	1998	$0.12
Meknes (lowest in Morocco)	2000	$0.20
Algiers	1998	$0.21
Amman	1999	$0.23
Sana'a	1999	$0.25
Khan Younis (Gaza)	1998	$0.29
Tunisia (countrywide)	1999	$0.32
Lebanon (average)	1998	$0.32
Marrakech	2000	$0.39
Canada	1999	$0.41
United States	1999	$0.50
Safi (highest in Morocco)	2000	$0.53
Spain	1999	$0.54
Ramallah	1994	$0.91
United Kingdom	1999	$1.15
France	1999	$1.17
Netherlands	1999	$1.19
Denmark	1999	$1.61
Germany	1999	$1.81

Table 18   Unaccounted-for Water

Description

Not all of the water put into a water system reaches customers or is paid for by water users. This water is typically called "unaccounted-for water," but it is measured and defined in a variety of ways. Many in the water industry consider all water that is not metered and billed to customer accounts to be unaccounted-for water. High rates of unaccounted-for water result in financial losses and poor performance of the water agency.

The most common definition is the difference between water supplied to a system and water sold expressed as a percentage of net water supplied. Unaccounted-for water can include unmeasured water put to beneficial use as well as water losses from the system. Unmeasured beneficial uses include firefighting, main flushing, and process water used at waste-treatment plants. Examples of water losses include illegal connections, accounting procedure errors, reservoir seepage and leakage, reservoir overflow, leaks, theft, and evaporation. As an indication of the importance of this issue, the amount of unaccounted-for water in Nairobi (Kenya's largest city) is estimated to be enough to supply all of Mombassa (Kenya's second largest city).

Where unaccounted-for water is 50 percent or more of total water supply, the World Health Organization (WHO) recommends that water agencies try to reach a target of 25 percent. But WHO also notes that significantly lower levels can and should be achieved. Once illegal connections, leakage, and other factors are reduced and adequate metering, billing, and collections procedures are maintained, unaccounted-for water can be as low as 8 to 10 percent.

The American Water Works Association similarly notes that sound water-system management can reduce unaccounted-for water to 10 percent and recommends the reduction of unaccounted-for water through universal metering and accounting of water use, routine meter testing and repair, and distribution-system leak detection and repair (see http://www.awwa.org/govtaff/watcopap.htm).

Limitations

No comprehensive collection of data on unaccounted-for water has been put together. The lack of standardized definitions contributes to the problem of measuring lost water and revenue. The data presented here come from a variety of sources and a variety of time periods, though an effort was made to restrict this table to data from the mid-1990s and later. We urge more comprehensive efforts be made to collect and standardize information on unaccounted-for water, as well as on methods for reducing unaccounted-for water to low levels.

Sources

a: Saghir, J., M. Schiffler, and M. Woldu, 1999, *World Bank Urban Water and Sanitation in the Middle East and North Africa Region: The Way Forward,* the World Bank, Middle East and North Africa Region Infrastructure Development Group, December, available at http://www.worldbank.org/wbi/mdf/mdf3/papers/finance/Saghir.pdf.

b: Saghir, J., World Bank compilation.

c: Http://www.sydneywater.com.au/html/tsr/performanceindicators/esd2.html.

d: World Bank benchmarking data, http://www.worldbank.org/watsan/topics/bench/bench_network_iup.html.

e: World Health Organization, 2000, *Global Water Supply and Sanitation Assessment 2000 Report*, available at http://www.who.int/water_sanitation_health/Globassessment/Global4-4.htm.

f: Asian Development Bank, 1997, *Second Water Utilities Data Book*, Manila, Philippines, available at http://www.adb.org/Documents/News/1997/nr1997111.asp.

g: Makuro, M., 2000, *Nairobi's Response to the Water Crisis*, UNCHS (Habitat) United Nations Centre for Human Settlements, Vol. 6, No. 3, available at http://www.unchs.org/unchs/english/hdv6n3/nairobi_response.htm.

h: Kenny, J.F., 2000, *Public Water-Supply Use in Kansas, 1987–97*, USGS Fact Sheet 187-99, January, United States Geological Survey, available at http://ks.water.usgs.gov/Kansas/pubs/fact-sheets/fs.187-99.html#HDR2.

i: AWWA recommends the use of audits to reduce unaccounted-for water, http://www.awwa.org/govtaff/watcopap.htm.

j: Barbados Water, 1999, *Managing Water Resources in an Integrated and Participatory Way Report of the First Stakeholder Meeting in Barbados*, September 29–30, 1999, available at http://www.commonwealthknowledge.net/Thanni/wwevh.htm.

k: League of Women Voters, 1999, *Drinking Water Supply in the Washington D.C. Metropolitan Area: Prospects and Options for the 21st Century*, available at http://www.dcwatch.com/lwvdc/lwv9903b.htm.

l: Adelson, N., 2000, *Water Woes: Private Investment Plugs Leaks in Water Sector, Business Mexico*, available at http://www.mexconnect.com/mex_/travel/bzm/bzmwaterwoes.html.

m: Trung, D.Q., R. Snow, L. Doukas, N. Thanh, and N. Trung, 1998, Water-loss reduction program in Vietnam, 24th WECD Conference, Water and Sanitation for All, Islamabad, Pakistan, available at http://www.lboro.ac.uk/departments/cv/wedc/papers/24/S/trung.pdf.

n: Esmay, J., 1998, *Roundtable on Municipal Water*, Vancouver, Canada, available at http://www.idrc.ca/industry/canada_e7.html.

o: Qamber, M., 2000, *Water Demand Management in State of Bahrain, Bahrain Ministry of Electricity & Water*, available at http://www.emro.who.int/ceha/AmmanConference-WaterDemandManagement.pdf.

p: Http://www.metro.seoul.kr/eng/smg/agenda/2-3.html.

TABLE 18   Unaccounted-for Water

Location		Percent	Source
Africa (large city average)	1990s	39	e
Algiers, Algeria	1990s	51	a
Amman, Jordan	1990s	52	a
Asia (large city average)	1990s	35 to 42	e, f
Bahrain	1993	36	o
Bahrain	2000	24	o
Barbados	1996	43	j
Buenos Aires, Argentina	1993	43	n
Buenos Aires, Argentina	1996	31	n
Canada (average)	1990s	15	b
Casablanca, Morocco	1990s	34	a, b
Damascus, Syria	1995	64	a
Dubai, United Arab Emirates	1990s	15	a
Gaza	1995	47	a
Gaza	1999	31	a
Haiphong, Vietnam	1998	70	m
Hanoi, Vietnam	1995	63	f
Hebron	1990s	48	a
Johor Bahru, Malaysia	1995	21	f
Kansas, United States (average)	1997	15	h
Kansas, United States (range)	1997	3 to 65	h
Lae, Papua New Guinea	1995	61	f
Latin America and Caribbean (large city average)		42	e
Lebanon	average	40	a
Male, Maldives	1995	10	f
Mandalay, Myanmar	1995	60	f
Mexico City, Mexico	1997	37	l
Mexico City, Mexico	1999	32	l
Nairobi, Kenya	2000	50	g
Nicosia, Cyprus	1990s	16	b
North America (large city average)	1990s	15	e
Oran, Algeria	1990s	42	a
Penang, Malaysia	1995	20	f
Phnom Penh, Vietnam	1995	61	f
Poland (medium utility range)	1990s	19 to 51	d
Rabat, Morocco	1990s	18	b
Ramallah	1990s	25	a
Rarotonga, Cook Islands	1995	70	f
Sana'a, Yemen	1990s	50	a
Seoul, South Korea	1996	35	p
Singapore	1990s	11	b
Singapore	1995	6	f
Sydney, Australia	1990s	13.4	c
Tamir, Yemen	1990s	28	b
Teheran, Iran	1990s	35	a
Tunisia (large utility range)	1990s	8 to 21	a, d
United Kingdom (small utility range)	1990s	14 to 30	d
United States (average)	1990s	12	b
Vietnam (average)	1998	50	m
Washington, D.C., area suppliers	1999	10 to 28	k

Table 19   United States Population and Water Withdrawals, 1900 to 1995

Description

This table and the accompanying figure show the population of the United States and total water withdrawals for all purposes from 1900 to 1995, the last year for which water-use data was available. United States population during this period increased from 76 million to 267 million, nearly a four-fold increase. Water withdrawals rose from 56 cubic kilometers per year, peaking around 1980 at over 600 cubic kilometers, a ten-fold increase, and then dropped nearly 10 percent from that level by 1995.

Limitations

The United States is one of the few countries with a consistent system for regularly reporting water use. These data are compiled from state surveys and the quality of water-use data varies from location to location. Data from the earliest years should be considered approximations.

Sources

Water-use data from U.S. Geological Survey, 1997, *Estimated Use of Water in the United States in 1995,* U.S. Geological Survey, Reston, Virginia, and personal communication, available at http://water.usgs.gov/watuse/pdf1995/html. Population data are from the U.S. Census Bureau, http://www.census.gov.

TABLE 19   **United States Population and Water Withdrawals, 1900 to 1995**

	1900	1910	1920	1930	1940	1945	1950	1955	1960	1965	1970	1975	1980	1985	1990	1995
Population (millions)	76	92	105	125	132	140	151	164	179	194	206	216	230	242	252	267
Total Water Withdrawals (km³/yr)	56	93	125	152	188	224	250	330	370	430	510	580	610	551	564	554

United States Population and Water Withdrawals, 1990 to 1995

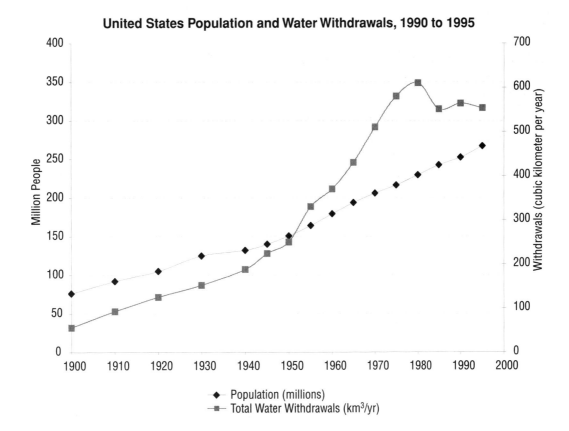

- ◆ Population (millions)
- ■ Total Water Withdrawals (km³/yr)

Table 20   United States GNP and Water Withdrawals, 1900 to 1996

Description

For many years, a fundamental assumption in the water planning community was that increases in water use were necessary to provide for increases in economic activity. The data in this table and accompanying figure show that this is no longer the case—a point made more explicitly in Chapter 1 of this book. The data here include the gross national product (GNP) of the United States from 1900 to 1996 (in 1996 dollars) together with annual water withdrawals for the same period. These two curves moved together until the late 1970s at which point GNP continued to rise while water use leveled off and even declined. The change in water use is a result of improving water-use efficiency and productivity in all U.S. sectors along with a change in the structure of those sectors away from water-intensive activities. The Pacific Institute is trying to collect a broad set of these comparisons for other countries (see Tables 21 and 22) and will present more of them in later volumes of *The World's Water.*

Limitations

Gross national or domestic product is an imperfect, albeit standard, measure of well-being or economic activity. As a result, GNP data are available for many countries and provide a benchmark against which other activities can be measured. Actual annual water withdrawal data are rarely collected for any country, even the United States. Good data are available every five years going back through most of the twentieth century, though data for earlier decades in the century are less reliable. For this graph, annual data were extrapolated from the five-year data collected and presented in Table 19.

Sources

Water-use data from U.S. Geological Survey, 1997, *Estimated Use of Water in the United States in 1995,* U.S. Geological Survey, Reston, Virginia, and personal communication, available at http://water.usgs.gov/watuse/pdf1995/html. U.S. GNP data are from the U.S. Department of Commerce, Bureau of Economic Statistics http://www.bea.doc.gov.

TABLE 20   United States GNP and Water Withdrawals, 1900 to 1996

Year	1996 ($)	Total Water Withdrawals km³/yr
1900	373	56
1901	416	60
1902	419	63
1903	440	67
1904	435	71
1905	467	75
1906	521	78
1907	530	82
1908	486	86
1909	566	89
1910	582	93
1911	597	96
1912	631	99
1913	637	103
1914	609	106
1915	604	109
1916	651	112
1917	656	115
1918	736	119
1919	710	122
1920	679	125
1921	620	128
1922	718	130
1923	805	133
1924	803	136
1925	870	139
1926	921	141
1927	920	144
1928	926	147
1929	878	149
1930	799	152
1931	748	156
1932	648	159
1933	639	163
1934	709	166
1935	773	170
1936	873	174
1937	920	177
1938	887	181
1939	959	184
1940	1,040	188
1941	1,218	195
1942	1,446	202
1943	1,682	210
1944	1,816	217
1945	1,795	224
1946	1,600	229
1947	1,583	234
1948	1,651	240

TABLE 20   *Continued*

Year	1996 ($)	Total Water Withdrawals km³/yr
1949	1,638	245
1950	1,784	250
1951	1,921	266
1952	1,993	282
1953	2,084	298
1954	2,070	314
1955	2,218	330
1956	2,262	338
1957	2,305	346
1958	2,280	354
1959	2,448	362
1960	2,508	370
1961	2,566	382
1962	2,723	394
1963	2,840	406
1964	3,005	418
1965	3,197	430
1966	3,402	446
1967	3,489	462
1968	3,652	478
1969	3,761	494
1970	3,765	510
1971	3,892	524
1972	4,106	538
1973	4,352	552
1974	4,330	566
1975	4,301	580
1976	4,538	586
1977	4,754	592
1978	5,006	598
1979	5,165	604
1980	5,146	610
1981	5,255	598
1982	5,137	586
1983	5,338	575
1984	5,705	563
1985	5,891	551
1986	6,061	554
1987	6,234	556
1988	6,477	559
1989	6,694	561
1990	6,784	564
1991	6,715	562
1992	6,892	560
1993	7,060	558
1994	7,293	556
1995	7,436	554
1996	7,638	552

Table 21   Hong Kong GDP and Water Withdrawals, 1961 to 2000

Description

For many years, a fundamental assumption in the water planning community was that increases in water use were necessary to provide for increases in economic activity. The data in this table and accompanying figure show that this is no longer the case—a point made more explicitly in Chapter 1 of this book. The data here include the gross domestic product (GDP) of Hong Kong from 1961 to 2000 (in 1990 Hong Kong dollars) together with water withdrawals (in million cubic meters per year) for the same period. These two curves moved together until the late 1980s/early 1990s, at which point GDP continued to rise while water use leveled off and even declined.

The change in water use is a result of improving water-use efficiency and productivity in all sectors along with a change in the structure of those sectors away from water-intensive activities. Water use may have also been affected by the transition of Hong Kong to China, although its economic productivity and population continued to rise during this period. The Pacific Institute is trying to collect a broad set of these comparisons for other countries (see Tables 20 and 22) and will present more of them in later volumes of *The World's Water*.

Limitations

Gross national or domestic product is an imperfect, albeit standard, measure of well-being or economic activity. As a result, GDP data are available for many countries and provide a benchmark against which other activities can be measures. Actual annual water withdrawal data are rarely collected for any country. Good water-use data are available for Hong Kong going back only to 1961.

Source

Data from Professor David Yongqin Chen, Chinese University of Hong Kong, 2001, personal communication to Peter H. Gleick.

TABLE 21   Hong Kong GDP and Water Withdrawals, 1961 to 2000

Year	GDP (billion 1990HK$)	Water Used (million cubic meters/year)
1961	61.75	111
1962	70.49	135
1963	81.56	118
1964	88.55	119
1965	101.36	182
1966	103.11	199
1967	104.86	181
1968	108.35	225
1969	120.59	259
1970	131.66	277
1971	140.98	301
1972	155.54	325
1973	174.77	358
1974	178.84	349
1975	179.43	361
1976	208.55	405
1977	233.02	387
1978	252.83	412
1979	281.95	467
1980	310.50	508
1981	339.04	507
1982	348.36	519
1983	368.17	592
1984	404.87	627
1985	406.62	637
1986	450.41	703
1987	508.76	750
1988	549.30	808
1989	563.37	845
1990	582.55	873
1991	612.02	884
1992	650.35	889
1993	690.22	915
1994	727.51	923
1995	755.83	919
1996	789.75	928
1997	829.02	913
1998	785.07	916
1999	808.83	911
2000	893.40	924

Data from David Yongqin Chen, Chinese University of Hong Kong, 2001.

Hong Kong GDP and Water Use

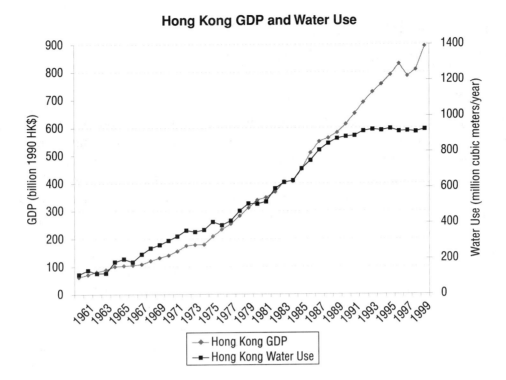

Source: David Chen (2001), Peter Gleick (2001)

Table 22   China GDP and Water Withdrawals, 1952 to 2000

Description

For many years, a fundamental assumption in the water planning community was that increases in water use were necessary to provide for increases in economic activity. The data in this table and accompanying figure show that this is no longer the case—a point made more explicitly in Chapter 1 of this book. The data here include an index of gross domestic product (GDP) of China from 1952 to 2000 (indexed against 1952 = 100) together with water withdrawals (in billion cubic meters per year) for the same period. The economic curve has followed an exponential growth typical of other national economic data sets; the water-use curve has also substantially increased over the past half century, but not in an exponential way. In recent years, total water withdrawals have actually leveled off and possibly even declined from the mid-1990s.

China is substantially water-limited in parts of the country, though in recent years there have also been improvements in the efficiency of water use as industries have modernized and as the government has slowly begun to address severe water pollution problems. The Pacific Institute is trying to collect a broad set of these comparisons for other countries (see Tables 20 and 21) and will present more of them in later volumes of *The World's Water*.

Limitations

Gross national or domestic product is an imperfect, albeit standard, measure of well-being or economic activity. As a result, GDP data are available for many countries and provide a benchmark against which other activities can be measured. The GDP data here were generated through a joint effort of the State Statistical Bureau of the People's Republic of China and the Institute of Economic Research of Hitotsubashi in Japan. Annual water withdrawal data are rarely collected for any country. For China, selected points are available back to 1952 but should be regarded with caution.

Sources

China's GDP data from "The Historical National Accounts of the People's Republic of China 1952–1995," Joint SSBC-Hitotsubashi University Team, Ministry of Education and Sciences of Japan, September 1997, available at http://www.ier.hit-u.ac.jp/COE/Japanese/online_data/china/china.htm.
China's water data come from various sources, including:
  • Research Center of Environment and Development, Chinese Academy of Social Sciences, 2001, *China Environment and Development Review*, 86–89, Social Sciences Documentation Publishing House, Beijing (Book B).
  • Year 2000 water-use data from Table 2, *The World's Water 2002–2003*.
  • The International Institute for Applied Systems Analysis, Laxenburg, Austria, available at http://www.iiasa.ac.at/Research/LUC/ChinaFood/data/water/wat_11.htm.

Year	China GDP Index (Real)	Water Use billion m³
1952	100.0	100
1953	115.6	
1954	120.5	
1955	128.6	
1956	147.9	
1957	155.5	
1958	188.6	
1959	205.2	200
1960	204.6	
1961	148.7	
1962	140.4	
1963	154.7	
1964	183.0	
1965	214.2	
1966	237.1	
1967	223.6	
1968	214.4	
1969	250.6	300
1970	299.3	
1971	320.6	
1972	332.4	
1973	358.6	
1974	366.9	
1975	398.8	
1976	392.4	
1977	422.2	
1978	471.6	
1979	507.5	
1980	547.1	443.68
1981	575.5	
1982	629.0	
1983	698.9	
1984	805.8	
1985	912.1	
1986	989.7	
1987	1103.5	
1988	1228.2	
1989	1279.8	
1990	1333.5	500
1991	1454.9	
1992	1660.0	
1993	1877.5	522
1994	2114.0	
1995	2304.3	
1996	2530.1	
1997	2745.2	563
1998	2959.3	543.5
1999	3169.4	
2000	3422.9	549.76

Note: "Gross Domestic Expenditure" (GDE) is also called "GDP by Expenditure Approach" in China.

Water Units, Data Conversions, and Constants

Water experts, managers, scientists, and educators work with a bewildering array of different units and data. These vary with the field of work: engineers may use different water units than hydrologists; urban water agencies may use different units than reservoir operators; academics may use different units than water managers. But they also vary with regions: water agencies in England may use different units than water agencies in France or Africa; hydrologists in the eastern United States often use different units than hydrologists in the western United States And they vary over time: today's water agency in California may sell water by the acre-foot, but its predecessor a century ago may have sold miner's inches or some other now arcane measure.

These differences are of more than academic interest. Unless a common "language" is used, or a dictionary of translations is available, errors can be made or misunderstandings can ensue. In some disciplines, unit errors can be more than embarrassing; they can be expensive, or deadly. In September 1999, the $125 million Mars Climate Orbiter spacecraft was sent crashing into the face of Mars instead of into its proper safe orbit above the surface because one of the computer programs controlling a portion of the navigational analysis used English units incompatible with the metric units used in all the other systems. The failure to translate English units into metric units was described in the findings of the preliminary investigation as the principal cause of mission failure.

This table is a comprehensive list of water units, data conversions, and constants related to water volumes, flows, pressures, and much more. Most of these units and conversions were compiled by Kent Anderson and initially published in P.H. Gleick, 1993, *Water in Crisis: A Guide to the World's Fresh Water Resources*, Oxford University Press, New York.

Water Units, Data Conversions, and Constants

Prefix (Metric)	Abbreviation	Multiple	Prefix (Metric)	Abbreviation	Multiple
deka-	da	10	deci-	d	0.1
hecto-	h	100	centi-	c	0.01
kilo-	k	1000	milli-	m	0.001
mega-	M	10^6	micro-	µ	10^{-6}
giga-	G	10^9	nano-	n	10^{-9}
tera-	T	10^{12}	pico-	P	10^{-12}
peta-	P	10^{15}	femto-	f	10^{-15}
exa-	E	10^{18}	atto-	a	10^{-18}

LENGTH (L)

1 micron (µ)	$= 1 \times 10^{-3}$ mm	**10 hectometers**	= 1 kilometer
	$= 1 \times 10^{-6}$ m	**1 mil**	= 0.0254 mm
	$= 3.3937 \times 10^{-5}$ in		$= 1 \times 10^{-3}$ in
1 millimeter (mm)	= 0.1 cm	**1 inch (in)**	= 25.4 mm
	$= 1 \times 10^{-3}$ m		= 2.54 cm
	= 0.03937 in		= 0.08333 ft
1 centimeter (cm)	= 10 mm		= 0.0278 yd
	= 0.01 m	**1 foot (ft)**	= 30.48 cm
	$= 1 \times 10^{-5}$ km		= 0.3048 m
	= 0.3937 in		$= 3.048 \times 10^{-4}$ km
	= 0.03281 ft		= 12 in
	= 0.01094 yd		= 0.3333 yd
1 meter (m)	= 1000 mm		$= 1.89 \times 10^{-4}$ mi
	= 100 cm	**1 yard (yd)**	= 91.44 cm
	$= 1 \times 10^{-3}$ km		= 0.9144 m
	= 39.37 in		$= 9.144 \times 10^{-4}$ km
	= 3.281 ft		= 36 in
	= 1.094 yd		= 3 ft
	$= 6.21 \times 10^{-4}$ mi		$= 5.68 \times 10^{-4}$ mi
1 kilometer (km)	$= 1 \times 10^5$ cm	**1 mile (mi)**	= 1609.3 m
	= 1000 m		= 1.609 km
	= 3280.8 ft		= 5280 ft
	= 1093.6 yd		= 1760 yd
	= 0.621 mi	**1 fathom (nautical)**	= 6 ft
10 millimeters	= 1 centimeter	**1 league (nautical)**	= 5.556 km
10 centimeters	= 1 decimeter		= 3 nautical miles
10 decimeters (dm)	= 1 meter	**1 league (land)**	= 4.828 km
			= 5280 yd
10 meters	= 1 dekameter		= 3 mi
10 dekameters (dam)	= 1 hectometer	**1 international nautical mile**	= 1.852 km
			= 6076.1 ft
			= 1.151 mi

(continues)

Water Units, Data Conversions, and Constants *(continued)*

AREA (L2)

1 square centimeter	$= 1 \times 10^{-4}\,\text{m}^2$	**1 square foot (ft²)**	$= 929.0\ \text{cm}^2$
(cm²)	$= 0.1550\ \text{in}^2$		$= 0.0929\ \text{m}^2$
	$= 1.076 \times 10^{-3}\ \text{ft}^2$		$= 144\ \text{in}^2$
	$= 1.196 \times 10^{-4}\ \text{yd}^2$		$= 0.1111\ \text{yd}^2$
1 square meter	$= 1 \times 10^{-4}\ \text{hectare}$		$= 2.296 \times 10^{-5}\ \text{acre}$
(m²)	$= 1 \times 10^{-6}\ \text{km}^2$		$= 3.587 \times 10^{-8}\ \text{mi}^2$
	$= 1\ \text{centare}$	**1 square yard (yd²)**	$= 0.8361\ \text{m}^2$
	(French)		$= 8.361 \times 10^{-5}$
	$= 0.01\ \text{are}$		hectare
	$= 1550.0\ \text{in}^2$		$= 1296\ \text{in}^2$
	$= 10.76\ \text{ft}^2$		$= 9\ \text{ft}^2$
	$= 1.196\ \text{yd}^2$		$= 2.066 \times 10^{-4}\ \text{acres}$
	$= 2.471 \times 10^{-4}\ \text{acre}$		$= 3.228 \times 10^{-7}\ \text{mi}^2$
1 are	$= 100\ \text{m}^2$	**1 acre**	$= 4046.9\ \text{m}^2$
1 hectare (ha)	$= 1 \times 10^4\,\text{m}^2$		$= 0.40469\ \text{ha}$
	$= 100\ \text{are}$		$= 4.0469 \times 10^{-3}\ \text{km}^2$
	$= 0.01\ \text{km}^2$		$= 43,560\ \text{ft}^2$
	$= 1.076 \times 10^5\ \text{ft}^2$		$= 4840\ \text{yd}^2$
	$= 1.196 \times 10^4\ \text{yd}^2$		$= 1.5625 \times 10^{-3}\ \text{mi}^2$
	$= 2.471\ \text{acres}$	**1 square mile (mi²)**	$= 2.590 \times 10^6\ \text{m}^2$
	$= 3.861 \times 10^{-3}\ \text{mi}^2$		$= 259.0\ \text{hectares}$
1 square kilometer	$= 1 \times 10^6\,\text{m}^2$		$= 2.590\ \text{km}^2$
(km²)	$= 100\ \text{hectares}$		$= 2.788 \times 10^7\ \text{ft}^2$
	$= 1.076 \times 10^7\ \text{ft}^2$		$= 3.098 \times 10^6\ \text{yd}^2$
	$= 1.196 \times 10^6\ \text{yd}^2$		$= 640\ \text{acres}$
	$= 247.1\ \text{acres}$		$= 1\ \text{section (of land)}$
	$= 0.3861\ \text{mi}^2$	**1 feddan (Egyptian)**	$= 4200\ \text{m}^2$
1 square inch (in²)	$= 6.452\ \text{cm}^2$		$= 0.42\ \text{ha}$
	$= 6.452 \times 10^{-4}\ \text{m}^2$		$= 1.038\ \text{acres}$
	$= 6.944 \times 10^{-3}\ \text{ft}^2$		
	$= 7.716 \times 10^{-4}\ \text{yd}^2$		

Water Units, Data Conversions, and Constants *(continued)*

VOLUME (L³)

1 cubic centimeter	$= 1 \times 10^{-3}$ liter	**1 cubic foot (ft³)**	$= 2.832 \times 10^4$ cm³
(cm³)	$= 1 \times 10^{-6}$ m³		$= 28.32$ liters
	$= 0.06102$ in³		$= 0.02832$ m³
	$= 2.642 \times 10^{-4}$ gal		$= 1728$ in³
	$= 3.531 \times 10^{-3}$ ft³		$= 7.481$ gal
1 liter (1)	$= 1000$ cm³		$= 0.03704$ yd³
	$= 1 \times 10^{-3}$ m³	**1 cubic yard (yd³)**	$= 0.7646$ m³
	$= 61.02$ in³		$= 6.198 \times 10^{-4}$
	$= 0.2642$ gal		acre-ft
	$= 0.03531$ ft³		$= 46656$ in³
1 cubic meter (m³)	$= 1 \times 10^6$ cm³		$= 27$ ft³
	$= 1000$ liter	**1 acre-foot**	$= 1233.48$ m³
	$= 1 \times 10^{-9}$ km³	**(acre-ft or AF)**	$= 3.259 \times 10^5$ gal
	$= 264.2$ gal		$= 43560$ ft³
	$= 35.31$ ft³	**1 Imperial gallon**	$= 4.546$ liters
	$= 6.29$ bbl		$= 277.4$ in³
	$= 1.3078$ yd³		$= 1.201$ gal
	$= 8.107 \times 10^{-4}$		$= 0.16055$ ft³
	acre-ft	**1 cfs-day**	$= 1.98$ acre-feet
1 cubic decameter	$= 1000$ m³		$= 0.0372$ in-mi²
(dam³)	$= 1 \times 10^6$ liter	**1 inch-mi²**	$= 1.738 \times 10^7$ gal
	$= 1 \times 10^{-6}$ km³		$= 2.323 \times 10^6$ ft³
	$= 2.642 \times 10^5$ gal		$= 53.3$ acre-ft
	$= 3.531 \times 10^4$ ft³		$= 26.9$ cfs-days
	$= 1.3078 \times 10^3$ yd³	**1 barrel (of oil)**	$= 159$ liter
	$= 0.8107$ acre-ft	**(bbl)**	$= 0.159$ m³
1 cubic hectometer	$= 1 \times 10^6$ m³		$= 42$ gal
(ha³)	$= 1 \times 10^3$ dam³		$= 5.6$ ft³
	$= 1 \times 10^9$ liter	**1 million gallons**	$= 3.069$ acre-ft
	$= 2.642 \times 10^8$ gal	**1 pint (pt)**	$= 0.473$ liter
	$= 3.531 \times 10^7$ ft³		$= 28.875$ in³
	$= 1.3078 \times 10^6$ yd³		$= 0.5$ qt
	$= 810.7$ acre-ft		$= 16$ fluid ounces
1 cubic kilometer	$= 1 \times 10^{12}$ liter		$= 32$ tablespoons
(km³)	$= 1 \times 10^9$ m³		$= 96$ teaspoons
	$= 1 \times 10^6$ dam³	**1 quart (qt)**	$= 0.946$ liter
	$= 10Q0$ ha³		$= 57.75$ in³
	$= 8.107 \times 10^5$		$= 2$ pt
	acre-ft		$= 0.25$ gal
	$= 0.24$ mi³	**1 morgen-foot**	$= 2610.7$ m³
1 cubic inch (in³)	$= 16.39$ cm³	**(S. Africa)**	
	$= 0.01639$ liter	**1 board-foot**	$= 2359.8$ cm³
	$= 4.329 \times 10^{-3}$ gal		$= 144$ in³
	$= 5.787 \times 10^{-4}$ ft²		$= 0.0833$ ft³
1 gallon (gal)	$= 3.785$ liters	**1 cord**	$= 128$ ft³
	$= 3.785 \times 10^{-3}$ m³		$= 0.453$ m³
	$= 231$ in³		
	$= 0.1337$ ft³		
	$= 4.951 \times 10^{-3}$ yd³		

(continues)

Water Units, Data Conversions, and Constants *(continued)*

VOLUME/AREA (L³/L²)

1 inch of rain	$= 5.610$ gal/yd^2	1 box of rain	$= 3,154.0$ lesh
	$= 2.715 \times 10^4$ gal/acre		

MASS (M)

1 gram (g or gm)	$= 0.001$ kg	1 ounce (oz)	$= 28.35$ g
	$= 15.43$ gr		$= 437.5$ gr
	$= 0.03527$ oz		$= 0.0625$ lb
	$= 2.205 \times 10^{-3}$ lb	1 pound (lb)	$= 453.6$ g
1 kilogram (kg)	$= 1000$ g		$= 0.45359237$ kg
	$= 0.001$ tonne		$= 7000$ gr
	$= 35.27$ oz		$= 16$ oz
	$= 2.205$ lb	1 short ton (ton)	$= 907.2$ kg
1 metric ton (tonne or te or MT)	$= 1000$ kg		$= 0.9072$ tonne
	$= 2204.6$ lb,		$= 2000$ lb
	$= 1.102$ ton	1 long ton	$= 1016.0$ kg
	$= 0.9842$ long ton		$= 1.016$ tonne
1 dalton (atomic mass unit)	$= 1.6604 \times 10^{-24}$ g	1 long ton	$= 2240$ lb
			$= 1.12$ ton
1 grain (gr)	$= 2.286 \times 10^{-3}$ oz	1 stone (British)	$= 6.35$ kg
	$= 1.429 \times 10^{-4}$ lb		$= 14$ lb

TIME (T)

1 second (s or sec)	$= 0.01667$ min	1 day (d)	$= 24$ hr
	$= 2.7778 \times 10^{-4}$ hr		$= 86400$ s
1 minute (min)	$= 60$ s	1 year (yr or y)	$= 365$ d
	$= 0.01667$ hr		$= 8760$ hr
1 hour (hr or h)	$= 60$ min		$= 3.15 \times 10^7$ s
	$= 3600$ s		

DENSITY (M/L³)

1 kilogram per cubic meter (kg/m³)	$= 10^{-3}$ g/cm^3	1 metric ton per cubic meter (te/m³)	$= 1.0$ specific gravity
	$= 0.062$ lb/ft^3		$= $ density of H$_2$O at 4°C
1 gram per cubic centimeter (g/cm³)	$= 1000$ kg/m^3		$= 8.35$ lb/gal
	$= 62.43$ lb/ft^3	1 pound per cubic foot (lb/ft³)	$= 16.02$ kg/m^3

Water Units, Data Conversions, and Constants *(continued)*

VELOCITY (L/T)

1 meter per second (m/s)	= 3.6 km/hr = 2.237 mph = 3.28 ft/s	**1 foot per second (ft/s)**	= 0.68 mph = 0.3048 m/s
1 kilometer per hour (km/h or kph)	= 0.62 mph = 0.278 m/s	**velocity of light in vacuum (c)**	= 2.9979×10^8 m/s = 186,000 mi/s
1 mile per hour (mph or mi/h)	= 1.609 km/h = 0.45 m/s = 1.47 ft/s	**1 knot**	= 1.852 km/h = 1 nautical mile/hour = 1.151 mph = 1.688 ft/s

VELOCITY OF SOUND IN WATER AND SEAWATER
(assuming atmospheric pressure and sea water salinity of 35,000 ppm)

Temp, °C	Pure water, (meters/sec)	Sea water, (meters/sec)
0	1,400	1,445
10	1,445	1,485
20	1,480	1,520
30	1,505	1,545

FLOW RATE (L³/T)

1 liter per second (1/sec)	= 0.001 m³/sec = 86.4 m³/day = 15.9 gpm = 0.0228 mgd = 0.0353 cfs = 0.0700 AF/day	**1 cubic decameters per day (dam³/day)**	= 11.57 1/sec = 1.157×10^{-2} m³/sec
1 cubic meter per second (m³/sec)	= 1000 1/sec = 8.64×10^4 m³/day = 1.59×10^4 gpm = 22.8 mgd = 35.3 cfs = 70.0 AF/day	**1 cubic decameters per day (dam³/day)**	= 1000 m³/day = 1.83×10^6 gpm = 0.264 mgd = 0.409 cfs = 0.811 AF/day
1 cubic meter per day (m³/day)	= 0.01157 1/sec = 1.157×10^{-5} m³/sec = 0.183 gpm = 2.64×10^{-4} mgd = 4.09×10^{-4} cfs = 8.11×10^{-4} AF/day	**1 gallon per minute (gpm)**	= 0.0631 1/sec = 6.31×10^{-5} m³/sec = 1.44×10^{-3} mgd = 2.23×10^{-3} cfs = 4.42×10^{-3} AF/day
		1 million gallons per day (mgd)	= 43.8 1/sec = 0.0438 m³/sec = 3785 m³/day = 694 gpm = 1.55 cfs = 3.07 AF/day

(continues)

Water Units, Data Conversions, and Constants *(continued)*

FLOW RATE (L³/T) (continued)

1 cubic foot per second (cfs)	= 28.3 1/sec = 0.0283 m³/ sec = 2447 m³/day = 449 gpm	**1 miner's inch**	0.02 cfs (in Idaho, Kansas, Nebraska, New Mexico, North Dakota, South
1 cubic foot per second (cfs)	= 0.646 mgd = 1.98 AF/day		Dakota, and Utah) = 0.026 cfs
1 acre-foot per day (AF/day)	= 14.3 1/sec = 0.0143 m³/sec = 1233.48 m³/day		(in Colorado) = 0.028 cfs (in British Columbia)
	= 226 gpm	**1 weir**	= 0.02 garcia
	= 0.326 mgd	**1 quinaria**	= 0.47–0.48 1/sec
	= 0.504 cfs	**(ancient Rome)**	
1 miner's inch	= 0.025 cfs (in Arizona, California, Montana, and Oregon: flow of water through 1 in² aperture under 6-inch head)		

ACCELERATION (L/T²)

standard acceleration of gravity	= 9.8 m/s² = 32 ft/s²

FORCE (ML/T² = Mass × Acceleration)

1 newton (N)	= kg-m/s² = 10⁵ dynes = 0.1020 kg force = 0.2248 lb force	**1dyne**	= g·cm/s² = 10⁻⁵ N
		1 pound force	= lb mass × acceler- ation of gravity = 4.448 N

Water Units, Data Conversions, and Constants *(continued)*

PRESSURE (M/L^2 = Force/Area)		
1 pascal (Pa)	= N/m^2	
1 bar	= 1 × 10^5 Pa	
	= 1 × 10^6 dyne/cm^2	
	= 1019.7 g/cm^2	
	= 10.197 te/m^2	
	= 0.9869 atmosphere	
	= 14.50 lb/in^2	
	= 1000 millibars	
1 atmosphere (atm)	= standard pressure	
	= 760 mm of mercury at 0°C	
	= 1013.25 millibars	
	= 1033 g/cm^2	
	= 1.033 kg/cm^2	
	= 14.7 lb/in^2	
	= 2116 lb/ft^2	
	= 33.95 feet of water at 62°F	
	= 29.92 inches of mercury at 32°F	

1 kilogram per sq. centimeter (kg/cm^2)	= 14.22 lb/in^2
1 inch of water at 62°F	= 0.0361 lb/in^2
	= 5.196 lb/ft^3
1 inch of water at 62°F	= 0.0735 inch of mercury at 62°F
1 foot of water at 62°F	= 0.433 lb/in^2
	= 62.36 lb/ft^2
	= 0.833 inch of mercury at 62°F
	= 2.950 × 10^{-2} atmosphere
1 pound per sq. inch (psi or lb/in^2)	= 2.309 feet of water at 62°F
	= 2.036 inches of mercury at 32°F
	= 0.06804 atmosphere
	= 0.07031 kg/cm^2
1 inch of mercury at 32°F	= 0.4192 lb/in^2
	= 1.133 feet of water at 32°F

TEMPERATURE	
degrees Celsius or Centigrade (°C)	= (°F–32) × 5/9
	= K–273.16
Kelvins (K)	= 273.16 + °C
	= 273.16 + ((°F- 32) × 5/9)

degrees Fahrenheit (°F)	= 32 + (°C x 1.8)
	= 32 + ((°K–273.16) × 1.8)

(continues)

Water Units, Data Conversions, and Constants *(continued)*

ENERGY(ML^2/T^2 Force × Distance)

1 joule (J)	= 10^7 ergs	1 kilowatt-hour	= 3.6 × 10^6 J
	= N·m	(kWh)	= 3412 Btu
	= W·s		= 859.1 kcal
	= kg·m^2/s^2	1 quad	= 10^{15} Btu
	= 0.239 calories		= 1.055 × 10^{18}J
	= 9.48 × 10^{-4} Btu		= 293 × 10^9 kWh
1 calorie (cal)	= 4.184 J		= 0.001 Q
	= 3.97 × 10^{-3} Btu		= 33.45 GWy
	(raises 1 g H$_2$O	1 Q	= 1000 quads
	1°C)		≈ 10^{21} J
1 British thermal	= 1055 J	1 foot-pound (ft-lb)	= 1.356 J
unit (Btu)	= 252 cal (raises		= 0.324 cal
	1 lb H$_2$O 1°F)	1 therm	= 10^5 Btu
	= 2.93 × 10^{-4} kWh	1 electron-volt (eV)	= 1.602 × 10^{-19} J
1 erg	= 10^{-7} J	1 kiloton of TNT	= 4.2 × 10^{12} J
	= g·cm^2/s^2	1 10^6 te oil equiv.	= 7.33 × 10^6 bbl oil
	= dyne·cm	(Mtoe)	= 45 × 10^{15} J
1 kilocalorie (kcal)	= 1000 cal		= 0.0425 quad
	= 1 Calorie (food)		

POWER (ML^2/T^3 rate of flow of energy)

1 watt (W)	= J/s	1 horsepower	= 0.178 kcal/s
	= 3600 J/hr	(H.P. or hp)	= 6535 kWh/yr
	= 3.412 Btu/hr		= 33,000 ft-lb/min
1 TW	= 10^{12} W		= 550 ft-lb/sec
	= 31.5 × 10^{18} J		= 8760 H.P.-hr/yr
	= 30 quad/yr	H.P. input	= 1.34 × kW input
1 kilowatt (kW)	= 1000W		to motor
	= 1.341 horsepower		= horsepower
	= 0.239 kcal/s		input to motor
	= 3412 Btu/hr	Water H.P.	= H.P. required to
10^6 bbl (oil) /day	≈ 2 quads/yr		lift water at a
(Mb/d)	≈ 70 GW		definite rate to
1 quad/yr	= 33.45 GW		a given distance
	≈ 0.5 Mb/d		assuming 100%
1 horsepower	= 745.7W		efficiency
(H.R or hp)	= 0.7457 kW		= gpm × total head
			(in feet)/3960

Water Units, Data Conversions, and Constants *(continued)*

EXPRESSIONS OF HARDNESS[a]

1 grain per gallon	= 1 grain $CaCO_3$ per U.S. gallon	**1 French degree**	= 1 part $CaCO_3$ per 100,000 parts water
1 part per million	= 1 part $CaCO_3$ per 1,000,000 parts water	**1 German degree**	= 1 part CaO per 100,000 parts water
1 English, or Clark, degree	= 1 grain $CaCO_3$ per Imperial gallon		

CONVERSIONS OF HARDNESS

1 grain per U.S. gallon	= 17.1 ppm, as $CaCO_3$	**1 French degree**	= 10 ppm, as $CaCO_3$
1 English degree	= 14.3 ppm, as $CaCO_3$	**1 German degree**	= 17.9 ppm, as $CaCO_3$

WEIGHT OF WATER

1 cubic inch	= 0.0361 lb	**1 imperial gallon**	= 10.0 lb
1 cubic foot	= 62.4 lb	**1 cubic meter**	= 1 tonne
1 gallon	= 8.34 lb		

DENSITY OF WATER[a]

Temperature		Density
°C	°F	gm/cm³
0	32	0.99987
1.667	35	0.99996
4.000	39.2	1.00000
4.444	40	0.99999
10.000	50	0.99975
15.556	60	0.99907
21.111	70	0.99802
26.667	80	0.99669
32.222	90	0.99510
37.778	100	0.99318
48.889	120	0.98870
60.000	140	0.98338
71.111	160	0.97729
82.222	180	0.97056
93.333	200	0.96333
100.000	212	0.95865

Note: Density of Sea Water: approximately 1.025 gm/cm³ at 15°C.

[a]*Source:* F. van der Leeden, F.L. Troise, and D.K. Todd, 1990. *The Water Encyclopedia*, 2d edition. Lewis Publishers, Inc., Chelsea, Michigan.

Index